Cisco IP Routing Protocols: Troubleshooting Techniques

LIMITED WARRANTY AND DISCLAIMER OF LIABILITY

Cisco IP Routing Protocols: Troubleshooting Techniques

V. Anand

K. Chakrabarty

CHARLES RIVER MEDIA, INC.
Hingham, Massachusetts

Acquisitions Editor: James Walsh
Cover Design: The Printed Image

CHARLES RIVER MEDIA, INC.
10 Downer Avenue
Hingham, Massachusetts 02043
781-740-0400
781-740-8816 (FAX)
info@charlesriver.com
www.charlesriver.com

This book is printed on acid-free paper.

V. Anand and K. Chakrabarty *Cisco IP Routing Protocols: Troubleshooting Techniques.*
ISBN: 1-58450-341-6

Library of Congress Cataloging-in-Publication Data

Anand, Vijay.
Cisco IP routing protocols : troubleshooting techniques / Vijay Anand and Koel Chakrabarty.— 1st ed.
p. cm.
ISBN 1-58450-341-6 (pbk. with cd-rom : alk. paper)
1. TCP/IP (Computer network protocol) 2. Routers (Computer networks) I. Chakrabarty, Koel. II. Title.

TK5105.585.A48 2004
004.6'2—dc22

2004006559

Printed in the United States of America
04 7 6 5 4 3 2 First Edition

CHARLES RIVER MEDIA titles are available for site license or bulk purchase by institutions, user groups, corporations, etc. For additional information, please contact the Special Sales Department at 781-740-0400.

Requests for replacement of a defective CD-ROM must be accompanied by the original disc, your mailing address, telephone number, date of purchase and purchase price. Please state the nature of the problem, and send the information to CHARLES RIVER MEDIA, INC., 10 Downer Avenue, Hingham, Massachusetts 02043. CRM's sole obligation to the purchaser is to replace the disc, based on defective materials or faulty workmanship, but not on the operation or functionality of the product.

Contents

1 Introduction to Troubleshooting 1
In This Chapter 1
Challenges and Issues of Complex Networks 1
Cisco Hierarchical Approach 5
The Access Layer 5
The Distribution Layer 6
The Core Layer 6
The Layered Troubleshooting Approach 7
The Layered Architecture of the OSI Model 8
The Layered Architecture of the TCP/IP Model 13
Mapping the TCP/IP and OSI Models 15
Using the Layered Approach 15
Problem Resolution Model Approach 19
Problem Definition 21
Facts Collection 21
Consideration of Possibilities 22
Action Plan Definition 22
Action Plan Implementation 22
Observe and Review Findings 22
Review Problem Solving Cycle 23
Isolation of Error 23
Resolve the Problem 23
Case Study 23
Summary 24
Points to Remember 24

2 Protocols and Their Characteristics 27
In This Chapter 27
Basic Protocol Behavior 27
Types of Connectionless and Connection-Oriented Protocols 28
Connectionless Behavior 30
Connection-Oriented Behavior 30
Connection-Oriented Behavior Versus Connectionless Behavior 32
Protocol Characteristics 34
Protocols of the Data-link Layer 36
Protocols of the Network Layer 41
Protocols of the Transport Layer 51
Protocols of the Session, Presentation, and Application Layers 53
Summary 59
Points to Remember 59

3 Diagnostic Mechanisms 61
In This Chapter 61
Cisco Network Management Tools 61
CiscoView 62
CiscoWorks 62
Traffic Director RMON 63
VLAN Director Switch Management 65
WAN Manager 65
NetSys Network Management Suite 67
NetSys Baseliner 4.0 67
NetSys SLM Suite 67
Cisco Diagnostic Commands 68
The Show Command 68
The Ping Command 79
The Debug Command 82
The Trace Command 86

Cisco Discovery Protocol (CDP) 87
Summary 90
Points to Remember 90

4 Troubleshooting Tools 93
In This Chapter 93
Connectivity and Cable Testers 94
Low Spectrum Cable Testers 94
High Spectrum Cable Testers 95
Digital Interface Testers 96
Network and Protocol Analyzers 97
Network Monitoring and Management System 98
Enterprise Management Systems 103
Management Solutions for LANs 105
Management Solutions for WANs 106
Management Solutions for VPN/Security 106
CiscoWorks QoS Policy Manager 107
Modeling and Simulation Tools 107
Traffic Generators 108
Cisco IOS Diagnostic Commands 109
The Show Command 109
The Debug Command 111
The Ping Commands 115
The Trace Command 117
Summary 118
Points to Remember 118

5 Troubleshooting TCP/IP 121
In This Chapter 121
TCP/IP Router Diagnostic Tools 121
Ping 122
Trace 127

Show Commands 128
Debug Commands 137
Troubleshooting Techniques 138
Problem Isolation in TCP/IP Networks 139
TCP/IP Problems and Symptoms 141
Summary 146
Points to Remember 147

6 Troubleshooting RIP Environments 149
In This Chapter 149
Features of RIP 149
Problem Isolation in RIP Environments 151
Misconfiguration 154
Configuration Problems 161
RIP Routes Missing from Routing Table 162
RIP Is Not Installing All Possible Equal-Cost Paths 167
Classless Routing 171
RIPv2 Cannot Reach Discontiguous Networks 177
Compatibility Between RIPv1 and RIPv2 181
Timer Problem 183
Looping 187
Hop Count 191
Split Horizon 191
Route Poisoning 194
Triggered Updates 195
Hold-down Timers 196
Summary 197
Points to Remember 197

7 Troubleshooting IGRP Routing Environments 199
In This Chapter 199
Features of IGRP 199

Problem Isolation in IGRP 201
Misconfiguration in IGRP 207
IGRP Routes Missing from Routing Table 208
IGRP Does Not Install All Possible Equal Cost Paths 213
Misconfigured ASN 216
Timer Problem in IGRP 221
Case Study 224
Summary 226
Points to Remember 226

8 Troubleshooting EIGRP Routing Environments 229

In This Chapter 229
Features of EIGRP 229
Problem Isolation in EIGRP 231
Misconfiguration in EIGRP 240
Network Not Declared in EIGRP 241
Same AS Not Defined in EIGRP Routers 244
EIGRP Neighbor Formation Problem 248
Observation 248
Problem Isolation 249
EIGRP Route Problem 252
Observation 252
Problem Isolation 253
EIGRP Metric Problem 254
Observation 255
Problem Isolation 255
Stuck in Active State 258
Observation 258
Problem Isolation 259
Redistribution Problem 260
Observation 261
Problem Isolation 261

Case Study 262
EIGRP Route Problem 263
EIGRP Redistribution Problem 264
Summary 265
Points to Remember 265

9 Troubleshooting OSPF Routing Environments 267
In This Chapter 267
OSPF Terminology 267
Resolution of Problems in OSPF 268
Show Commands 268
Debug Commands 275
Problems with Assigning Priority 279
OSPF Neighbor States 281
INIT State 282
EXCHANGE or EXSTART State 283
2-Way State 283
Nothing at All State 284
Load State 286
OSPF Routing Table 286
NBMA Networks 287
Configure Interface as Point-to-Point Links 288
Define OSPF Network as Point-to-Multipoint Links 289
OSPF Stub Areas 290
Redistribution in OSPF 291
Case Study 294
B2 Is Not Visible as a Neighbor of B1 294
Route 10.6.1.1/27 Is Unavailable in Routing Table of B4 297
Summary 297
Points to Remember 297

10 Troubleshooting IS-IS Routing Environments 299
In This Chapter 299
Features of IS-IS 299
Problem Isolation in IS-IS 302
Misconfiguration Problems in IS-IS 312
L1 Router Problem 317
L2 Router Problem 320
Redistribution Problem 321
Case Study 323
Redistribution of Routing Information 324
R1 and R2 Configured with Same System ID 325
Summary 328
Points to Remember 328

11 Troubleshooting BGP for Routing Environments 329
In This Chapter 329
Problem Isolation in BGP 329
BGP Neighbor Relationship 341
BGP Route Advertisement 345
Missing Routes 347
Misconfiguration Problems 349
Incorrectly Defined Numbers 349
Undeclared Networks 350
Attribute Problems 351
Route Dampening 353
Redistribution Problems 355
BGP Communities 357
BGP Multihoming and Loadsharing 358
Case Study 360
Summary 367
Points to Remember 367

12 Troubleshooting Redistribution Routing Environments 369
In This Chapter 369
EIGRP and OSPF Redistribution Environment 370
Network 192.168.30.0/27 Unavailable at B7 373
Network 192.168.50.0/24 Unavailable at B3 376
RIP and OSPF Redistribution Environment 378
Network 10.10.12.0/27 Unreachable from RIP Domain 380
Network 172.16.12. 0/24 Unreachable from RIP Domain 381
Static Routing and the OSPF Redistribution Environment 382
Static Routing and the RIP Redistribution Environment 385
IS-IS and OSPF Redistribution Environment 389
Case Study 393
Nonoptimal Path from Network 192.168.10.0/24 to Network 172.16.1.0/24 397
Route Unavailable for Network 172.16.10.0/25 from RIP 398
Summary 399
Points to Remember 399

Appendix: About the CD-ROM 401

Index 403

1 Introduction to Troubleshooting

IN THIS CHAPTER

- Challenges and Issues of Complex Networks
- Cisco Hierarchical Approach
- The Layered Troubleshooting Approach
- Problem Resolution Model Approach

CHALLENGES AND ISSUES OF COMPLEX NETWORKS

The advent of globalization has led to the need for interconnecting isolated LANs into a single network. Enterprises are now looking at sharing data and resources among hosts spread across different geographical locations. This has led to the concept of WANs, which connect an organization's standalone LANs, located in different cities across the globe. As compared to LANs, WANs span larger geographical areas, which increases the complexity of the connected networks. In addition to WANs, wireless networks have been developed that connect multiple LANs and WANs using a satellite link.

The interconnectivity between LANs, WANs, and wireless networks has led to the development of complex networks. Thus, a complex network consists of multiple networks, connected with each other, using different types of links and a multi-protocol environment. Figure 1.1 depicts a complex network.

Complex networks encounter more problems than independent LANs and WANs because of their multi-protocol architecture. LANs and WANs in a complex network are connected through different protocols, such as Transmission Control Protocol/Internet Protocol (TCP/IP), Open Shortest Path First (OSPF), and Interior Gateway Routing Protocol (IGRP).

A multi-protocol architecture requires maintenance and monitoring of multiple protocols, simultaneously. Complex networks span large geographical and

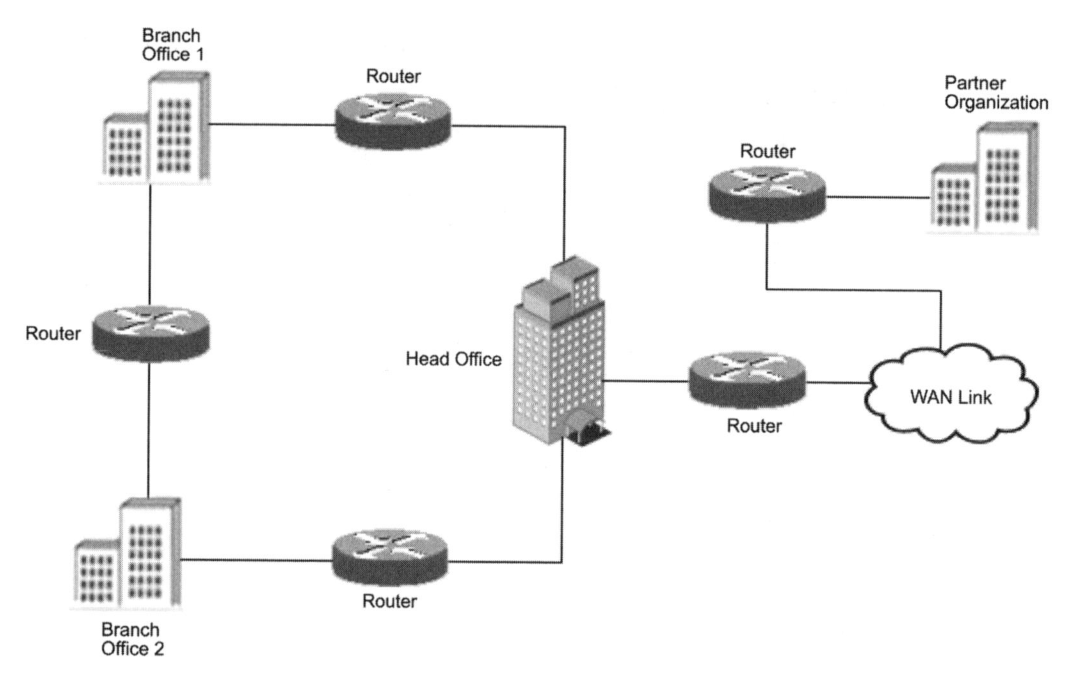

FIGURE 1.1 A complex network.

organizational areas, and as a result, they need to be robust and reliable. However, even the best-known technologies are prone to errors and malfunction.

Complex networks use high-end infrastructure and technology, and any kind of malfunctioning results in excessive financial and productivity losses. Therefore, troubleshooting mechanisms are required to minimize such losses. The functioning and problems of a complex network can be explained with the help of the following example.

Spacesoft Inc. is an organization that deals in research and development of aeronautical engineering. Their space shuttle launch station is located in Virginia, and the development center is located in California. In addition, Spacesoft has its space-suite manufacturing unit at Atlanta, Georgia, which is connected with the development center and the space shuttle launch station. The three offices have individual networks that are interconnected with each other through a WAN link and Cisco 4000 series routers. Figure 1.2 displays the network architecture of Spacesoft Inc.

Spacesoft is in the process of launching a new satellite, WAVE-series I, that will provide weather forecasts for the North American region. The Spacesoft team at the launching station is conducting last minutes checks for WAVE-series I, a report of which is being simultaneously analyzed and documented at the development center.

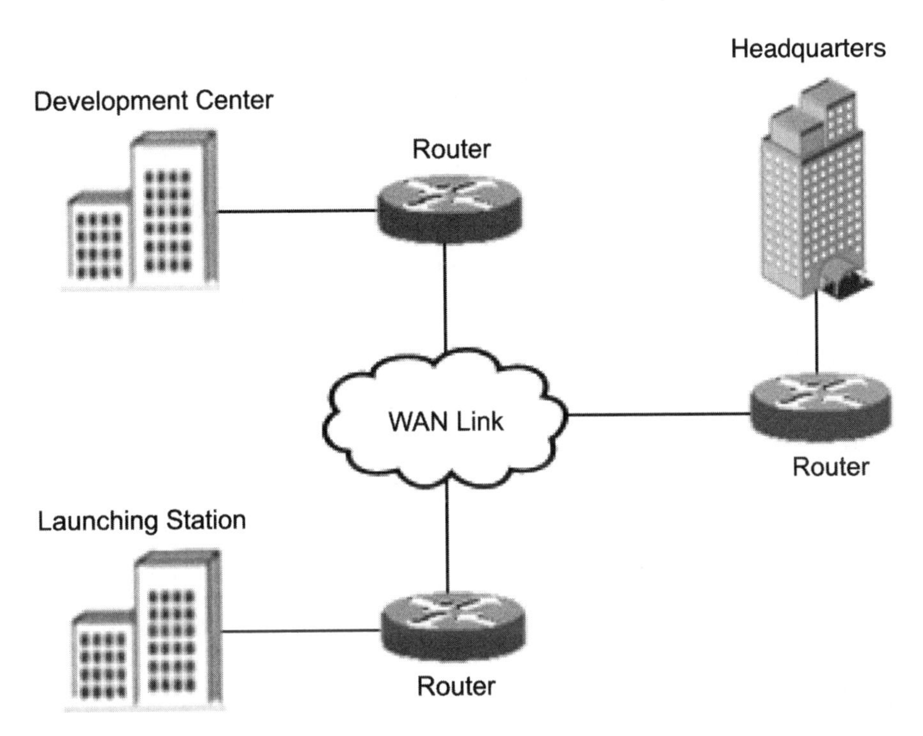

FIGURE 1.2 The network architecture of Spacesoft Inc.

During this process, the WAN link between the launching station and the development center suddenly becomes nonfunctional. This renders the launching station in a noncommunicating stage until the WAN link is restored. The troubleshooting team at the launching station takes nearly 30 minutes to restore the link. After the link is restored, the development center informs the launching station that data sent 15 minutes prior to the breakdown is lost. As a result, the engineers at the launching station have to redo the checks, which delays the launching of the satellite.

A 30-minute delay results in significant financial losses to Spacesoft Inc. This motivates the management of Spacesoft to formulate a well-defined troubleshooting methodology. As a result, they are prepared for any contingency that may arise in the future.

The problem that occurred at Spacesoft is an example of the types of possible network problems and their impact. The following are some common areas in which problems may occur in complex networks, which can lead to finance and data losses:

Backbone network: Enables high-speed data transfer among hosts in a network. This backbone network consists of multiple protocols that need to be monitored constantly. Any problem in the backbone network makes the entire network nonfunctional, leading to excessive financial and productivity losses.

Transmission media: Provides data transmission services through various media, such as WANs, satellite links, infrared links, and fiber optic cables. Such links can become nonfunctional due to several reasons, such as high-level electrical interference, malfunctioning of routers, and breakdown of cables. The delay caused by nonfunctional links can be disastrous for mission-critical projects, such as space shuttle launches, online banking transactions, and telemedicine systems.

Multiple networks and multiple protocols: Enables communication and data transfer between networks. As a result, incompatibility issues within protocols can render the entire network or a part of it nonfunctional. As a result, organizations with complex networks have teams of network engineers that monitor their networks, 24/7. This ensures that any problem is identified and rectified without loss of time.

Server: Enables multiple clients to access data from a central location and provides connectivity to other servers. As a result, a nonfunctional server can render the entire or a part of the network inaccessible.

Internal and external security mechanisms: Ensure security of the data transmitted across a network. Examples of security mechanisms are firewalls, encryption techniques, and hash algorithms. Problems in the functioning of these security mechanisms also impact network performance. For example, if there is a malfunction in a hash algorithm used for encryption, the data is corrupted. A problem like this requires immediate troubleshooting, because security of data is an important factor for measuring network performance and efficiency.

Because such problems can hamper the smooth functioning of complex networks, typically organizations keep a separate team of network engineers to troubleshoot network problems.

Troubleshooting any network requires a systematic approach, because a small error in the troubleshooting process may lead to significant financial and productivity losses. To provide a systematic methodology to troubleshoot network problems, you can use the following troubleshooting approaches:

- Cisco hierarchical approach
- Layered approach
- Problem resolution approach

CISCO HIERARCHICAL APPROACH

Cisco uses the hierarchical model to design and troubleshoot a network. This model is an efficient design topology that allows you to divide the network into three distinct layers: core, access, and distributed. These layers are a logical and not physical segmentation of the network. The Cisco hierarchical model divides the network into three layers, which reduces the complexity of internetworks that consist of multiple networks, communication subnets, and routers. This makes it possible to maintain each part of the network without disturbing the entire network.

Using the hierarchical model, you can design, maintain, and troubleshoot a network with respect to the following factors:

- Number of network users
- Number of running applications
- Number of subnetworks involved in making the complex network
- Number of servers in the network and the estimated load that each server can effectively handle

Figure 1.3 displays the Cisco hierarchical model.

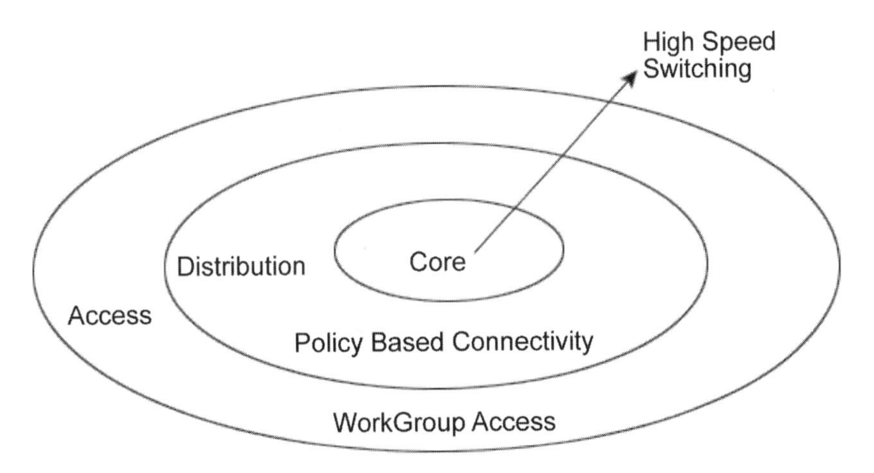

FIGURE 1.3 The Cisco hierarchical model.

The Access Layer

The access layer enables user interaction with the network. It is also called the *desktop layer*. End user workstations and local resources such as the printer are placed at this layer. Routers serve as gatekeepers at the entry and exit of this layer, and ensure that local server traffic is not forwarded to the wider network. The access layer controls user and workgroup access to internetwork resources. Other functions performed on this layer include sharing and switching of bandwidth, MAC-layer filtering, and micro segmentation.

Therefore, troubleshooting the access layer would require identifying problems of the local servers or network resources. In addition, problems at the access layer can include switch malfunction. This can be corrected by using backup switches for the main switches.

The Distribution Layer

Next in the Cisco hierarchical model is the distribution layer. This layer enables routing of data to the destination nodes and establishes WAN links between subnetworks. In addition, this layer performs data filtering functions to ensure security of data and the network. It is also called the *workgroup layer*.

The distribution layer serves as a bridge between the access layer and the core services of the network because it determines which data packets and user requests should be allowed access to core services of the network with respect to the associated user request. The distribution layer also routes data packets through the fastest route to access the required core service of the network or the required destination node.

This layer implements network policies and controls network traffic and data movement. In addition, it performs complex CPU-intensive calculations pertaining to routing, filtering, inter-VLAN routing, Access Control Lists (ACLs), address or area aggregation, security, and it identifies alternate paths to access the core.

To prevent network congestion, the distribution layer segregates the network into domains to distribute the load effectively over the entire network. In addition, the distribution layer ensures compatibility among different types of networks, such as Ethernet and Token Ring. The distribution layer ensures that the data transfer between different types of networks is not hampered because of compatibility issues between the networks.

Problems at the distribution layer are related to network congestion and the nonfunctioning WAN links between the subnetworks. Troubleshooting the problems of the distribution layer involves restoring the WAN link and using appropriate filtering mechanisms or an appropriate security mechanism.

The Core Layer

The core layer is a high-speed switching backbone that transports large volumes of traffic reliably and quickly. The traffic transported across the core is between end users and enterprise services, such as e-mail, video-conferencing, and dial-up access to the network. The links in the core layer are point-to-point.

NOTE

The services provided by the network are common to all the users of the network and are called enterprise services.

The core layer is responsible for providing high-speed transport. As a result, there is no room for latency and complex routing decisions pertaining to filters and access lists. Therefore, protocols such as OSPF and BGP, which have fast convergence time, are implemented at this layer. Quality of Service (QoS) may be implemented at this layer to ensure higher priority to traffic that may be lost or delayed in congestion. The core layer should have a high degree of redundancy.

This is the most important layer of a complex network and needs to be equipped with extensive troubleshooting support. This is because, if there is a malfunction in the backbone, the entire network is rendered nonfunctional.

THE LAYERED TROUBLESHOOTING APPROACH

Networks are based on the Open System Interconnection (OSI) model of networks, which is a layered architecture used to design networks compatible with all types of operating systems. The OSI model contains seven layers, which are organized in the order of their role for facilitating data transfer. In addition to the OSI model, another model that is widely used as a standard for internetwork communication is the Transmission Control Protocol/Internet Protocol (TCP/IP) model. This model has a four-layered architecture, and each layer corresponds to one or more layers of the OSI model. The functions of these layers are similar to those performed by the layers of the OSI model.

Because most networks are based on the layered architecture of OSI or the TCP/IP model, the troubleshooting approach applied to such networks is called the *layered troubleshooting approach*. Using the layered approach, you can isolate and troubleshoot problems pertaining to a specific layer. As a result, other layers are not affected, and troubleshooting is carried out on only that area of the network where the problem has occurred; the rest of the network functions without any interruption. In addition, the layered troubleshooting approach allows for easy and quick identification of the type of error or problem.

The layered troubleshooting approach provides various advantages as compared to the general troubleshooting methods. The following are the main advantages of the layered troubleshooting approach:

- Sequential analysis and identification of the problem
- Easy identification of the possible problem areas, because the function of each layer is predefined
- Facilitation of improvement of network performance

To understand the troubleshooting of networks based on the OSI or TCP/IP models, you need to understand the layered architecture of both these models.

The Layered Architecture of the OSI Model

The OSI model is a collection of seven layers:

- Physical
- Data-link
- Network
- Transport
- Session
- Presentation
- Application

Each layer performs a specific set of functions that are required to transfer data between hosts. In addition, to facilitate smooth data transfer, each layer provides services to the layer above and below it. Figure 1.4 shows a simple

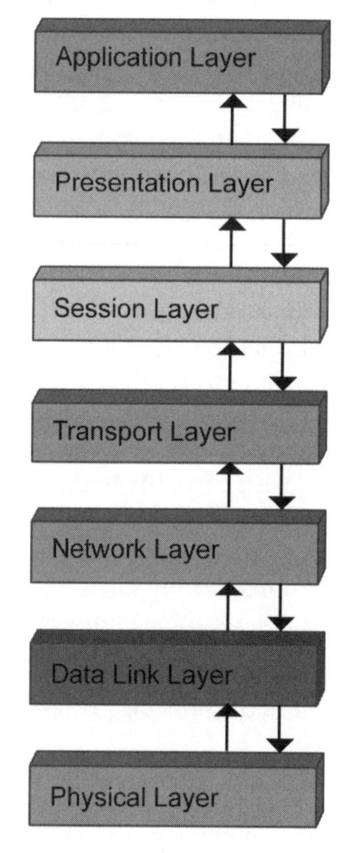

FIGURE 1.4 Position of layers in the OSI model.

diagrammatic representation of the position and appearance of these layers in the OSI model.

Each layer is connected to the layer above and below it through an interface, which enables the layers to interact and communicate with each other. For example, the interface between the Application and Presentation layers is called the *application-presentation interface.* The OSI model states that data flowing from source to destination must always pass through each layer of the model at both ends. Figure 1.5 displays the flow of data from source to destination.

In Figure 1.5, consider A as the source and B as the destination. According to the OSI model, any data starting from the Application layer of source A will travel up to the Physical layer. After reaching the Physical layer of destination B, the data moves up to the Application layer of B. At each layer, the data or message is modified according to the functionality of each layer. During the entire process, each layer of source A will communicate only with the corresponding layer of destination B. This means that the Presentation layer of A will communicate with the Presentation layer of B, and the Session layer of A will communicate with the Session layer of B, and so on.

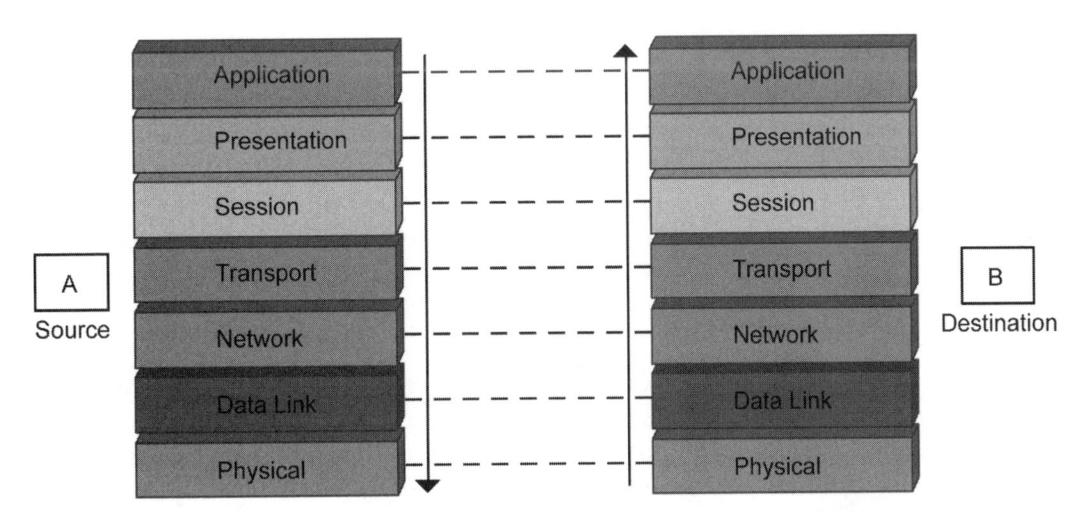

FIGURE 1.5 The data movement from source to destination in the OSI model.

The benefits of a layered protocol model are:

- Easy to troubleshoot, because each layer performs a specific function
- Interoperability of products from different vendors
- Ease in modifying programming interfaces
- Simplified troubleshooting
- Enhanced compatibility between diverse systems
- Future upgrade of a single layer does not affect other layers

Physical Layer

The Physical layer transmits bit streams of data over a physical medium from one host to another. The Physical layer encodes bit streams of 1s and 0s to signals before transmitting data. In addition, the Physical layer decides the line configuration and topology (ring, star, or mesh) of hosts connected in the network and defines the transmission mode (simplex, half-duplex, or full-duplex) between hosts.

Data-Link Layer

The Data-link layer enables node-to-node delivery of data in a network by converting bit streams received from the Network layer into manageable units. These units are called *frames*. A header is added to each frame to generate the physical address of the source and destination nodes. To maintain consistent rates of data transmission between the source and destination, the Data-link layer applies the flow control mechanism. This mechanism is a precautionary measure to ensure that the bit stream sent by a fast source node does not flood the destination node.

Network Layer

The Network layer manages internetworking and communication across multiple networks connected through routers and gateways. The Network layer handles the delivery of data from the source to the destination host. In addition, it facilitates delivery of data packets across multiple networks by generating the logical address of the data packet to be delivered. This provides easy identification of a data packet when it is transmitted across networks.

Transport Layer

The Transport layer enables delivery of data packets to the specific application running on the destination host. The Transport layer divides a data packet into data segments of uniform size, to which it assigns sequence numbers. Each sequence number identifies the segment and the associated data packet. The sequence number can identify the data packets that are lost or corrupted during transmission. The Transport layer also applies flow, connection, and error control mechanisms to a data packet.

The Transport layer implements two types of connection controls: connectionless and connection-oriented. In connectionless control, the Transport layer treats each data segment as a separate message, which is delivered to the Transport layer of the destination network. However, in connection-oriented control, the Transport layer of the source network first establishes a connection with the Transport layer of the destination network, then transmits the data segments. Thereafter, the connection between the two Transport layers is terminated.

Session Layer

The Session layer is the network dialog controller that establishes, maintains, and synchronizes interaction among networks. The layer first establishes a session between the source and the destination networks, which may be either in half-duplex or full-duplex mode. The Session layer then adds checkpoints to the data being transmitted.

In half-duplex transmission, you can either transmit or receive data at a single instance, whereas in full-duplex transmission, you can simultaneously transmit and receive data.

Presentation Layer

The Presentation layer is responsible for the syntax and semantics of the data being transferred. The Presentation layer is concerned with all aspects pertaining to the presentation of the data packets being transmitted. It translates the data being transferred into a format understandable to the destination node, encrypts the data packets to prevent data corruption and hacking, and compresses the data packets for smooth transmission.

Application Layer

The Application layer is the uppermost layer of the OSI Model. This layer handles high-level protocols such as HTTP, FTP, and SMTP. The Application layer uses these protocols to communicate with the application on the destination system. In addition, this layer handles flow control and error recovery.

Table 1.1 lists the different protocols running on each layer of the OSI model.

TABLE 1.1 Protocols Used on OSI Layers

Layer	Standards/Protocols/Applications
Application	DNS
	FTP and TFTP
	BOOTP
	SNMP and SMTP
	MIME
	NFS
	FINGER
	TELNET
	NCP
	APPC
	AFP
	SMB

(continued)

TABLE 1.1 *(continued)*

Layer	Standards/Protocols/Applications
Presentation	PICT TIFF MIDI MPEG
Session	NetBIOS NFS RPC Mail Slots DNA SCP Names Pipes
Transport	TCP SPX NetBIOS/NetBEUI ATP ARP, RARP NWLink
Network	IGMP IPX NetBEUI OSI DDP IP ARP RARP ICMP RIP OSFP IGMP DECnet X.25
Data-link	HDLC SDLC LAPB PPP ISDN SLIP
Physical	IEEE 802 IEEE 802.2 EIA/TIA-232 EIA-530 ISDN RS232 ATM

The Layered Architecture of the TCP/IP Model

TCP/IP is the most widely used networking protocol on the Internet. Both TCP and IP work together to facilitate safe and fast delivery of data through a network. To increase the transmission speed of large blocks of data, the data is split into smaller data fragments. These fragments are called *data packets.* TCP/IP plays a pivotal role in this context. TCP affixes a header to the data packet, which contains information pertaining to the destination address, source address, and length of the data. The role of IP is to ensure that once the data packets reach the destination address, they are reassembled into the original data block, and sent for use by the intended application. TCP/IP is a condensed version of the OSI model and has fewer layers. Figure 1.6 displays the layered architecture of the TCP/IP model.

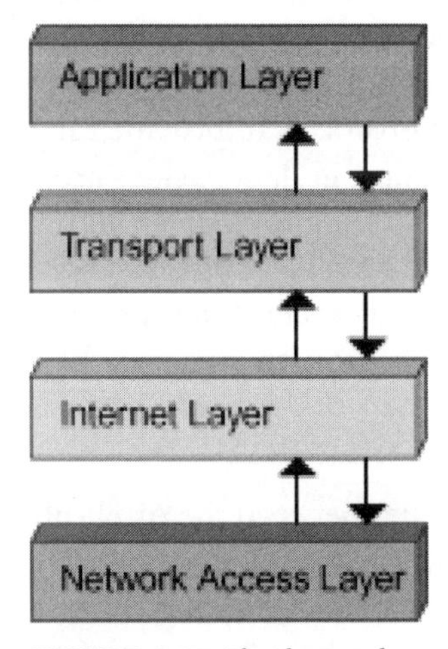

FIGURE 1.6 The layered architecture of the TCP/IP model.

As shown, the TCP/IP model consists of four layers:

- Network Access layer
- Internet layer
- Transport layer
- Application layer

Network Access Layer

The Network Access layer is placed at the bottom of the TCP/IP model hierarchy. The functions provided by the Network Access layer include encapsulating IP datagrams into frames and mapping IP addresses with physical devices.

All the processes in the Network Access layer are carried out by software applications and drivers customized to suit individual parts of hardware. Configuration often involves selecting the required driver for loading and selecting TCP/IP as the protocol.

Internet Layer

The Internet layer lies immediately below the Transport layer and above the Network Access layer of the TCP/IP model. The operating protocol of this layer is Internet Protocol (IP). The Internet Protocol builds the foundation of the packet delivery system, which serves as the basis for the entire concept of TCP/IP networking. This protocol manages connections over networks when data packets are transferred from the source to the destination.

IP is a connectionless protocol, which means it does not provide features such as a source-to-destination control of communication flow. IP relies on other layers and their associated protocols to provide this feature. Even functions such as error detection and correction in data packets are executed by other layers. In this context, IP is sometimes thought to be an unreliable protocol, though this does not imply that IP is not to be relied upon to deliver data via a network. It means that IP itself does not execute error checking and correcting functions. All information that flows through the TCP/IP networks uses the Internet Protocol.

Transport Layer

The Transport layer lies between the Application and Internet layers of the TCP/IP model. The two primary protocols associated with this layer are the Transmission Control Protocol (TCP) and the User Datagram Protocol (UDP).

TCP is a connection-based protocol, which enables error detection and correction in the data packets. It also ensures reliable delivery of data packets from source to destination. UDP is a connectionless and faster protocol as compared to TCP. It has low overhead costs and time associated with it, because it provides quick transfer of data. However, it does not provide the error detection and correction features of TCP.

The selection of protocols in the Transport layer is dependent on end user needs. TCP is used if a reliable connection session with two-way communication of data is of paramount importance. UDP is used to develop applications that are low on overheads.

Application Layer

The Application layer is the top layer of the TCP/IP reference model. It includes both the Presentation and Session layers of the OSI model. The word *application* is used to define any process that takes place above the Transport layer. The Application layer encompasses all processes that use the Transport layer protocols and transfer data to the Internet layer.

Mapping the TCP/IP and OSI Models

The TCP/IP suite of protocols works in Layers 3 and 4 of the OSI model. TCP/IP is made up of two protocols: TCP and IP. TCP/IP is the protocol used for communication over the Internet. Besides working in Layers 3 and 4, TCP/IP also has specifications for applications such as mail and FTP. Figure 1.7 depicts the mapping of the OSI and TCP/IP models.

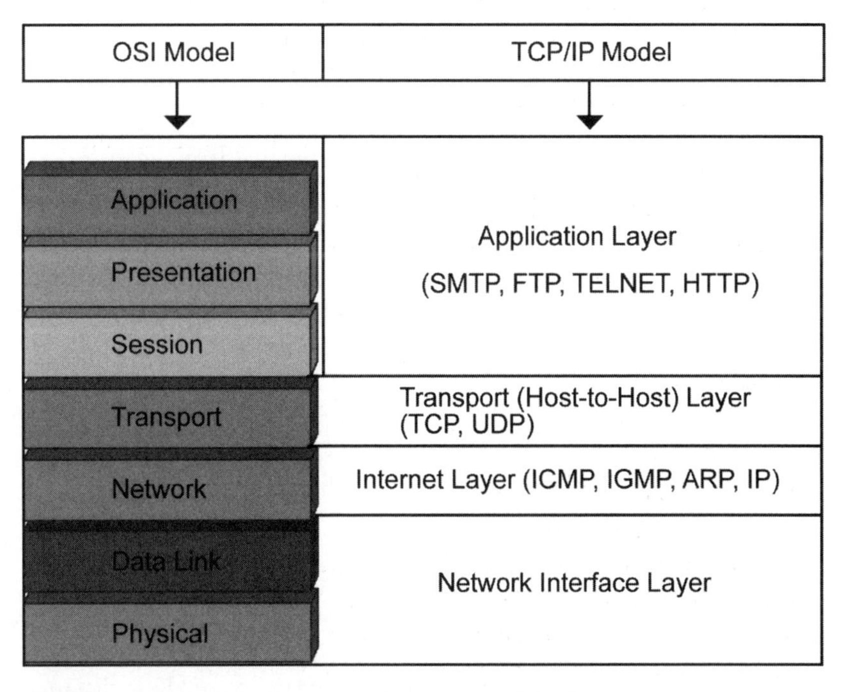

FIGURE 1.7 Mapping the OSI and TCP/IP models.

Using the Layered Approach

Using the layered approach, you can troubleshoot problems pertaining to each layer separately. Let us understand the different problems that occur on each layer of the OSI model and their possible solutions.

Troubleshooting Problems on the Physical Layer

The Physical layer of the OSI model is concerned with the transmission of the stream of data in bits. Data transmission between networks takes place through the

Physical layer. This transmission takes place using various transmission media, such as:

Magnetic media: Physically transmits data from one host to another. Magnetic transmission media provides maximum bandwidth for data transmission and can be reused multiple times to transmit data. But magnetic media are prone to destruction, both accidental or due to natural calamity. Examples of magnetic media are floppy disks and cassette tapes.

Twisted pair cables: Pairs of copper wires that are twisted to reduce electrical interference from other twisted pair cables placed alongside. Twisted pair can transmit data across long distances with the help of repeaters. In addition, the bandwidth provided by twisted pair depends upon the thickness of the cable and the distance it must travel. The most common application of twisted pair cables is the telephone system.

Coaxial cable: Copper wire cable protected from changes in temperature by an insulating material and a conducting material. Coaxial cable provides high bandwidth and can facilitate high-speed data transmission for longer distances than twisted pair cables.

Fiber optics: Uses light signals to transmit data over long distances without using repeaters.

Coaxial cables are of two types: baseband and broadband. Baseband coaxial cable is of 50 ohm and transmits digital data; broadband cable is of 75 ohm and transmits analog data. (Ohm is the unit for resistance.)

Problems of the Physical layer are related to the choice of the transmission media, which depends on the amount, distance, and rate of data transmission.

If data needs to be transmitted in small blocks, over a short distance, and frequency of data transmission is less, twisted pair or coaxial cable can be used. However, if large blocks of data are to be transmitted over a long distance and with high frequency, optical fiber should be used.

Apart from this, data can be transmitted across the networks using wireless media such as radio, infrared, and microwave transmission. These transmission media provide connectivity to mobile users. Unlike static transmission media, wireless media provide long-distance communication and data transmission, which spans continents across the globe.

The Physical layer also deals with the frequency and bandwidth problems of the data being transmitted. The Physical layer applies multiplexing, such as time-division and frequency-division multiplexing, to overcome problems related to optimizing bandwidth of the transmission media.

Troubleshooting Problems on Data-Link Layer

The Data-link layer of the OSI model provides the Network layer with a well-defined interface, groups the bits transmitted to the Physical layer into frames, identifies transmission errors, and applies flow control mechanisms that prevent a slow receiver from being flooded with data sent by a fast transmitter. For providing all these services, the layer uses mechanisms, such as Cyclic Redundancy Check (CRC), for error detection and flow control.

However, the main function of the Data-link layer is to ensure delivery of data from the Network layer of the source host to the Network layer of the destination host. While performing these functions, it is possible for the data to become corrupted or lost. To counter this problem, the Data-link layer uses the following protocols:

Unrestricted Simplex Protocol: Assumes that the data is transmitted in only one direction and Network layers of the source and destination networks are functional. This protocol is unreliable because there is always some loss of data during transmission.

Simplex Stop-and-Wait Protocol: Assumes that the receiving Data-link layer stores the data frames until these are transmitted to the Network layer. In addition, this is a simplex protocol, which assumes that the communication channel is error free. The disadvantage of implementing this protocol is that it does not provide the recipient node any mechanism to prevent overflow of data.

Sliding Window Protocol: Uses a single channel to transmit control and data frames to keep track of the data frames being transmitted. It was introduced to provide bi-directional data communication and transmission.

The Data-link layer also uses various error-detection techniques, such as CRC and bit stuffing, to detect transmission errors that corrupt the data frames. In addition, the Data-link layer applies a flow control mechanism to prevent a fast sender from flooding a slow receiver with data frames.

Troubleshooting Problems on the Network Layer

The main function of the Network layer is to deliver data packets from the source to the destination network using routers, switches, and bridges. The problems that the Network layer might encounter include:

- Overloading of a specific transmission route
- Inappropriate subnet topology
- Dependency of the Network layer on the subnet topology and its number
- Network congestion

To counter these problems, the subnet topology should be independent of the Network layer. Similarly, routing algorithms should be used to ensure proper routing of the data packets from the source to the destination network. Examples of routing algorithms include flooding, shortest path routing, and flow based routing.

In the flooding routing algorithm, each data packet is sent to every router. This results in the creation of multiple data packets. In the shortest path routing algorithm, a graph is created to identify the shortest path between the source and destination.

Both the flooding and the shortest path routing algorithm are based on the topology of subnets but do not take into account the load on each router. As a result, these routing algorithms do not provide accurate results. To ensure accurate results with respect to selecting the appropriate router, the anticipated load on the router needs to be considered. This is done by using the flow-based routing algorithm, which identifies the shortest route between the source and destination networks with minimum data transmission load.

Network congestion hampers smooth flow of data and may corrupt the data packets being transmitted. The Network layer applies congestion control algorithms to counter the problems of network congestion. These congestion control algorithms monitor the network to detect the probable areas where network congestion can occur and apply corrective action to reduce the congestion.

Troubleshooting Problems on the Transport Layer

The Transport layer is the core of the OSI model, because it serves as a bridge between the lower and upper layers to ensure reliable data transmission. The Transport layer can encounter the following problems:

- Unreliable connection between source and destination networks
- Unprotected transmission of data packets
- Data transmission delay
- Priority of data packets to be transmitted
- Problems with error detection and correction
- Flooding the slow receiver with data packets

To overcome the above-mentioned problems, the Transport layer uses the three-way handshake protocol. In this protocol, the source host sends a connection request to the destination host. If the destination host is ready to receive data packets, it sends a connection acknowledgment signal to the source host. After receiving the connection acknowledgment signal from the destination host, the source host sends the data packet to the destination host.

On receipt of the data packet, the destination host sends an acknowledgment signal to the source host. If the transmitted data packet is lost during transmission, the destination host does not send an acknowledgment signal to the source host.

In such a situation, the source host resends the data packet after a stipulated period. Using this protocol, the Transport layer provides retransmission of the lost or corrupted data packets.

Troubleshooting Problems on the Application Layer

The Application layer generally encounters problems pertaining to data security during transmission. To protect data from unauthorized access and hacking during transmission, the Application layer applies various encryption techniques. The types of encryption techniques are private key encryption and public key encryption.

To implement private key encryption, the Network layer uses algorithms such as Data Encryption Standard (DES). In the DES algorithm, the data contained in the data packets is encrypted in blocks of 64 bits of cipher data using a 56-bit key. The sender and the receiver agree on this 56-bit key before transmitting the data packets. After the data is transmitted, the receiver uses this 56-bit key to decrypt the encrypted data.

Similarly, to implement public key encryption, the Network layer uses the Rivest-Shamir-Adleman (RSA) algorithm. Using the RSA algorithm, the sender and receiver use a pair of keys to encrypt and decrypt the data contained in the data packets. The key pair contains a public key and a private key. The sender uses the public key of the key pair to encrypt the data, and the receiver uses the private key to decrypt the data.

PROBLEM RESOLUTION MODEL APPROACH

The problem resolution model contains a set of nine well-defined steps that enable you to identify a problem and efficiently implement a solution with minimum effort and time. Using this model, the entire task is divided into a series of manageable steps. The problem resolution model is a Cisco-proprietary solution used for troubleshooting networks. This model divides the troubleshooting process into the following steps:

- Problem definition
- Facts collection
- Consideration of possibilities
- Action plan definition
- Action plan implementation
- Observe and review findings
- Review problem solving cycle
- Isolation of error
- Resolve the problem

Figure 1.8 depicts Cisco's problem resolution model.

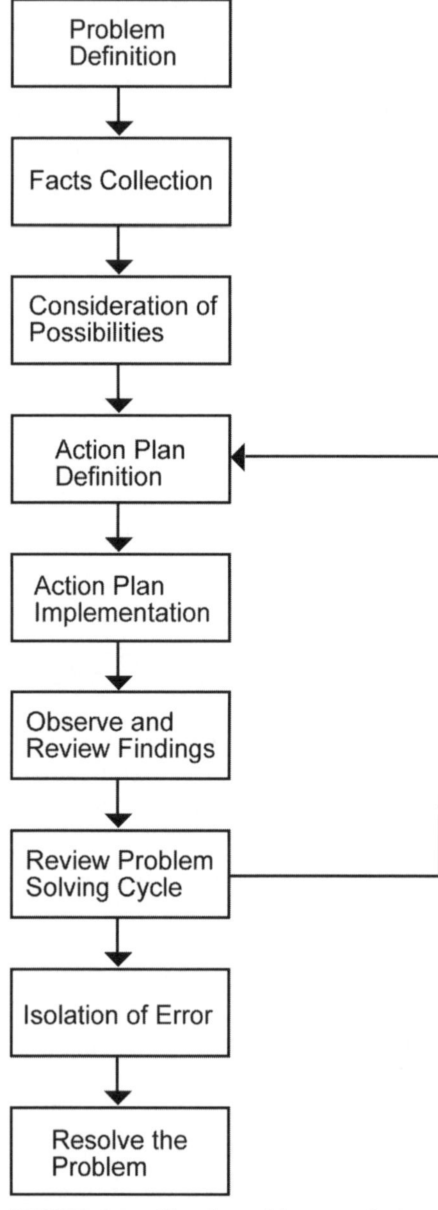

FIGURE 1.8 Cisco's problem resolution model.

Problem Definition

In this step, you ascertain the domain where the problem exists and define the cause and scope of the problem. This enables you to create an action plan in the subsequent steps of the Cisco problem resolution model. Figure 1.9 depicts a scenario in which Intranet A and Intranet B are connected over a high-speed backbone. Computer 1 and Computer 2 are part of Intranet A, and Computer 3 and Computer 4 are a part of Intranet B.

In Figure 1.9, a problem arises when Computer 1 is unable to send data to Computer 4 due to a break in any of the connection links between: Computer 1 and Intranet A, Intranet A and Intranet B, or Computer 4 and Intranet B. As a result, Computer 1 cannot establish a link to Computer 4. To resolve the problem, you need to gather some facts pertaining to the problem. The output of problem definition serves as an input to the next step: facts collection.

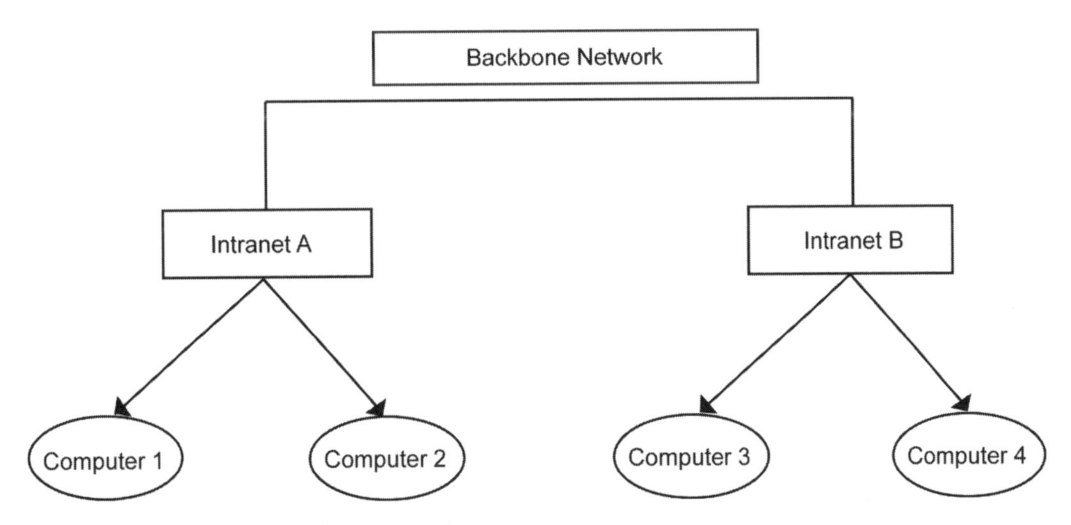

FIGURE 1.9 Diagram of an internetwork.

Facts Collection

The problem resolution model initiates a fact collection process after the completion of problem definition. In this process, you gather all information pertaining to the problem, using the following parameters:

- Possible causes
- Logs of the events before and after the problem occurred
- Scope of the problem to estimate the probable effect on other areas of the network

Taking the example of Intranets A and B, following are some basic facts that enable an optimal process of resolving the problem:

- Is Computer 1 able to link to Computer 2?
- Is Computer 3 able to link to Computer 4?
- Is Intranet A able to link to Intranet B?

Consideration of Possibilities

After you have collected all the facts, you analyze and select the one relevant to the problem. Using the example of Intranets A and B, you can list the different possibilities, such as the link between the Intranets or the links between hosts of an Intranet are dysfunctional. This step is useful to iteratively eliminate the possibilities that are not relevant to the problem in contention.

Action Plan Definition

In this stage, you need to document the steps to be taken to solve the problem. Here, you select the most likely cause for the problem and create an action plan. An action plan contains a systematic procedure that lists the steps for resolving the error that has occurred. This step entails all the inputs of the facts collection step and consideration of possibilities. For example, with Intranets A and B, the probability of a breakdown of the link between the two Intranets is the highest.

The action plan should not be defined in such a way that its implementation leads to other cascading problems. When defining the action plan, you should perform an appropriate risk analysis.

Action Plan Implementation

Action plan implementation is a step-by-step process of rectifying the problem by making changes. These changes are implemented such that any repercussions to a particular change can be easily traced. In addition, the actions performed for resolving the problem should be reversible in case of any subsequent problem. You also need to take incremental backups and maintain a log of the entire implementation process.

Observe and Review Findings

In this step, the final solution implemented for the problem is observed and reviewed. You need to analyze the changes done during implementation, any additional impact on the system due to the implementation of the action plan, and the alternate solutions considered during the definition step.

This step is a review process for the previous steps of the model. The facts collection at the second step should be reviewed. You need to ensure that the information gathered for solving the problem is optimally utilized.

Review Problem Solving Cycle

The process of reviewing all the steps implemented and observing the results so far is called *review problem solving cycle*. The objective of this step is to distinctly identify the actual causes for the problem. In addition, it enables you to enhance the problem definition model and eliminate irrelevant changes.

Isolation of Error

This step isolates the error occurring in the system. Based on the inputs from the previous steps, you need to isolate the actual problem or error from the system. You also need to ensure that the entire process is logged and the error is timestamped to serve as a reference, in case the error occurs again.

Resolve the Problem

In this step, the identified problem is resolved using the solutions identified in the previous steps. After the problem is resolved, it is tested to confirm that the identified problem is rectified and the network is working properly. Besides testing the rectified problem, the entire process of resolving the problem is documented for future reference.

Case Study

Davis has recently joined Blue Moon Computers as an assistant network administrator. Blue Moon Computers deals with manufacturing hardware devices such as monitors, printers, and scanners. The head office of Blue Moon Computers is situated at Denver, with its branches in Chicago and New York. Each branch office has an individual network, which is connected to the head office, where these networks are maintained and administered.

James, an employee of the Chicago branch, contacts Davis and tells him that he is unable to transfer a file to the New York branch. Davis consults his supervisor, Norman, who suggests that Davis collect all the relevant information from James, identify the possible causes, and resolve the problem.

When Davis explores the causes of the problem, he discovers that a connection could not be established between Host B of the Chicago network and Host S of the New York network. After defining the problem, Davis should:

1. Collect facts and details about the possible causes of the problem. The causes could be that the router is overloaded and may not be taking any more requests to transfer data, or a connection is being established but the data being transmitted is corrupted during transmission.
2. Identify the most likely cause of the problem. In this case, the most likely cause of the problem is that the router is overloaded and is unable to take any more requests to establish a connection and transfer data.
3. Design an action plan to provide a solution to the problem. A probable solution could be that a priority algorithm must be applied to all the requests being sent to the router. This enables the router to prioritize the routing of data.
4. Implement a priority algorithm, such as the round-robin algorithm to prioritize the requests.
5. Review the results of implementing the action plan.
6. Identify whether there are any other related problems that have occurred due to the implementation of the priority algorithm.

If reviewing the results does not indicate any other errors or problems, the problem being solved is isolated, and its solution process is documented for future reference.

SUMMARY

In this chapter, you learned about the challenges and issues of complex networks. You also learned about the need for a well-defined troubleshooting approach to counter these problems. There are primarily three approaches that you can use to effectively troubleshooting networks.

In the next chapter, you will learn about the behaviors and characteristics of protocols.

All code listings, figures, and tables presented in this book can be found on the book's companion CD-ROM.

POINTS TO REMEMBER

- A complex network consists of multiple networks, connected with each other, using different types of links and a multi-protocol environment.
- Because of their multi-protocol architecture, complex networks have more problems as compared to independent LANs and WANs.

- Most complex networks have problems pertaining to the backbone network, transmission media, multiple networks and protocols, servers, and internal and external security mechanisms.
- There are three troubleshooting methodologies: Cisco hierarchical approach, layered approach and the problem resolution approach.
- In the hierarchical approach, you divide the network into three layers—Core Distribution and Access. This makes it possible to troubleshoot problems on each layer, without disturbing the entire network.
- In the layered troubleshooting approach, you isolate and troubleshoot the problem pertaining to a specific layer
- The Data-link layer enables node-to-node delivery of data in a network by converting bit streams received from the Network layer into manageable units.
- Problems at the Data-link layer usually pertain to the data loss or corruption. To counter this problem, protocols such as Unrestricted Simplex Protocol, Simplex Stop-and-Wait Protocol, and Sliding Window Protocol are used.
- Problems at the Network layer usually pertain to overloading of a specific transmission route, inappropriate subnet topology, dependency of the Network layer on the subnet topology and its number, and network congestion.
- To counter problems at the Network layer, the subnet topology should be independent of the Network layer. Similarly, routing algorithms should be used to ensure proper routing of data packets from the source to the destination network.
- In the problem resolution approach, you divide the entire troubleshooting process into a set of seven well-defined tasks: problem definition, facts collection, consideration of possibilities, action plan, definition, action plan implementation, observe and review findings, review problem solving cycle, isolation of error, resolve the problem.

2 Protocols and Their Characteristics

IN THIS CHAPTER

- Basic Protocol Behavior
- Protocol Characteristics

A protocol is a set of rules defined to govern data transmission between applications residing on different systems. Protocols form the basis of all networks, and their function is to provide smooth transfer of data from source to destination. In addition to data transfer, protocols detect errors during data transmission, apply flow control, and facilitate retransmission of data packets corrupted during transmission. The protocols use acknowledgment signals to track data packets sent by the source node and received by the destination node.

The benefit of using a protocol for data transmission is that there are fewer chances of discrepancies such as loss of data during transmission. In addition, if a discrepancy occurs, the fault can be easily located and resolved.

BASIC PROTOCOL BEHAVIOR

A network based on the OSI model is segregated into layers. Each layer has its own protocol, which performs a specific set of functions to enable data transfer from source to destination. The characteristic of most protocols can be divided into two categories: connection-oriented and connectionless.

Connection-oriented protocols are also called *reliable* protocols, because the sequence of data packets is the same at the source and destination nodes. On receipt of data packets, the destination node sends an acknowledgment to the source node. If any data packet is lost or corrupted during transmission, connection-oriented protocols demand retransmission of the missing data packet. However, a few connection-oriented protocols, such as Frame Relay (FR), are unreliable, because they require virtual connectivity for sending data packets and do not have to depend on any reliability mechanism. In connection-oriented protocols:

- The path taken by data packets from source to destination is established before actual data transfer.
- The resources required for a connection are reserved in advance.
- Resource reservation of a connection is active throughout the life of that connection.
- The allocated resources are available for use after the connection is terminated.

On the other hand, connectionless protocols are unreliable, because data packets sent using these protocols might not be received in the same sequence. The receiver does not send any acknowledgment on receipt of data packets. In addition, connectionless protocols do not demand retransmission of missing or corrupted data packets. Therefore, there is no guarantee that all the data packets have actually reached the destination. This is similar to sending mail without receiving a return receipt.

For reliability, connectionless protocols depend on protocols residing in the higher layers of the OSI model. The higher layer protocols arrange the data packets in the appropriate sequence, handle timeout counters, and demand retransmission of missing data packets. For example, in the TCP/IP protocol suite, IP is a Layer 3 connectionless protocol, and TCP is a Layer 4 connection-oriented protocol. However, there are a few exceptions. For example, Open Shortest Path First (OSPF) is a reliable connectionless protocol, which demands acknowledgments from its neighbors when it multicasts routing updates. In connectionless protocols:

- Each message is independent of other messages.
- The destination node manages nonsequential arrival of data packets.
- The protocols provide dynamic flow through the network. This is because connectionless protocols are always in the data transfer phase with no explicit setup or the release phase (like connection-oriented protocols).
- The protocols provide alternate paths for data transfer.

Types of Connectionless and Connection-Oriented Protocols

Connectionless protocols transfer data packets independently without establishing a logical connection between the source and destination nodes. Therefore, the protocol sends data packets with all the information about the destination node. This information includes the destination address and the services required by the data packet.

Some connectionless protocols are:

IEEE 802.2: Provides three types of services, of which Type 1 and Type 3 are connectionless. The Type 1 service is an unacknowledged connectionless service, whereas Type 3 provides acknowledged connectionless services. Type 3 service is also referred to as a Logical Link Control (LLC). (Type 2 service of IEEE 802.2 is a connection-oriented service.)

Connectionless Network Protocol (CLNP): Provides connectionless service at the Network layer.

IP: Functions at the Network layer of the TCP/IP protocol suite and is connectionless.

UDP: Resides at the Transport Layer of the TCP/IP protocol suite, and provides connectionless services.

Fast Sequenced Transport (FST): Functions above the Network layer and is a connectionless transport protocol. The FST protocol uses the IP header to implement sequencing of the data packets being transferred.

Protocols using connection-oriented data transfer services establish a logical connection between the source and destination nodes. This logical connection consists of three phases: connection establishment, data transfer, and connection termination.

The main connection-oriented protocols are:

Type 2 service of IEEE 802.2: Is widely used in LAN networks, and also called the LLC2 connection-oriented data-link layer protocol.

ATM: Transfers data packets using the concept of Virtual Circuits (VC).

TP0, TP1, TP2, TP3, and TP4: Are a part of the Connection-Oriented Network Services (CONS) of the OSI model. For this, the OSI uses the X.25 packet-level protocol to provide connection-oriented data transfer.

TCP: Provides reliable and connection-oriented data transfer. It specifies the data format of the data packets being transferred and controls the entire data transfer operation.

A VC provides a private connection between any two nodes in an ATM network to facilitate data transfer between the nodes. A VC consists of a Virtual Path Identifier (VPI) and a Virtual Channel Identifier (VCI).

However, most networks use a combination of connectionless and connection-oriented protocols. This ensures that advantages of both the connectionless and connection-oriented services are used by networks during data transfer.

Like protocols, the different layers in a protocol display two types of behaviors to the above layers:

- Connectionless
- Connection-oriented

Connectionless Behavior

In connectionless behavior, there is no dedicated setup between communicating layers. Each data packet is independent of other data packets placed before and after it. This process is known as connectionless data transfer. For example, consider a courier service that forwards packets from one branch office to another. When branch office A sends a set of packets to branch office B, each packet is treated as a separate entity. Office A sends packets nonsequentially; it is the responsibility of office B to arrange them in the correct sequence. Even if two packets are sent to the same destination, they are considered different entities. The first packet is processed and sent to the destination first, followed by the second packet. However, there might be a situation in which the second packet reaches the destination prior to the first packet. This happens if there is some error in delivery. After office B receives the packets, it arranges them in a sequence, detects the missing packet, and demands retransmission.

In the same manner, a layer in a protocol sends packets to the layer above it, nonsequentially. After the upper layer receives the data packets, it arranges them in the correct sequence, detects the missing packets, and demands retransmission from the lower layer.

Connectionless data transfer offers two important advantages:

- Easy implementation
- Decreased network traffic as compared to connection-oriented data transfer

While troubleshooting connectionless data transfer, you should check the areas where:

- Data packet receipts are not acknowledged.
- Data packet errors are not reported to the sender.
- Data packets are not in correct sequence.
- Flow control is missing.

Connection-Oriented Behavior

In connection-oriented behavior, there is a dedicated setup between the communicating layers. Data is transmitted sequentially in streams of packets or frames over a single path through the network. Examples of connection-oriented protocols are TCP and ATM.

Connection-oriented behavior is similar to a telephone system. To make a telephone call, you:

1. Pick up the receiver and dial the destination number.
2. Converse with the person at the destination.
3. Disconnect the call when information exchange is complete.

Similarly, in a connection-oriented behavior:

1. The connection is established.
2. Data is transferred.
3. The connection is terminated.

The essential aspect of connection-oriented behavior is that it acts like a tube when the source node sends data packets from one end and the destination node receives them at the other end, in the same order. This process, also known as a *handshake*, is depicted in Figure 2.1.

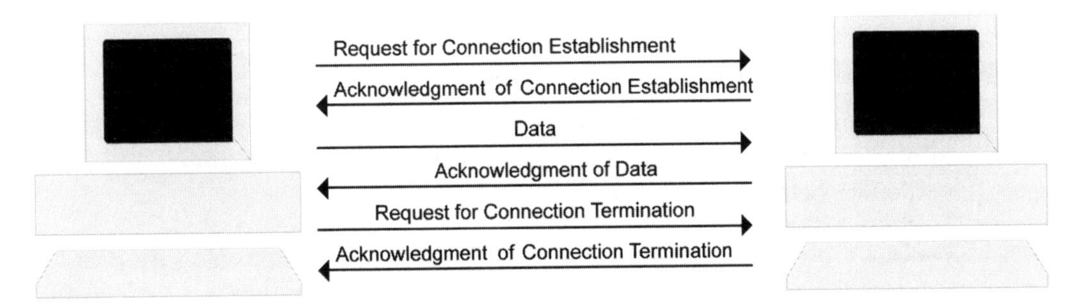

FIGURE 2.1 The handshake process in a connection-oriented behavior.

The services offered by connection-oriented protocols are:

Flow Control: Implemented by specifying definite space for data packets. There may be situations in which the source node sends the data at a transmission speed that is higher than what the destination node can handle. This leads to network congestion. To avoid this, connection-oriented protocols use the Windowing technique, in which the receiver specifies the space available for a data packet. The sender sends the data packet pertaining to the specified space and waits for an acknowledgment. The acknowledgment ensures that the recipient has received the data packet and can receive more.

Sequenced Delivery: Ensured by assigning a connection identifier and a sequence number to each data packet. In addition, if data packets are not received in the correct sequence, the receiver uses a unique sequential numbering scheme. This scheme is decided during the Request for Connection (RFC) establishment phase.

Error Control: Ensured when the receiver sends a request for retransmission, if the data packet received is damaged.

The receiver checks whether the data packet received is damaged or not, with respect to the number of bits received.

While troubleshooting connection-oriented data transfer, you need to monitor the:

- Number of data packet retransmissions in the network. If there is an excessive number of retransmissions, you need to inspect why the upper layer protocols are requesting them.
- Parameters of protocols such as level of incrementing and controlling of sequence numbers.

Connection-Oriented Behavior Versus Connectionless Behavior

There are many differences between connection-oriented and connectionless protocols. Table 2.1 lists the major differences between connection-oriented and connectionless behavior.

TABLE 2.1 Differences between Connection-Oriented and Connectionless Behavior

Connectionless Behavior	Connection-Oriented Behavior
Data packets are sent non-sequentially; that is, a packet that is sent later reaches the destination first.	Data packets are sent sequentially, by assigning a connection identifier and a sequence number.
Unreliable and nonsequential delivery.	Reliable and in-order delivery.
Simple to implement and has comparatively low requirement for network traffic.	Comparatively complex to implement and has high requirement for network traffic.
No dedicated link between the communicating layers.	A dedicated link is established between the communicating layers.

TABLE 2.1 *(continued)*

Connectionless Behavior	Connection-Oriented Behavior
No acknowledgment is sent by the destination node on receipt of data packets.	An acknowledgment is sent by the destination node on receipt of each data packet.
Does not allocate resources to connections.	Allocates resources to connections at switches.
A route packet is based on the destination address, which is in every packet.	A route packet is based on an identifier (connection ID is in every packet).
Per packet routing; that is, data packets are rerouted if a failure occurs.	Per connection routing; that is, in case of failures, connection should be re-established.
Examples of connectionless protocols are Ethernet, Frame Relay, and IP.	Examples of connection-oriented protocols are Token Ring, X.25, and TCP.

Table 2.2 lists the examples of connection-oriented and connectionless protocols.

TABLE 2.2 Examples of Connection-Oriented and Connectionless Protocols

Connectionless Protocols	Connection-Oriented Protocols
UDP	ICMP
XDR	DVMRP
NetBIOS	RSVP
DNS	SLIP
DHCP	CSLIP
NTP	IP
BOOTP	X.25
TFTP	TCP
SNMP	ISO-DE
Remote Unix Services	PPTP
	LDAP

(continued)

TABLE 2.2 *(continued)*

Connectionless Protocols	Connection-Oriented Protocols
	RPC
	LPP
	HTTP
	FTP
	SMTP
	Telnet

PROTOCOL CHARACTERISTICS

To enable communication over heterogeneous systems, a common standard such as the OSI reference model is used. This model contains seven layers, and each layer performs a specific set of functions. Each layer communicates with the corresponding layer of the other system. These protocols are components of layers of a protocol suite. A single layer may include one or more protocols. For example, Layer 2 of the OSI model includes more than one protocol—PPP and ATM. Table 2.3 lists the seven layers of the OSI Model.

TABLE 2.3 The Seven Layers of the OSI Model

Layer	Function
Application	Is concerned with the user interface and the application program interface, such as FTP or Telnet. It is also concerned with the management of the data transfer process.
Presentation	Is concerned with the data format received from the Application layer, and the format required by the destination node.
Session	Establishes a logical session between the source and destination nodes of the network. In addition, this layer maintains and synchronizes the established session until data is transferred to the destination node.
Transport	Provides end-to-end connectivity between the source and the destination nodes using acknowledgments and flow control mechanisms.

TABLE 2.3 *(continued)*

Layer	Function
Network	Provides logical and physical addresses to data packets being transferred and routes them along the best route available.
Data-link	Facilitates data transfer through the communication channel by consolidating data into frames. This layer also checks for transmission errors and applies mechanisms for correcting it.
Physical	Provides the actual mechanical and electrical interface for data transmission. Data transmission at this layer is in bits and bytes.

The most commonly used protocol suite of the Transport and Network layers is TCP/IP. Using the packet-switching technology, the TCP/IP protocol suite enables data transfer between dissimilar networks. This was developed by the Defense Advanced Research Projects Agency (DARPA) and Stanford University. Together, these two institutions developed the Internet Protocol (IP) suite, which consists of the TCP and IP protocols.

The TCP/IP protocol suite provides communication among complex networks. For example, the network of an organization ABC is connected through a WAN link with the network of its partner, XYZ. ABC uses the TCP/IP protocol suite, whereas XYZ is using NetWare protocols. To enable data transfer and communication between these two networks, TCP/IP provides data translation and compatibility features.

Table 2.4 displays protocols of the TCP/IP protocol suite that are applicable to various layers of the OSI layers.

TABLE 2.4 Protocols for OSI Layers

OSI Layer	Protocols
Network layer	IP and ICMP
Transport layer	TCP and UDP
Session, Presentation, and Application layers	RIP, IGRP, and OSPF

Other protocols, such as FTP, TELNET, SMTP, and SNMP are also applicable to the Session, Presentation, and Application layers.

Protocols of the Data-link Layer

The Data-link layer arranges raw data transmitted by the Physical layer into data frames, transmits them sequentially, and processes acknowledgment frames. The protocols used by the Data-link layer are:

- Point-to-Point Protocol (PPP)
- Synchronous Data Link Control (SDLC)
- High-Level Data Link Control (HDLC)

Point-to-Point Protocol

PPP manages error detection, permits authentication, supports multiple protocols, negotiates IP addresses at the time of connection, and solves the problem of assigning and managing IP addresses. PPP consists of:

Framing Flag: Defines the beginning and end of each frame.

Link Control Protocol (LCP): Establishes, configures, and tests the data-link connection.

Network Control Protocols (NCP): Establishes and configures different Network layer protocols by having a different NCP for every Network layer.

To establish communication over a PPP link:

1. The initiating node sends LCP packets to configure the data link.
2. The Network layer receives a series of NCP data packets when the link is established.
3. Packets from the Network layer are sent over the data link after the Network layer protocols are configured.

Figure 2.2 depicts a PPP frame format.

Flag	Addresses	Control	Protocol	Data	Checksum	Flag
(1 Byte)	(1 Byte)	(1 Byte)	(2 Bytes)	(Variable)	(2 or 4 Bytes)	(1 Byte)

FIGURE 2.2 The PPP frame format.

Table 2.5 lists functions of each field of the PPP format.

TABLE 2.5 Fields of PPP Frame

Field	Description
Flag	Indicates the beginning or end of each frame. All the PPP frames begin with the binary value 01111110.
Address	Contains the binary value 11111111, which is a common broadcast address. Because PPP does not allocate a separate node address, it supports only one connection between two nodes.
Control	Contains the binary value 00000011, which indicates an un-numbered frame, because PPP does not offer reliable transmission using sequence numbers and acknowledgments.
Protocol	Identifies the protocol in the Data field. For example, a 0 bit indicates that the protocol is a Network layer protocol, such as IP and IPX.
Data	Contains the data for transfer. This field is of variable length, which is decided using LCP during the communication line setup. However, the default length of this field is 1500 bytes.
Checksum	Checks for any errors in data transfer. This field is normally 2 bytes. However, 4 bytes can be used for more efficient error detection.

Figure 2.3 illustrates the different phases of establishing and terminating a PPP link.

Table 2.6 lists the different phases of a PPP link connection.

TABLE 2.6 Phases of the PPP Link

Phase	Description
Link Dead	The communication line is dead; that is, no Physical layer carrier is available. In addition, no Physical layer connection exists.
Link Established	The Physical layer connection is established, that is, negotiation of an LCP option begins. If it is successful, the link moves to the Link Authenticated phase.

(continued)

TABLE 2.6 *(continued)*

Phase	Description
Link Authenticated	The two communicating nodes check the identity of the other.
Link Networked and Opened	The suitable NCP protocol is invoked to configure the Network layer. If successful, data transfer takes place.
Link Terminated	In this phase, the communication link is terminated, after data transfer is complete.

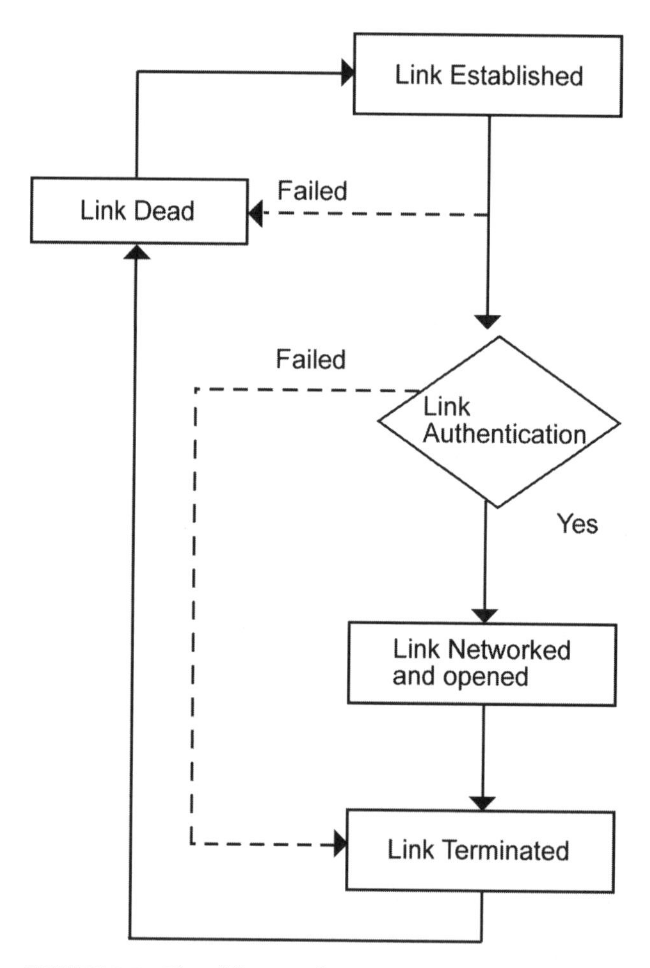

FIGURE 2.3 The different phases of a point-to-point link connection.

Synchronous Data Link Control (SDLC)

SDLC is a synchronous bit-oriented operation protocol, which can be used with links and topologies such as:

- Point-to-point link
- Half-duplex transmission
- Full-duplex transmission
- Circuit-switched networks
- Packet-switched networks

SDLC has two types of network nodes: primary and secondary. The primary node manages the operation of one or more nodes, also known as *secondary nodes*. It also establishes and terminates links with the secondary node. The primary node requests that the secondary nodes send the data in a predetermined sequence. The primary nodes manage the secondary nodes. The secondary nodes can send data to a primary node only after getting permission from the primary node, as depicted in Figure 2.4.

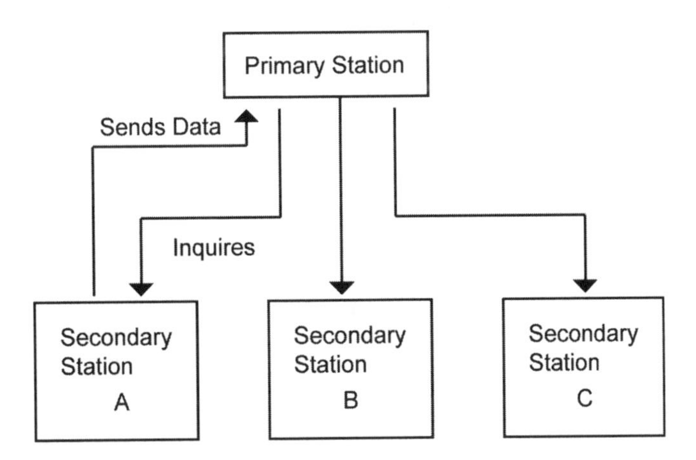

FIGURE 2.4 Functions of primary and secondary nodes.

Table 2.7 shows the four main configurations in which primary and secondary nodes can be configured.

TABLE 2.7 Configuration of Primary and Secondary Nodes

Configuration	Description
Point-to-point	Connects two nodes, primary and secondary.
Multipoint	Connects one primary node with multiple secondary nodes.

(continued)

TABLE 2.7 *(continued)*

Configuration	Description
Hub Go-ahead	Involves two channels, incoming and outgoing. The primary node uses the outgoing channel to communicate with the secondary nodes. The secondary node uses the inbound channel to send data to the primary node.
Loop	Connects a primary node to the first and last secondary nodes. The intermediate nodes pass messages to one another for communicating with the primary node.

Figure 2.5 displays the SDLC frame format.

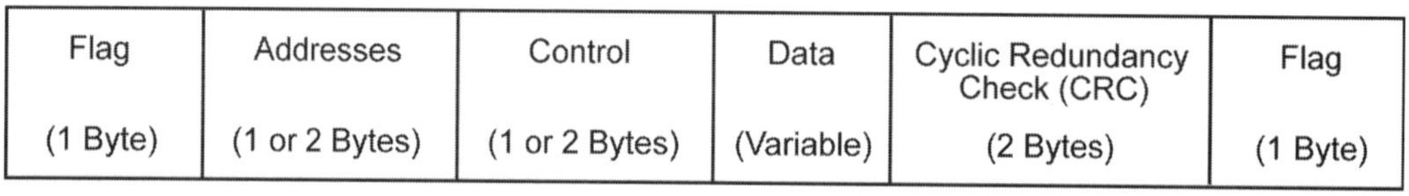

Flag	Addresses	Control	Data	Cyclic Redundancy Check (CRC)	Flag
(1 Byte)	(1 or 2 Bytes)	(1 or 2 Bytes)	(Variable)	(2 Bytes)	(1 Byte)

FIGURE 2.5 The frame format of the SDLC protocol.

Table 2.8 states the description of each field in the SDLC frame format.

TABLE 2.8 Fields of the SDLC Frame

Field	Description
Flag	Indicates the beginning or end of each SDLC frame.
Address	Contains the address of the secondary node involved in current communication.
Control	Uses three different formats: Information, Supervisory, and Unnumbered. These formats depend on type of SDLC frame used.
Data	Contains the data to be communicated and is of variable length.
CRC	Performs error detection during data transfer.

High-Level Data Link Control (HDLC)

HDLC is a multipurpose data link control protocol that can be used on both point-to-point and multipoint links. Like SDLC, it supports full-duplex transparent-mode transmission.

HDLC is similar to SDLC in that their frame formats are identical. In addition, both protocols support a full-duplex operation. Table 2.9 lists the differences between HDLC and SDLC.

TABLE 2.9 Differences Between SDLC and HDLC

SDLC	HDLC
Supports hub go-ahead and loop configurations.	Does not support hub go-ahead and loop configurations.
Does not have the 4-byte checksum option or CRC.	Contains 4-byte checksum option.
Supports only one transfer mode—the Normal Response Mode (NRM). In this mode, secondary nodes cannot communicate with the primary node until the primary node permits them.	Supports three transfer modes: NRM, Asynchronous Response Mode (ARM), and Asynchronous Balanced Mode (ABM). In ARM, secondary nodes can start communicating with the primary node without obtaining its permission. ABM introduces a combined node that can serve as a primary or secondary node with respect to the situation.

The HDLC frame is synchronous and relies on the Physical layer to provide a method of clocking and synchronizing transmission and receipt of frames. This protocol uses the technique of zero insertion/deletion, known as bit stuffing, to ensure that the bit pattern of the delimiter flag does not occur in the fields between flags. Figure 2.6 displays the details of an HDLC frame structure.

Flag	Address	Control	Information	FCS	Flag
(8 Bits)	(8 Bits)	(8 or 16 Bites)	(Variable)	(8 Bits)	(8 Bits)

FIGURE 2.6 The HDLC Frame Structure

Protocols of the Network Layer

IP is the primary protocol used on the Network layer. This protocol routes data packets from the source to the destination node in a network. Therefore, this protocol is also called a routing protocol. The IP protocol fragments data packets into datagrams as the data packets are transmitted to the destination node. Each datagram is 1500 bytes and is fragmented into smaller segments as it is transmitted. At the destination node, IP reassembles the fragmented datagrams to form the original data packet. In addition, IP identifies and detects errors that may occur during data transmission.

When data packets are fragmented into datagrams, each datagram has an associated header attached to it. This header contains vital information about the datagram, such as the source address, destination address, and protocol version. The header of a data packet is of 32 bits, which includes 20 bytes of fixed length fields. The remaining fields are of variable length. Figure 2.7 displays the header datagram of IP.

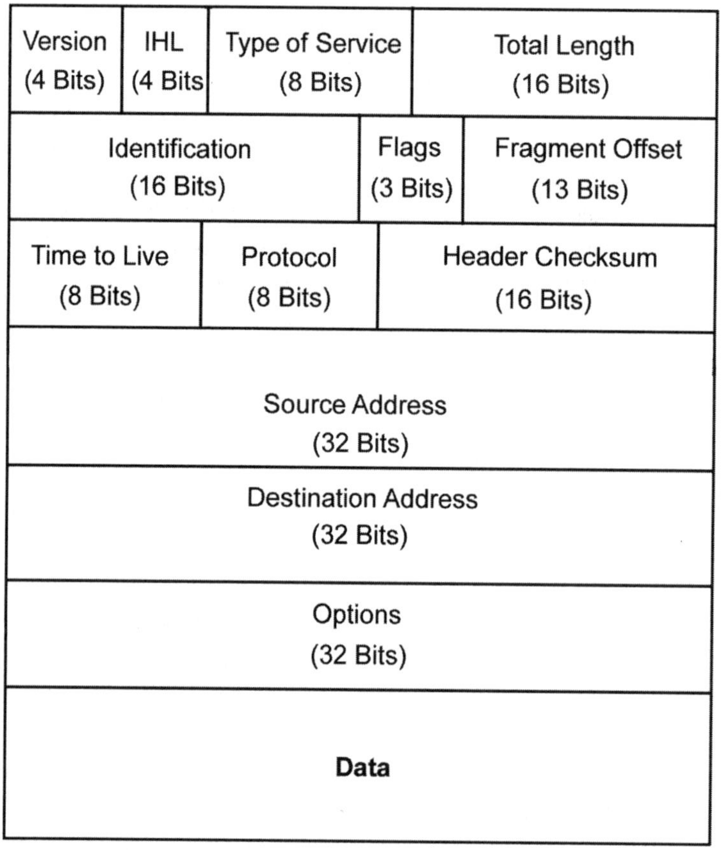

Version (4 Bits) | IHL (4 Bits | Type of Service (8 Bits) | Total Length (16 Bits)
Identification (16 Bits) | Flags (3 Bits) | Fragment Offset (13 Bits)
Time to Live (8 Bits) | Protocol (8 Bits) | Header Checksum (16 Bits)
Source Address (32 Bits)
Destination Address (32 Bits)
Options (32 Bits)
Data

FIGURE 2.7 The IP header datagram.

The fields of the IP header datagram are:

Version: Indicates the current version of the IP protocol. If the current version of IP is 4, it is called IPv4. The size of this field is 4 bits.

IHL: Indicates the length of the IP header. By default, the length of the IP header is 32 bits. The size of this field is 4 bits.

Type of Service: Indicates the manner in which the upper-layer protocol wants the datagram to be handled. Using this field, the priority level for a particular datagram can be specified. The size of this field is 8 bits.

Total Length: Specifies the total length of the data packet. The total length includes the length of the datagram header. The size of this field is 16 bits.

Identification: Specifies the identification number of the current datagram. This identification number enables reassembling of datagrams at the destination node. The size of this field is 16 bits.

Flags: Indicates the fragmentation and the total bits in the entire data packet. The size of this field is 3 bits. The two lower bits of the three bits control the fragmentation of the datagram. The first of these two bits indicates the possibility of the current packet being fragmented. If this bit is set to 1, it indicates that the data packet cannot be fragmented; if the bit is 0, the data packet can be fragmented. The second bit of the two lower order bits specifies whether the current packet is the last in a series of packets. If this bit is 0, it indicates that the current packet is the last of the series; if the bit is 1, there are more packets following the current packet. The third bit of the three bits is a high-order bit and is set to 0.

Fragment Offset: Indicates the exact location of the current datagram in the entire data packet. The size of this field is 13 bits.

Time-to-Live (TTL): Specifies the total time for which a datagram is active and valid. A counter is associated with this time, which gradually decrements to 0, at which point, the datagram is discarded and considered invalid. The size of this field is 8 bits.

Protocol: Specifies an integer value, which indicates that an upper-layer protocol has received the incoming data packets after the IP protocol processing is complete. For example, if the value in this field is 6, TCP will receive the processed incoming data packets.

Header Checksum: Detects errors in the header of the data packet. The size of this field is 16 bits.

Source Address: Specifies the address of the source host. The size of this field is 32 bits.

Destination Address: Specifies the address of the destination host. The size of this field is 32 bits.

Options: Provides additional information, such as security information or information related to a path adopted by the data packet. This field can specify the complete path that a data packet needs to follow. The size of this field is 32 bits.

Data: Consists of information related to the upper layers.

In addition to the header information, addressing schemes used at the Network layer are an important part of routing IP data packets.

Addressing at the Network Layer

An IP address is a unique 32-bit Layer 3 address that uniquely defines a host in the network and allows it to participate in a TCP/IP network. The IP address is a 32-bit series divided into four groups of eight bits (octets) each. This series is written in a decimal notation with numbers ranging from 0 to 255.

NOTE

The IP address is assigned by the network administrator and differs from a MAC address, which is allocated by the hardware manufacturer.

An IP address is made up of 32 bits. This address is divided into four bytes or octets. Table 2.10 shows a 32-bit IP address in a structured addressing scheme.

TABLE 2.10 The 32-bit IP Address

IP Address Notation	Address
Dotted decimal	192.168.12.1
Binary	11000000.10101000.00001100.00000001
Subnet mask in dotted decimal	255.255.255.0
Subnet mask in binary	11111111.11111111.11111111.00000000

Each network can again be subdivided into a number of subnets, depending on the requirement. Finally, each host within the same network or subnet should have a unique IP address, which identifies it in the entire network. This is also called the host address. In Table 2.10, the host address is 192.168.12.1. The network part of the IP address identifies the network to which the particular host belongs. In order to communicate, all hosts should have the same network address.

When a greater number of hosts and lesser number of networks are required, more bits are allocated to the hosts and less to the networks, and vice versa. For example, 192.168.2.4 is an IP address with a subnet mask of 255.255.0.0. The first two

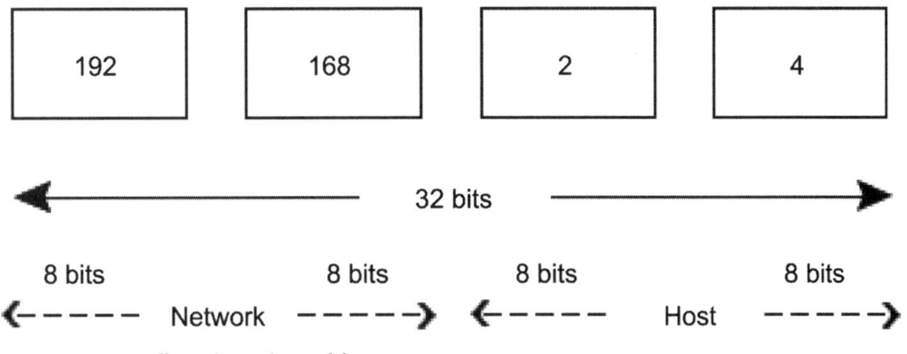

FIGURE 2.8 Allocation of IP addresses.

bytes or octets represent a network address, and the last two octets are the host address, as shown in Figure 2.8.

The number of bits assigned to the network and host portions depends upon the number of networks to be configured. In a public network, InterNIC assigns this network number. To identify the number of bits that determine the network portion, a subnet mask has to be used along with the IP address. The subnet mask determines the demarcation point between the network and the host portion of the IP address. This is explained later in the chapter.

Classes of IP Addressing

The requirement to allocate IP addresses to networks of varying sizes was addressed by introducing the concept of address classes. The hierarchical model of the IP addressing scheme gave rise to different classes of IP addressing. InterNIC assigns the classes of IP addresses to an internetwork with respect to the size of the network. This is to avoid any confusion during the allocation and distribution of IP addresses. There are five different address classes:

- Class A
- Class B
- Class C
- Class D
- Class E

Internetworks are divided into three sizes:

Big internetworks: Assigned a larger number of hosts and fewer networks. These organizations are allocated Class A address.

Medium internetworks: Assigned a requirement of hosts and subnetworks in between the big and small internetworks. These organizations are allocated Class B address.

Small internetworks: Assigned a smaller number of hosts and larger number of subnetworks. These organizations are allocated Class C address.

Table 2.11 lists the specifications and options associated with different classes.

TABLE 2.11 Class Requirement and Available Options

Class	Purpose	Maximum Networks	Maximum Hosts
A	Large organizations	127	16,777,214
B	Medium-size organizations	16,384	65,543
C	Small organizations	2,097,152	254

The Internet community defined a set of rules in the hierarchical IP addressing scheme. For addresses in Class A, the leading bits of the first octet should always start with 0. The leading bits should be 10 for Class B, 110 for Class C, 1110 for Class D, and 1111 for Class E. Table 2.12 shows the address ranges of different classes.

TABLE 2.12 Leading Bits and Address Ranges of Classes

Class	Leading Bit	Address - Range
A	0	1.0.0.0 -127.255.255.255
B	10	128.1.0.0 - 191.254.0.0
C	110	192.0.0.0 - 223.255.255.255
D	1110	224.0.0.0 - 239.255.255.255
E	1111	240.0.0.0 - 254.255.255.255

Class A

Class A addresses range from 1.0.0.0 to 126.0.0.0, where the first octet represents the network portion, and the last three octets represent the host. The Class A address format is used for large organizations with networks supporting a large number of end users. The maximum number of networks possible with Class A addressing is 127, and the maximum number of hosts per network number is 16,777,214. The highest order of the network bits is always the most significant and defines the class of the network. In the case of Class A networks, the highest order bit—the first bit of the first octet—is zero. Figure 2.9 depicts the Class A addressing format.

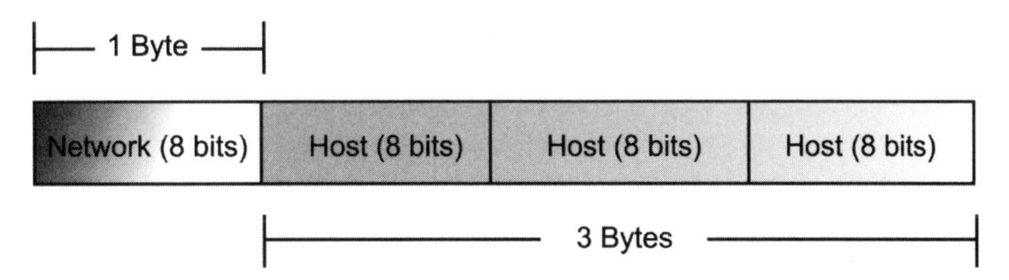

FIGURE 2.9 The addressing format of Class A.

Class B

Class B addresses range from 128.1.0.0 to 191.254.0.0, where the first two octets represent the network portion, and the other two octets represent the host. The Class B address format is used for networks of mid-sized organizations. The maximum number of networks possible with Class B addressing is 16,384, and the maximum number of hosts per network is 65,543. The highest order of the network bits is always 10. The first bit of the first octet is set to 1, and the second bit is set to 0. Figure 2.10 depicts the Class B addressing format.

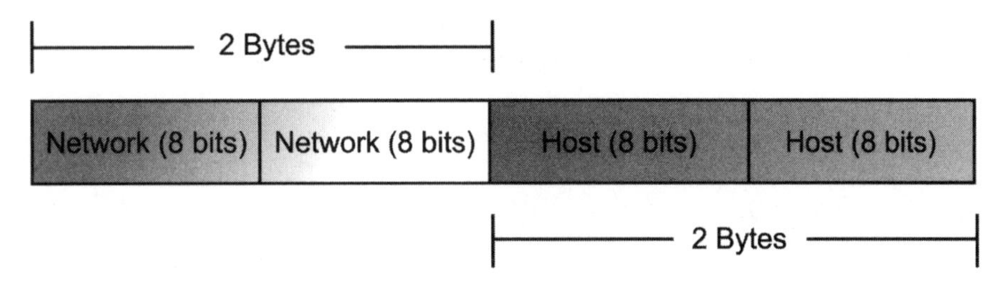

FIGURE 2.10 The addressing format of Class B.

Class C

Class C addresses range from 192.0.1.0 to 223.255.254.0, where the first three octets represent the network portion, and the last octet represents the host. The Class C addressing format is used for small organizations with networks supporting a large number of users. The maximum number of networks possible with Class C addressing is 2,097,152, and the maximum number of hosts per network number is 254. The highest order of the network bits is always 110. Figure 2.11 depicts the Class C addressing format.

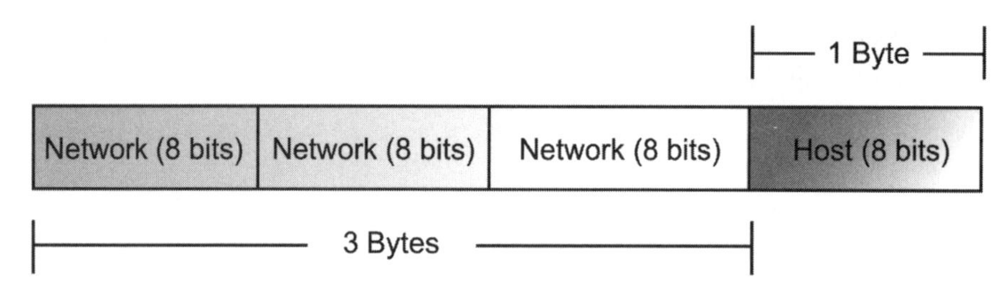

FIGURE 2.11 The addressing format of Class C.

Classes D and E

Unlike classes A, B, and C, classes D and E are not for commercial use. Class D addresses are used for multicast groups and range from 224.0.0.0 to 239.255.255.255. Class E addresses are used for experimental purposes and range from 240.0.0.0 and 254.255.255.255.

Table 2.13 shows a breakup of IP addresses of Class A, B, and C in binary format.

TABLE 2.13 IP Addresses in Binary Format

Network Number (Dotted decimal)	Network Number (Binary)
10.1.1.0	00001010.00000001.00000001.00000000 (Class A)
150.5.5.0	10010110.00000101.00000101.00000000 (Class B)
192.1.1.0	11000000.00000001.00000001.00000000 (Class C)

In Table 2.13, note that Class A address has 0 in the first bit of the first octet, Class B has 10 in the first two bits of the first octet, and Class C has 110 as the first three bits of the first octet. Table 2.14 lists the characteristics of all the five classes of IP addresses.

TABLE 2.14 Characteristics of Classes A, B, C, D, and E

Class	Format	Purpose	Leading Bit	Address Range	Maximum Networks	Maximum Hots
A	N.H.H.H	Large organizations	0	1.0.0.0 – 126.0.0.0	127	16,777,214
B	N.N.H.H	Medium-size organizations	10	128.1.0.0 – 191.254.0.0	16,384	65,543

TABLE 2.14 *(continued)*

Class	Format	Purpose	Leading Bit	Address Range	Maximum Networks	Maximum Hots
C	N.N.N.H	Small organizations	110	192.0.1.0 – 223.255.254.0	2,097,152	254
D	N/A	Multicast groups	1110	224.0.0.0 – 239.255.255.255	N/A	N/A
E	N/A	Experimental	1111	240.0.0.0 – 254.255.255.255	N/A	N/A

It is possible to connect networks with different classes of IP addresses. Figure 2.12 depicts a scenario in which the networks 10.1.1.0 (Class A), 150.5.5.0 (Class B), and 192.1.1.0 (Class C) are internetworked.

In Figure 2.12, the networks belong to three different classes of address: 10.1.1.0/8 (Class A), 150.5.5.0/16 (Class B), and 192.1.1.0/26 (Class C).

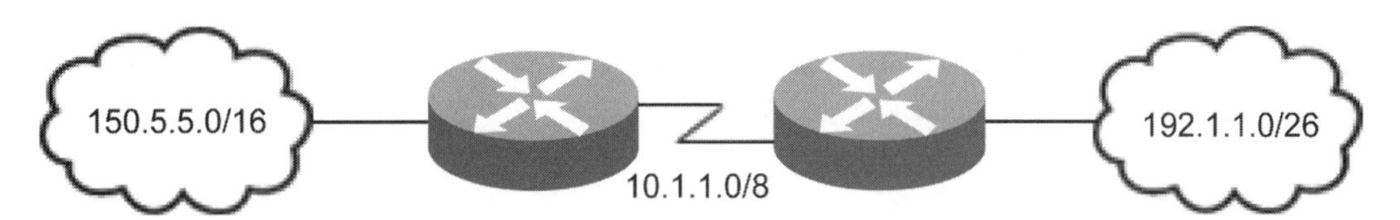

FIGURE 2.12 Internetworking of different classes of an IP network.

Routing at the Network Layer

The main function of the Network layer is to route data packets from the source to the destination nodes. IP performs this function by using the IP address of data packets. An IP address enables IP to identify the source and destination nodes. These source and destination nodes can be on different networks connected through routers.

For routing data packets on the appropriate path, the Network layer uses dynamic routing protocols, which contain routing software that decides the routes to be taken by each incoming data packet at specified intervals.

The Network layer uses the Internet Control Message Protocol (ICMP) and ICMP Router Discovery Protocol (IRDP) to route data packets on the appropriate path. In addition, protocols use the IP routing table, which specifies the destination address and the next-hop pair for the data packet. The next-hop pair specifies the address of an intermediate node, which is the next stop of the data packet from its current location to reach the specified destination address. The IP routing table does not specify the exact path of the data packet at the source node. It contains the address of the next intermediate node. The address of the next intermediate node is calculated by matching the destination address of the data packet with the address in the routing table of the current node. This type of routing process is known as *dynamic routing.*

Table 2.15 displays a sample routing table.

TABLE 2.15 Sample Routing Table

Destination Address	Next-Hop
36.10.0.0	45.32.22.1
79.3.0.0	35.32.10.9
17.12.3.0	45.32.22.2

From Table 2.15, you can infer that to reach subnet 10 of network 36, the next stop of the data packet is 45.32.22.1.

However, the IP routing table does not provide any mechanism to track whether data packets have reached the destination without any error, such as corruption of the data packet. To track errors related to data packets or routing failures, the Network layer uses ICMP.

ICMP

ICMP is a protocol used to track errors that corrupt data packets during transmission and routing failures that prevent data packets from reaching the destination node. In addition, ICMP reports transit errors to the source node.

To check for the destination node in a network, the ICMP protocol uses two commands:

ping: Performs the echo function that sends the data packet on a roundtrip between two hosts in a network. It provides details about packet forwarding and packet loss at each router and link in the path.

trace: Traces the path by sending router advertisement messages to identify the addresses of routers that are directly attached to the subnetworks, produces the command-line report output about each router that is crossed, and generates the Roundtrip Time (RTT) for each hop.

In addition, ICMP can also be used as a diagnostic tool to determine routing faults, security attacks, and misconfigurations within the network.

IRDP

IRDP uses router advertisements to identify addresses of routers that belong to directly attached subnets. Using IRDP, each router periodically broadcasts router advertisements to all hosts in a network. The hosts can use router advertisements to identify nearby routers and send requests to them.

Protocols of the Transport Layer

The Transport layer of the OSI model uses TCP and UDP protocols for transferring data through the network. TCP is responsible for the connection-oriented data transfer, whereas UDP provides connectionless data transfer.

TCP

TCP provides full-duplex and acknowledged services to the upper layers. In addition, TCP applies the flow-control mechanism to prevent a destination node from being flooded with data packets from a high-speed source node. TCP sends the data packets in the form of continuous and unstructured byte streams, which are uniquely identified by sequence numbers. Figure 2.13 displays a TCP packet format.

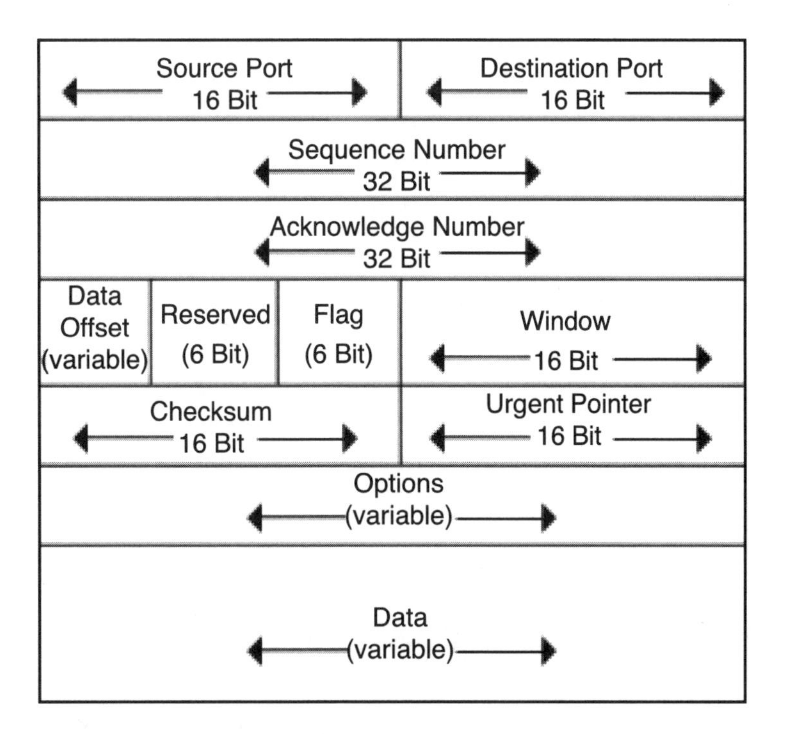

FIGURE 2.13 The header format of TCP.

The fields of the header format are:

Source Port: Identifies the source processes in the upper layers that receive TCP services. The size of this field is 16 bits.

Destination Port: Identifies destination processes in the upper layers that receive TCP services. The size of this field is 16 bits.

Sequence Number: Contains a sequence number assigned to the first byte of the data being sent. This sequence number uniquely identifies the data message being transmitted. The size of this field is 32 bits.

Acknowledgment Number: Contains the sequence number of the next byte of data being transmitted. The size of this field is 32 bits.

Data Offset: Indicates the number of 32-bit words in a TCP header. The size of this field is variable length.

Reserved: Is reserved for future use. The size of this field is 6 bits.

Flag: Specifies control information, such as establishing, terminating, and resetting a session. The size of this field is 6 bits.

Window: Specifies the size of the buffer space available at the source node to receive incoming data packets. The size of this field is variable length.

Checksum: Identifies if a data packet was corrupted during transmission. The size of this field is 16 bits.

Urgent Pointer: Contains a pointer to the last byte of any urgent data that is being transmitted to the destination node. The size of this field is 16 bits.

Options: Contains additional information about the data being transmitted to the destination node. The size of this field is variable length.

Data: Contains information about the upper layers. The size of this field is variable.

UDP

UDP is a simpler protocol than TCP, as it contains only four fields in its header. UDP is used where the source and the destination nodes do not require reliable and connection-oriented service. The fields of the UDP packet are:

Source Port: Identifies the source processes in the upper layers that receive UDP services. The size of the field is 16 bits.

Destination Port: Identifies the destination processes in the upper layers that receive UDP services. The size of the field is 16 bits.

Length: Specifies the total length of the UDP datagram, which includes a header and the user data. The size of the field is 16 bits.

Checksum: Identifies if a data packet was corrupted during transmission. A value of 0 in this field indicates that the checksum has not been calculated and used. This field is optional; its size is 16 bits.

Protocols of the Session, Presentation, and Application Layers

The Session, Presentation, and Application layers of the OSI model are concerned with the applications accessed by end users. Therefore, protocols supported by these layers perform several application-oriented functions, such as network management, file transfer and management, e-mail, and distributed network services.

Some protocols used by the Session, Presentation, and Application layers to ensure effective delivery of data packets to the destination node are:

- Routing Information Protocol
- Interior Gateway Routing Protocol
- Enhanced Interior Gateway Routing Protocol

Routing Information Protocol

The Routing Information Protocol (RIP) is suitable for small, homogenous networks and is relatively easy to implement. It has certain limitations that render it unsuitable for large and complex networks:

- Does not support classless routing (leads to waste of address space)
- Cannot be used in networks that require CIDR and VLSM
- Is not suitable for networks that require more than 15 hops
- Convergence is slow and is prone to routing loops
- Uses broadcasts for routing updates causing an increase in network traffic

RIP uses hop-count for its metric. A hop-count of 16 signifies an unreachable network. RIP sends request and response messages during routing updates. RIP uses UDP port 520 for all messages. RIP sends out request messages, using broadcasts when the RIP process begins.

Neighboring routers send response messages containing their routing tables, after receiving updates. On receiving the response message from its neighbor, a router checks to see whether the update is new or not. If the update is found to be new, it is entered in the routing table along with the address of the advertising router. If the entry already exists for that network, the updates are ignored unless one is received with a lower hop-count.

A hop count between 0 and 15 is considered valid. If the hop count is 16, it would indicate a networkthat is unreachable.

RIP uses the major classful network number for route summarization, because it does not carry subnet mask routing information. Subnet masks enable efficient use of the IP addressing scheme. The lack of subnet mask information carried by RIP means that a router should assume that the subnet mask it has been configured with is effective for all the subnets.

When multiple paths can be used to reach a destination network, the router makes a choice based on the reliability of the routing information source, known as its *administrative distance*. The higher the value of administrative distance, the lower its reliability. RIP has an administrative distance of 120.

RIP is defined in Request For Comment (RFC) 1058 and 1723.

The fields of the RIP header format are:

Command: Contains an integer with a value of 1 or 2. Value 1 indicates a request command. This command requests the responding network system to send all or part of its routing table to the destination node. Value 2 indicates a response command that responds to the request command and includes all or part of the routing table of the responding network system.

Version: Specifies the version of RIP, which can be either 1 or 2.

Address family identifier: Identifies the type of address family being used for the current RIP implementation. The size of this field is 16 bits.

Metric: Contains hop counts that a data packet needs to traverse before reaching the destination node. However, the maximum hop limit is 15. A destination that requires more than 15 hops is called unreachable.

In addition, RIP uses certain timers that regulate its performance. For example, the RIP routing update timer is set at 30 seconds, which ensures that every router sends its entire routing table to its neighbors.

Interior Gateway Routing Protocol

Interior Gateway Routing Protocol (IGRP) is a distance vector protocol, which is suitable for medium and large-sized networks. IGRP advertises the entire or a part of the routing table to neighbors via broadcasts. The metrics used in IGRP are a combination of delay, bandwidth, reliability, load, and Maximum Transmission Unit (MTU). IGRP has features such as split horizon, split horizon with poison reverse, route poisoning, hold-down timers, triggered updates, and count to infinity.

Being a distance vector protocol, IGRP advertises the entire or a part of the routing table to its neighbors every 90 seconds via broadcasts. If a route remains unreachable even after three consecutive updates, it is declared inaccessible. However, the inaccessible route is not removed from the routing table. It is only deleted after it fails to respond to seven consecutive routing updates.

The routes advertised by IGRP can be classified into:

Interior routes: For subnets connected to the interface of the router.

System routes: For networks from routers within the same AS. These networks do not include subnetworks.

Exterior routes: From networks outside ASs. These routes are used while defining a default gateway using the Gateway of Last Resort.

The message format of IGRP is depicted in Figure 2.14.

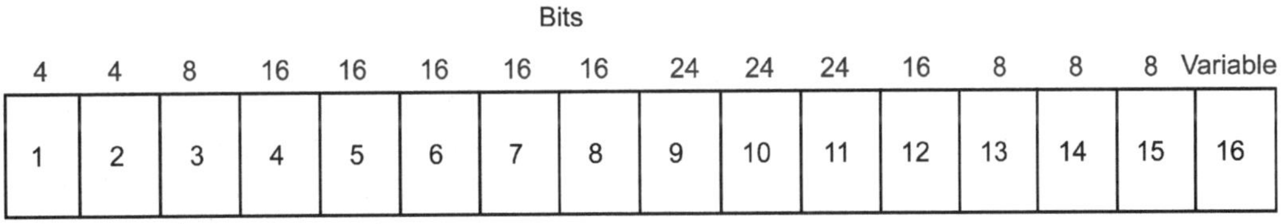

Bits

4	4	8	16	16	16	16	16	24	24	24	16	8	8	8	Variable
1	2	3	4	5	6	7	8	9	10	11	12	13	14	15	16

1- Version	9- Destination
2- opcode	10- Delay
3- Edition	11- Bandwidth
4- AS Number	12-MTU
5- No. of Interior routes	13-Reliability
6- No. of System routes	14- Load
7- No. of exterior routes	15- Hop Count
8- Checksum	16- Route Entries

FIGURE 2.14 The message format of IGRP.

The descriptions of the various fields of the message packet formats are listed in Table 2.16.

TABLE 2.16 IGRP Message Format Fields

Field	Description
Version	Sets the value to 0 x 01 always.
Opcode	Sets the value to 0 x 01 for a request and 0 x 02 for an update.

(continued)

TABLE 2.16 *(continued)*

Field	Description
Edition	Increments the counter by the sender such that the receiving router knows that it always keeps the latest version of the routing updates.
Autonomous System Number (ASN)	Specifies the IGRP process ID number.
Number of Interior Routes	Indicates the number of routes that are subnets of a directly connected network.
Number of System Routes	Indicates the number of routes for networks from routers within the same AS.
Number of Exterior Routes	Indicates the number of routes from networks outside the ASs.
Checksum	Is calculated from an algorithm based on the header and the entries.
Destination	Indicates the destination network.
Delay	Expresses delay as a multiple of 10 microseconds.
Bandwidth	Defines IGRP bandwidth.
MTU	Expresses MTU in bytes.
Reliability	Specifies error rates of the link.
Load	Specifies load on the link.
Hop Count	Is a number between 0 x 00 and 0 x FF

Enhanced Interior Gateway Routing Protocol

EIGRP is a Cisco proprietary routing protocol. This means that EIGRP does not support routers manufactured by any vendor other than Cisco. The characteristics of EIGRP are:

- Supports VLSM
- Achieves faster convergence than RIP
- Supports multiple routed protocols such as IP, IPX, and AppleTalk
- Requires less bandwidth
- Uses Diffusing Update Algorithm (DUAL)
- Performs load balancing over paths having unequal paths
- Finds alternate routes

EIGRP packets are encapsulated in IP with the protocol field set to 88. The source IP address of the packet is the interface generating this packet. The destination packet can be multicast or unicast depending on the packet type.

The maximum length of the packet is determined by the MTU. The value of MTU varies with respect to the technologies used; it has a default value of 1500 for Ethernet. The different fields of the EIGRP packet are shown in Figure 2.15.

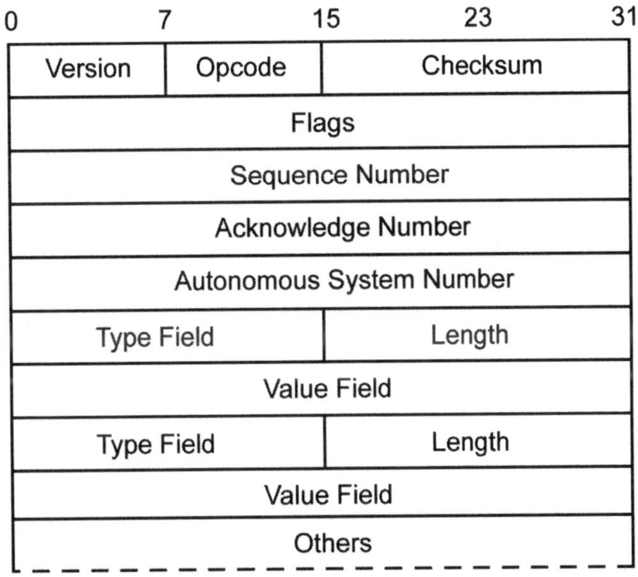

0 7 15 23 31

Version | Opcode | Checksum

Flags

Sequence Number

Acknowledge Number

Autonomous System Number

Type Field | Length

Value Field

Type Field | Length

Value Field

Others

FIGURE 2.15 The packet format of EIGRP.

The functions of the key fields of the EIGRP packet are:

Version: Indicates the version of the EIGRP process.

Opcode: Specifies the type of the EIGRP packet. Table 2.17 shows the EIGRP packet's Opcode types.

TABLE 2.17 Opcode Types

Opcode	Type
1	Update
3	Query
4	Reply
5	Hello
6	IPX SAP

Checksum: Specifies the checksum of the entire EIGRP packet, excluding the IP header.

Flags: Shows the flag type. An EIGRP packet has two flags. The first bit is called the init bit and is used in a new neighbor relationship. The second bit is the conditional receive bit and is used in a proprietary reliable multicast algorithm.

Sequence Number: Is a 32-bit sequence number. It is used to send messages reliably using RTP.

Acknowledgment Number: Is used to send messages reliably using RTP. It uses a 32-bit sequence number heard from a neighbor to which a packet has been sent.

ASN: Identifies the EIGRP process that is sending the packet. The destination to which the EIGRP packet is being sent will process the packet only if it has an EIGRP routing process with the same number; otherwise, the packet will be rejected.

Type and Length Value (TLV) Field: Follows the EIGRP header. Table 2.18 lists the protocol-specific TLVs.

TABLE 2.18 Protocol-specific TLV Types

Number	TLV Types
	General TLV types
0 x 0001	EIGRP parameters (hello/ hold-time)
0 x 0003	Sequence
0 x 0004	Software version
0 x 0005	Next multicast sequence
	TLV Types Specific to IP
0 x 0102	IP internal routes
0 x 0103	IP external routes
	TLV Types Specific to AppleTalk
0 x 0202	AppleTalk internal routes
0 x 0203	AppleTalk external routes
0 x 0204	AppleTalk cable configuration
	TLV Types Specific to IPX
0 x 0302	IPX internal routes
0 x 0303	IPX external routes

SUMMARY

In this chapter, you learned about protocol types and behaviors. You also learned about the characteristics of protocols residing on the Data-link, Transport, Network, Session, Presentation, and Application layers. In the next chapter, we will move on to the troubleshooting techniques for TCP/IP networks.

POINTS TO REMEMBER

- Connection-oriented protocols ensure that the sequence of data packets is the same at the source and destination nodes. These protocols are also called reliable protocols.
- Connectionless protocols do not send data packets in the same sequence at the source and destination nodes. These protocols are called unreliable protocols.
- Examples of connection-oriented protocols are Type 2 service of IEEE 802.2, ATM, TP0, TP1, TP2, TP3, and TP4, and TCP.
- Examples of connectionless protocols are Type 1 and 3 services of IEEE 802.2, CLNP, IP, and FST.
- The Data-link layer uses three types of protocols: PPP, SDLC, and HDLC.
- The primary protocol used on the Network layer is IP, which routes data packets from the source to the destination node in a network.
- The Transport layer of the OSI model uses TCP and UDP protocols for transferring data through the network. TCP is responsible for the connection-oriented data transfer, whereas UDP provides connectionless data transfer.
- The Session, Presentation, and Application layers are concerned with the applications accessed by end users. Examples of protocols residing on these layers are RIP, IGRP, and EIGRP.

3 Diagnostic Mechanisms

IN THIS CHAPTER

- Cisco Network Management Tools
- Cisco Diagnostic Commands

Managing networks is all about keeping the network running smoothly. Problems need to be detected and diagnosed proactively to prevent or at least minimize network disruptions. Cisco offers a host of management tools and diagnostic commands to provide diagnostic capabilities to network administrators. Cisco network management tools, such as CiscoWorks™ and CiscoView™, enable you to monitor the network continuously for performance, connectivity, resilience, security, and accessibility.

These tools can be selected as per the organization's specific requirements. For example, CiscoWorks is available for LAN management and IP telephony networks. You can select the most appropriate management tool for a network type and the most appropriate command options for a particular situation. The Cisco IOS commands include show, debug, trace, ping, and cdp. These commands run in different modes on the router console and provide a wide range of options that can be selected to perform specific diagnostic operations.

CISCO NETWORK MANAGEMENT TOOLS

The Cisco network management tools have been developed on the Cisco AVVID architecture guidelines. These tools provide centralized monitoring and management capabilities to enterprises working in different types of network environments such as WAN, wireless and mobile networks, or VLAN setups.

The basic parameters that determine network service levels are availability, resilience, responsiveness, and security. Using Cisco's network management tools, you can diagnose the specific causes of failure or disruptions on the network. These tools can be used independently or in combination, using a modular approach to meet the diagnostic requirements of the network. For example, CiscoView can be used as a diagnostic tool for the WAN management solutions provided by CiscoWorks.

CiscoView

CiscoView is a device management tool that uses a Graphic User Interface (GUI) to display the status of Cisco devices on the network. It provides real-time information about the status and configuration of Cisco devices on the network. CiscoView integrates with various other Cisco solutions such as CiscoWorks to provide centralized management of Cisco devices. The color codes and schemes of CiscoView enable you to view problems within the network, as well as their severity, through a single console.

The tool has been designed to diagnose and troubleshoot any errors in the network. Its major advantage is that it provides all the information about the network on a single interface, enabling you to manage large networks from a central location. This reduces the chances of network disruptions and increases system performance, because the entire network is constantly monitored.

CiscoWorks

CiscoWorks is a comprehensive application for managing networks of different sizes and types. It constitutes a family of products including LAN and WAN management solutions, IP telephony network management solutions, and mobile network management. These products manage various aspects of the network and provide troubleshooting capabilities.

The CiscoWorks suite is available in four versions, with varied capabilities and functionality:

CiscoWorks2000: A Web-based application suite that provides platform independent management capabilities for networks.

CiscoWorks Blue: A suite of applications used for managing highly complex networks for IBM SNA and IP environments.

CiscoWorks Switched Internetwork Solution (CWSI): A management application for switched networks that provides monitoring and analysis features. It can be integrated with SNMP management platforms, such as SunNet Manager, HP OpenView, and NetView. It also consists of LAN management applications such as VLAN Director, Traffic Director, and CiscoView.

CiscoWorks Windows: A management product for PC-based networks that provides network management capabilities for small and medium-sized networks.

Functions that can be performed using CiscoWorks include:

Device management: CiscoWorks allows you to create a database of the network inventory consisting of hardware, software, release levels of operation components, and the individuals who need to manage the inventory.

Device monitoring: Using this application, you can monitor the status of network devices. You can configure the interval after which the device state should to be checked. This information is logged using the Log Manager.

Security management: Authorizing user access can be easily configured for CiscoWorks in order to enhance security.

Inventory management: CiscoWorks provides a Sybase database, which can be used to store the status of software and hardware devices. The device information in the database is sorted by platform and software image. It invokes the Device Software Manager, which updates the status of the specific device.

Show commands: These commands display configuration information along with version details and interface states. They can be executed from the CiscoWorks console.

Configuration management: CiscoWorks simplifies the analysis of configuration files on local and remote Cisco system devices. You can compare the configuration files in the database or the currently active configuration running on the device with the configuration when the last database-to-device command was executed.

Configuration snap-in management: CiscoWorks supports the Global Command Scheduler application, which you can use to execute system commands on a device or a group of devices.

Traffic Director RMON

The Traffic Director Remote Monitor is a traffic management solution for switched networks. It performs traffic analysis and performance management functions for the embedded RMON agent within Catalyst switches and standalone Cisco SwitchProbe products. It determines the traffic utilization, broadcast levels, error rates, and number of collisions on particular ports. To persistently monitor the performance of Catalyst switches, you can configure predefined thresholds that generate alerts each time the threshold limit is exceeded.

Traffic Director performs advanced monitoring and troubleshooting functions and allows multilayer traffic analysis, threshold alerts, and remote packet capture. The thresholds are configured for Management Information Base (MIB) variables within the RMON agent. If these thresholds are violated, traps are sent to the required station, and notifications are sent to the technical support staff about the problem.

Traffic Director is an enterprise-level solution that also performs the following functions:

- Trend and pattern analysis of network traffic
- Protocol-level troubleshooting
- Trend reporting based on history data
- Configuration of thresholds for generating alerts

This application has been developed for providing high-end support for Catalyst LAN switches using the Cisco IOS RMON agents. A switch is managed by the Traffic Director as a special device with an embedded RMON (containing statistics, history, alarms, event-related tasks, and information) and Switched Port Analyzer (SPAN) that provides continuous support to RMON and RMON2 agents for all ports on the switch. In addition, this application can simultaneously recognize four SwitchProbes on a Fast Ethernet trunk or server link. This information can be integrated with traffic displays and viewed simultaneously for enhanced SwitchProbe management.

In addition, Traffic Director can manage VLANs and provide SQL-based trend reporting capabilities. You can use these trend reports to study traffic patterns and determine the busiest devices on the network. Using this information, you can perform load balancing and tuning tasks for enhancing performance. This application is highly comprehensive and provides a number of capabilities:

Seven-layer traffic analysis: You can monitor network traffic for all seven layers of the network. The RMON agent and SwitchProbe display network traffic for the entire enterprise from the Network, Data-link, Transport, and Application layers. It provides a multilayer traffic summary that enables you to assess network load and protocol distribution. You can view detailed information about each segment, ring, switch, or port. Using the real-time capabilities of the application, you can also perform high-end diagnostic functions for hosts, connections, or packets.

Protocol analysis: Traffic Director provides seven-layer decodes for the AppleTalk, DECnet, IP, ISO, Novell, SNA, Sun, NFS, Banyan VINES, and XNS protocol suites. As a result, Traffic Director provides centralized diagnostic and troubleshooting capabilities. The data is stored in a sniffer format, which is available for further analysis of protocol-related problems.

Distributed polling and threshold configuration: The Resource Monitor tool available with Traffic Director enables polling and monitoring of SNMP thresholds for large and remote networks.

VLAN Monitoring: You can analyze the activities performed within the VLAN setup along with trunk links, LAN segments, rings, and switch ports. The application provides a breakup of the statistics for each VLAN, which enables a drill-down analysis of network activities and events at the lowest level.

It also provides information pertaining to utilization, broadcast, multicast, and error rates for VLANs.

VLAN Director Switch Management

The VLAN Director Switch Management application provides management capabilities for VLAN ports. This application simplifies VLAN port assignment in the network and enables:

- Physical network representation and configuration management
- Access to device-specific configuration information
- Discrepancy reports in case of configuration conflicts
- Identification of erroneous device configurations for system level VLANs and troubleshooting
- Fast detection of switch port changes in VLANs
- Security implementation through user authentication and write protection

The GUI of the VLAN Director provides mapping capabilities for configuring logically defined workgroups. The VLAN Director provides drag-and-drop features for easy port configuration when configuring users for VLANs. In addition, it can be integrated with common SNMP management platforms such as HP OpenView™, Sun Solaris™, and Tivoli TME 10™ to optimize resource utilization and enhance reporting capabilities.

While VLANs enhance network manageability by segregating the network into logical segments with respect to network users and functionality, the VLAN Director makes it easy to configure and manage VLANs in large networks. It automates the configuration of interswitch VLAN links, reducing the errors that may occur during manual configuration. The interface makes administration of VLANs easier, especially when you need to add, remove, or modify VLANs.

WAN Manager

The Cisco WAN Manager (CWM) is a multi-protocol network management application based on SNMP. As the name suggests, the application has been designed for WAN management. CWM can integrate with other Cisco applications, such as Cisco Info Center and Cisco Provisioning Center. In addition, CWM supports the management of Cisco BPX Service Expansion Shelf (SES) and Cisco Advanced ATM Multiservice Portfolio (AAMP).

CWM is designed to provide exhaustive service management capabilities and process automation, and as a result, it ensures simplified management of complex networks. CWM is usually associated with service provider networks that demand management of high-level connections and a variety of services. This task becomes highly cumbersome when it needs to be done on a large scale, as in the case of WANs.

Using CWM, you can easily manage connections, detect faults, configure devices, and track network statistics. The network statistics can be stored in the SQL database and integrated with the currently active network management and operating systems. The functions performed by CWM are:

Fault management: CWM can detect, isolate, and rectify network faults. It can also generate reports about faulty network services. The fault management function of CWM correlates several services such as availability, reliability, survival, quality assurance, and alarms. These services combine to provide a comprehensive fault management solution.

Configuration management: CWM enables you to configure and control network elements, identify resources, collect information about resources such as CPU and memory, and manage connectivity between network elements. This involves monitoring the state of network elements, services, and resources.

Performance management: CWM enables performance management by collecting and reporting trends in network elements and services, including tasks such as quality assurance, management, control, and analysis.

Account management: This feature of CWM enables data collection for measuring the utilization of network resources and services. It also provides enterprise-control capabilities and allows management of the flow of funds within the enterprise, including tariff or pricing and usage measurement.

Security management: CWM detects and prevents unauthorized access to network resources. In addition, it recovers network elements in the event of a security breach.

Planning, modeling, and analysis: This is a management level function of CWM, which can anticipate potential network performance bottlenecks. This function includes simulation of networks and management systems, inventory, bandwidth capacity, usage, and cost analysis. Based on the data collected for these parameters, future network usage patterns can be defined.

CWM consists of various tools that perform the above-mentioned functions. For example, the network administrator provides information such as the node name and IP address of the node in your network. Other tools available with CWM are:

Statistics Collection Manager (SCM): Collects statistical data from the network nodes at predefined intervals. This data is stored in the database and is used for analysis. The SCM enables you to define network objects such as connections, ports, and trunks. In addition, it provides extensive error handling capabilities.

Informix Reporting Application: Handles history or trend data through the SQL architecture because of the large volume of data in the database and the error-prone process of retrieving the data using SNMP. The statistics reports generated by CWM are available through the Informix Wingz Report application. Based on this application, CWM can provide data on the rate of node use. The reports are generated with respect to certain predefined criteria, which can be a network object or report interval.

NetSys Network Management Suite

NetSys is a simulation tool that provides network planners, analysts, and managers with planning and analytical capabilities. This tool is especially useful when a new network design is being considered. NetSys is also used to display, debug, and validate the network configuration.

There are two versions of NetSys that cater to different network platforms and adopt specific functionality for simulation and planning: NetSys Baseliner (for Windows NT) and the NetSys SLM Suite.

NetSys Baseliner 4.0

The NetSys Baseliner is a tool that diagnoses network problems by simulating the current configuration of the network. It instantly creates a network model with the current network configuration and detects potential problems that might occur in the network. Based on this diagnosis, you can change the network configurations.

The Connectivity Baseliner creates a network map, while the Connectivity Solver analyzes the environment for various network device failures and their impact on performance, connectivity, and process flow.

This tool gathers the entire configuration data from a live network and uses these configurations to process Cisco IOS commands and build a network model. This model is then analyzed for errors and topology-related issues. Reports are generated, which can be used to reconfigure the problem areas in the model. After the configuration changes are tested, they can be applied to the live network.

NetSys SLM Suite

The NetSys SLM suite is a policy-based network management application, which establishes service levels while defining, monitoring, and assessing network connectivity, security, and performance policies.

The product consists of four modules that perform a specific set of functions:

Connectivity Manager: Monitors the existing network configuration and analyzes data for the availability of key network services. You can configure service levels for checking connectivity, reliability, and security of services. The View, Isolate, Solve, Test, and Apply (VISTA) methodology is adopted for diagnosing problems.

Performance Manager: Works in conjunction with the Connectivity Manager module to check the performance standards of network services. Using this module, you can define, monitor, and determine the optimum performance standards. Using the modeling feature of this module, you can develop accurate network models and then define performance standards during simulation.

LAN Service Manager: Provides LAN switching topology views for the Connectivity Manager module. This provides an integrated view of router/LAN switching networks and traffic paths. The LAN Service Manager monitors the LAN switch domain functionality and helps enhance the spanning tree configuration settings.

WAN Service Manager: Provides WAN switching analysis for SLM. The WAN Service Manager includes features such as integrated Layer 2/Layer 3 topologies and automated integrity checking. In addition, it provides a "What If" analysis capability for the WAN setup.

The modules interoperate and integrate depending on the requirements of the planners and analysts.

CISCO DIAGNOSTIC COMMANDS

Diagnostic commands are built into Cisco IOS and provide information for diagnosing network problems. Some of the diagnostic commands are basic commands used for determining the version of Cisco devices and their configuration. Depending on the problem, this information may be very important for troubleshooting. Usually, troubleshooting is initiated by using these commands. Such functions are performed by the show commands. A more advanced command available with Cisco IOS is the debug command. It provides drill-down information about various network problems. The data collected by these commands is so vast that it adversely impacts the performance of the router. Cisco recommends that this command be used only for a specific purpose.

The ping and trace commands are used to identify connectivity and accessibility problems. Cisco supports these commands for AppleTalk and IPX environments. The Cisco Discovery Protocol (CDP) is a Cisco-proprietary protocol used for building network maps to manage the network. It provides valuable information about Cisco devices at the lower layers of the OSI model. You will need to decide the specific commands and options that help diagnose the network problem.

The Show Command

The show command can be executed in user and privilege modes. Cisco IOS provides a range of show commands that display information about the rate of utilization of router resources, network interface status, and router configurations. You can isolate problems and determine the exact cause of performance slowdown or failure using show commands.

The Show Version Command

The show version command provides the IOS version and its internal name. Using the internal name and IOS capabilities, it displays the hardware configuration such as, processor type, memory size, and existing controllers. In addition, it displays other nonstandard software options such as the bootstrap software in use. It also displays the system uptime, the status of the system when it was last started, and the name of the image file. This information is useful if the router crashes and has to be reloaded, because the command will display the reason for router reload. At the end of the output of the show version command, the configregister value is displayed in hexadecimal format. The output of the show version command is displayed in Figure 3.1.

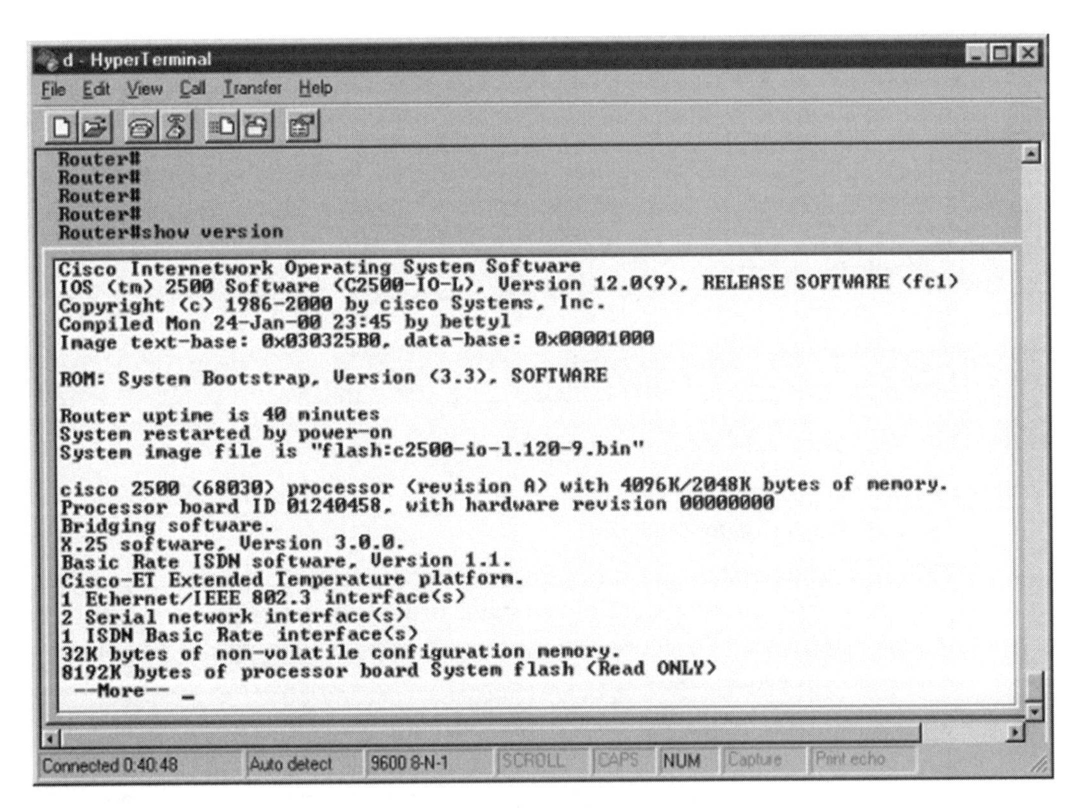

FIGURE 3.1 The output of the show version diagnostic command.

The Show Startup-config Command

The show startup-config command displays the router configuration stored in the NVRAM. This information is useful if the router configuration is changed during the session, and you need to determine the configuration of the router during bootup. The output of the show startup-config command is displayed in Figure 3.2.

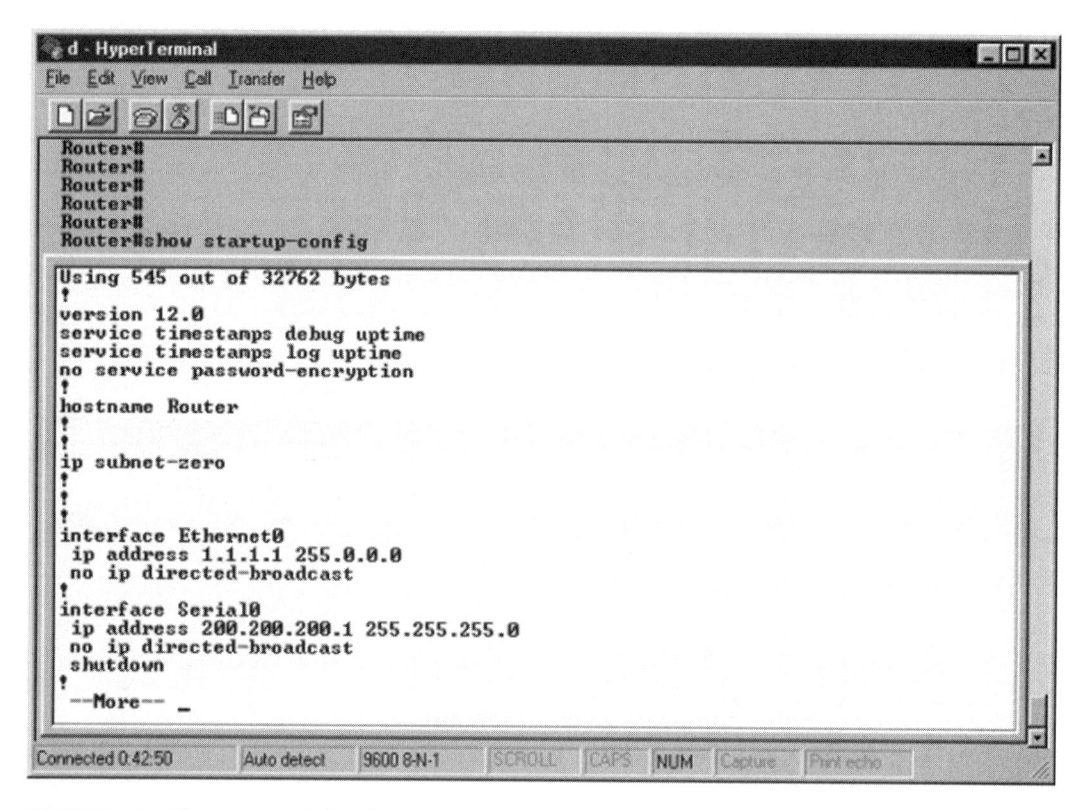

FIGURE 3.2 The output of the show startup-config diagnostic command.

The Show Running-config Command

The show running-config command displays the currently active router configuration. The show running-config command is used to isolate problems with the router or the reasons for a crash. The output of the show running-config command is displayed in Figure 3.3.

The Show Interfaces Command

The show interfaces command displays the status of all the interfaces configured for a router or access server. The output of the show interfaces command depends on the version and type of router being used. The output of this command is displayed in Figure 3.4.

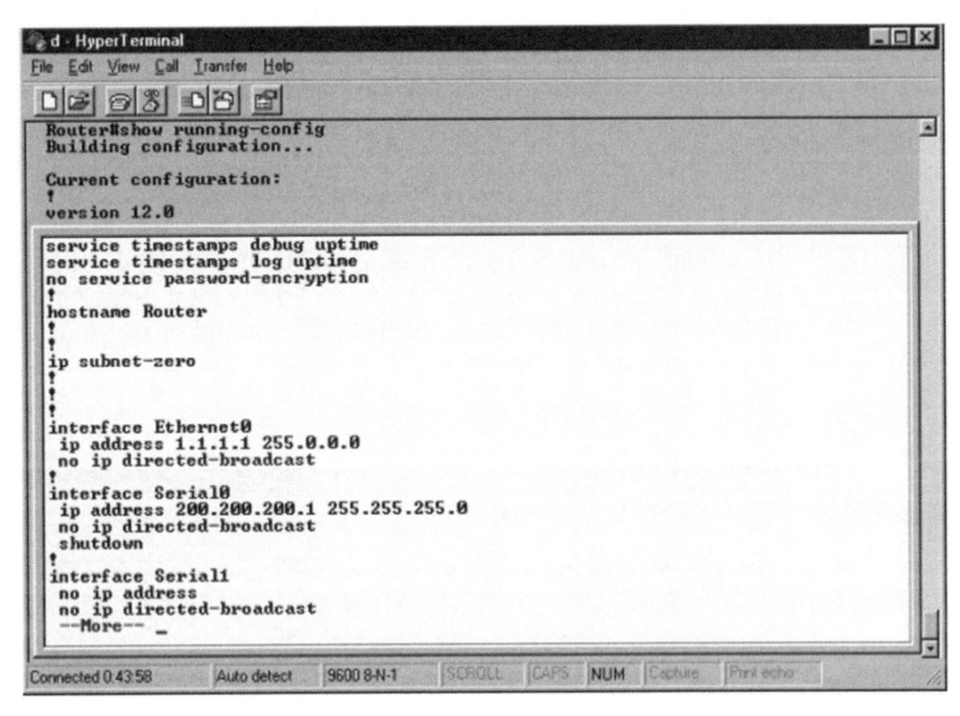

FIGURE 3.3 The output of the show running-config diagnostic command.

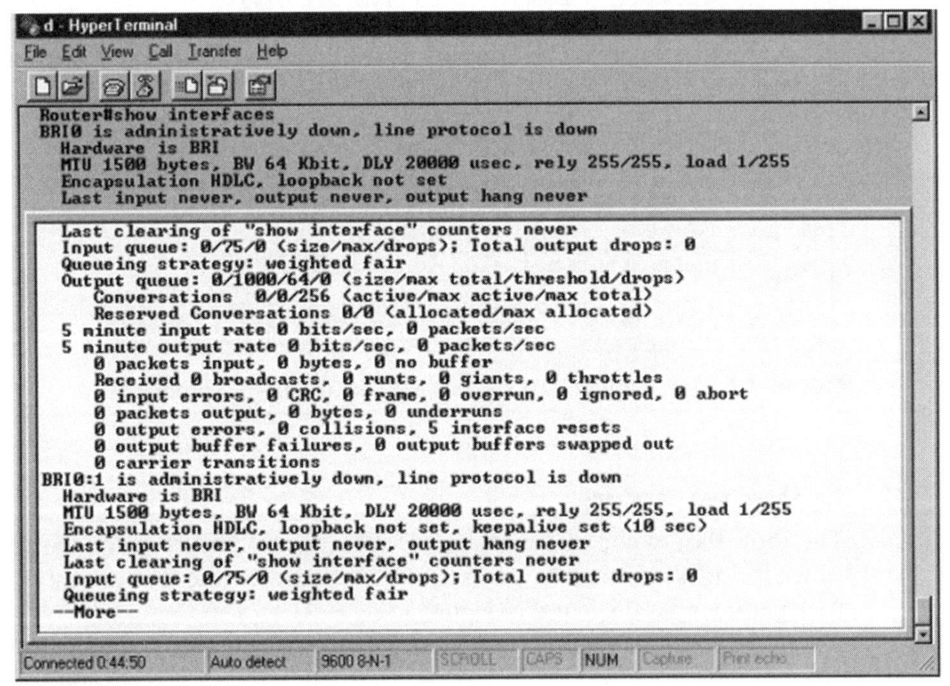

FIGURE 3.4 The output of the show interfaces diagnostic command.

The Show Controllers Command

The show controllers command displays the interface card controller statistics. Depending upon the type of interface of the network, the output displays various details including the microcode of the card. For example, the show controllers cxbus command is used with Cisco 7x00 series. In addition, it provides statistics about the Channel Interface Processor (CIP) including the hardware and microcode version, CPU utilization, free and total static RAM (SRAM), and the total free and dynamic RAM (DRAM). This information is used by Cisco technical support to determine the problems related to interface controllers. The output of the show controllers command is displayed in Figure 3.5.

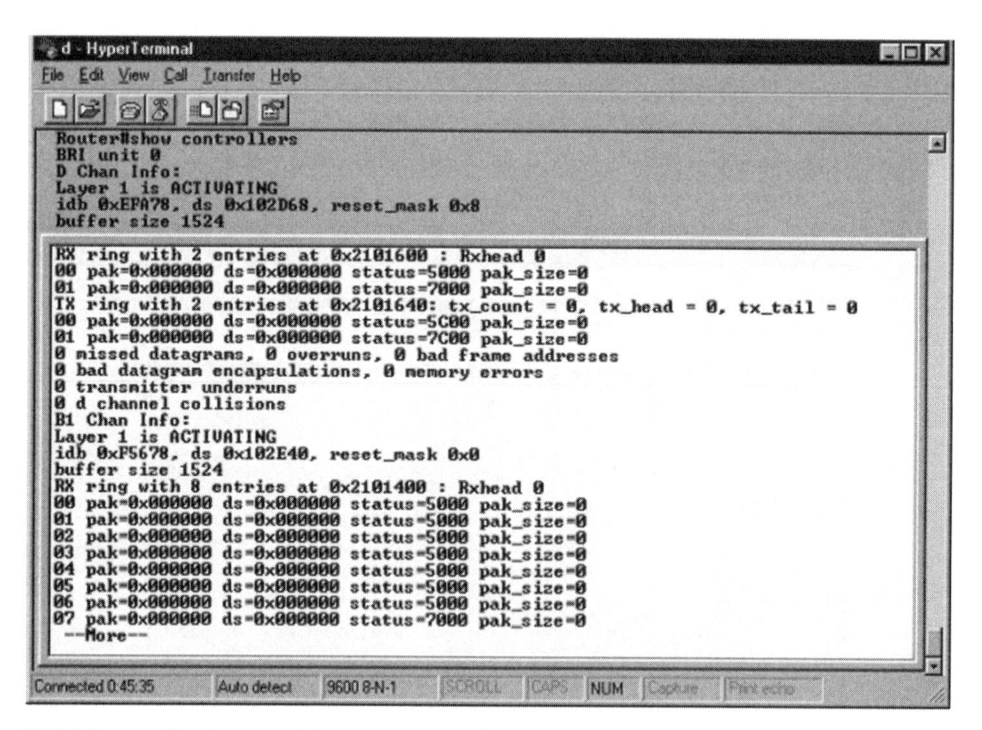

FIGURE 3.5 The output of the show controllers diagnostic command.

The Show Flash Command

The show flash command displays the content and layout of the flash memory, which includes information about the IOS software engine. The output of the show flash command is displayed in Figure 3.6.

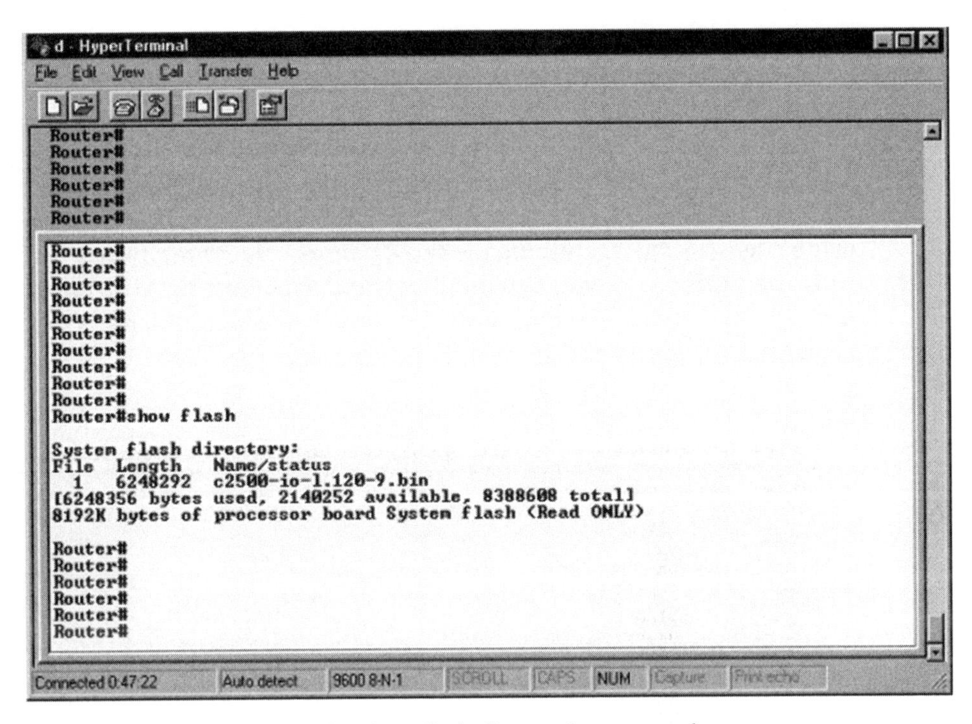

FIGURE 3.6 The output of the show flash diagnostic command.

The Show Buffers Command

Routers consist of system buffers that are allocated from the shared system memory to store packets during process switching. At times, some parameters associated with these buffers require tuning to synchronize process switching and maintain a standard system performance. These parameters apply to each buffer size:

Permanent: Specifies the minimum number of buffers that need to be allocated for the system. In the event of buffer de-allocation, the number of de-allocated buffers cannot be reduced below the value specified in this parameter.

Max Free: Determines the maximum number of buffers that have been allocated but are not in use. If the number of free buffers exceeds the value specified for the Max Free parameter, de-allocation is performed using the trim command.

Min Free: Specifies the minimum number of buffers that can be free within the system. This parameter is the opposite of Max Free. If the number of free buffers is more than the value specified for this parameter, buffer allocation is triggered using the create command.

Initial: Specifies the number of buffers to be allocated during router initialization. The value of this parameter usually exceeds the Permanent parameter value.

Buffers need to be allocated and free when the packets arrive. If free buffers are not available, the packets are dropped. The show buffers command is useful when the network experiences a large number of missing packets for particular buffer sizes. After running the show buffers command, if there is a problem with the buffer size, you can modify the Permanent or Min Free values to recover the system. The output of the show buffers command is displayed in Figure 3.7.

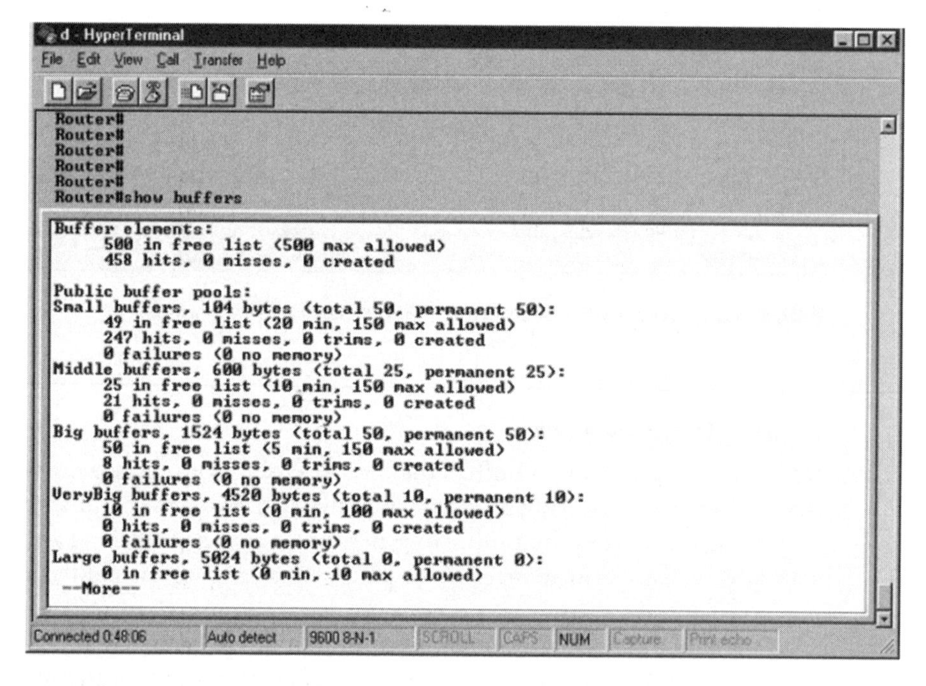

FIGURE 3.7 The output of the show buffers diagnostic command.

The Show Memory Command

The show memory command displays memory pool statistics and information about the activities of the system memory allocator. It displays a block-by-block list of the rate of memory usage. This command is useful when router performance is a problem area. The output of the show memory command is displayed in Figure 3.8.

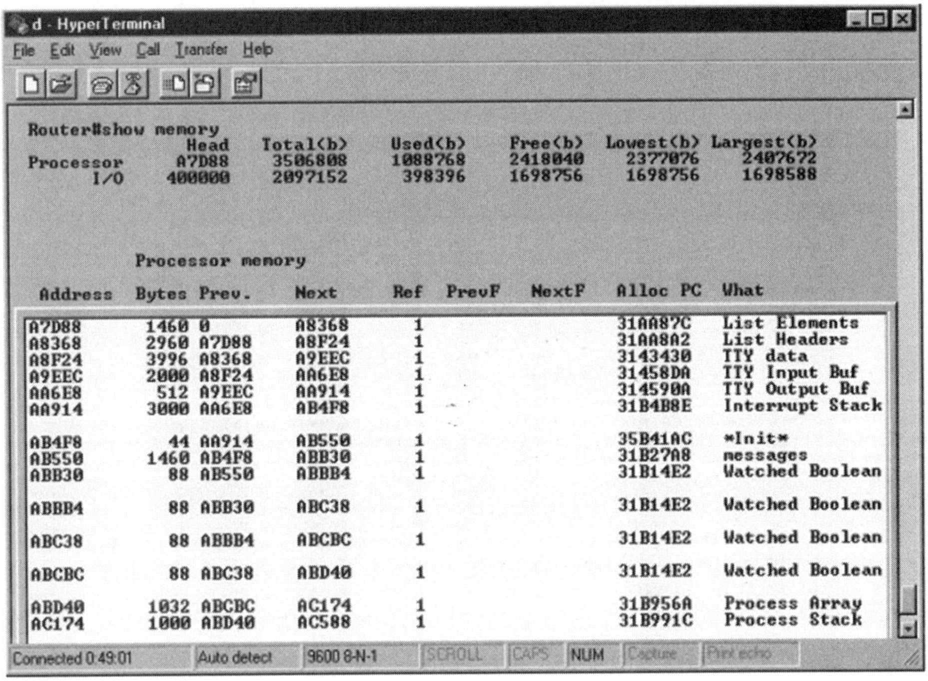

d - HyperTerminal

File Edit View Call Transfer Help

```
Router#show memory
                 Head     Total(b)    Used(b)     Free(b)   Lowest(b)  Largest(b)
Processor       A7D88     3506808     1088768     2418040    2377076    2407672
      I/O      400000     2097152      398396     1698756    1698756    1698588


          Processor memory

 Address    Bytes Prev.     Next      Ref   PrevF     NextF    Alloc PC  What
A7D88       1460 0          A8368      1                       31AA87C   List Elements
A8368       2960 A7D88      A8F24      1                       31AA8A2   List Headers
A8F24       3996 A8368      A9EEC      1                       3143430   TTY data
A9EEC       2000 A8F24      AA6E8      1                       31458DA   TTY Input Buf
AA6E8        512 A9EEC      AA914      1                       314590A   TTY Output Buf
AA914       3000 AA6E8      AB4F8      1                       31B4B8E   Interrupt Stack
AB4F8         44 AA914      AB550      1                       35B41AC   *Init*
AB550       1460 AB4F8      ABB30      1                       31B27A8   messages
ABB30         88 AB550      ABBB4      1                       31B14E2   Watched Boolean
ABBB4         88 ABB30      ABC38      1                       31B14E2   Watched Boolean
ABC38         88 ABBB4      ABCBC      1                       31B14E2   Watched Boolean
ABCBC         88 ABC38      ABD40      1                       31B14E2   Watched Boolean
ABD40       1032 ABCBC      AC174      1                       31B956A   Process Array
AC174       1000 ABD40      AC588      1                       31B991C   Process Stack
```

Connected 0:49:01 | Auto detect | 9600 8-N-1 | SCROLL | CAPS | NUM | Capture | Print echo

FIGURE 3.8 The output of the show memory diagnostic command.

The Show Process CPU Command

The show process cpu command displays the active processes on the router along with the corresponding process ID and status of the priority scheduler test. It displays the CPU utilization of the router, the CPU time used by the router, and the number of times it was invoked. While using this command, it is better to execute it several times with a lapse of a minute. This displays the trend or pattern followed by the active processes on the system. This information shows more reliable trends compared to when the command is executed just once. The output of the show process cpu command is displayed in Figure 3.9.

The Show Stack Command

The show stack command displays the status of stack utilization of processes and interrupt routines. It also displays the reasons for the last system reboot. It is mainly used when a system crash occurs, because it displays information about the failure type, failure program counter address, and the stack trace of the operand, which is stored by the ROM monitor. The commonly monitored error types displayed by the show stack command are:

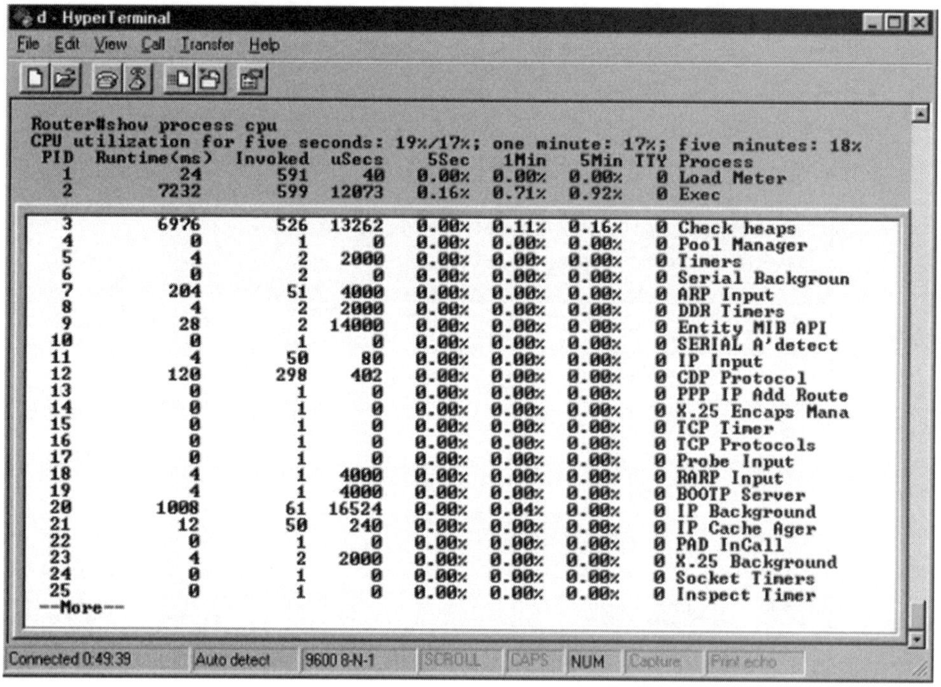

```
Router#show process cpu
CPU utilization for five seconds: 19%/17%; one minute: 17%; five minutes: 18%
 PID  Runtime(ms)   Invoked   uSecs    5Sec    1Min    5Min  TTY Process
   1           24       591      40   0.00%   0.00%   0.00%    0 Load Meter
   2         7232       599   12073   0.16%   0.71%   0.92%    0 Exec
   3         6976       526   13262   0.00%   0.11%   0.16%    0 Check heaps
   4            0         1       0   0.00%   0.00%   0.00%    0 Pool Manager
   5            4         2    2000   0.00%   0.00%   0.00%    0 Timers
   6            0         2       0   0.00%   0.00%   0.00%    0 Serial Backgroun
   7          204        51    4000   0.00%   0.00%   0.00%    0 ARP Input
   8            4         2    2000   0.00%   0.00%   0.00%    0 DDR Timers
   9           28         2   14000   0.00%   0.00%   0.00%    0 Entity MIB API
  10            0         1       0   0.00%   0.00%   0.00%    0 SERIAL A'detect
  11            4        50      80   0.00%   0.00%   0.00%    0 IP Input
  12          120       298     402   0.00%   0.00%   0.00%    0 CDP Protocol
  13            0         1       0   0.00%   0.00%   0.00%    0 PPP IP Add Route
  14            0         1       0   0.00%   0.00%   0.00%    0 X.25 Encaps Mana
  15            0         1       0   0.00%   0.00%   0.00%    0 TCP Timer
  16            0         1       0   0.00%   0.00%   0.00%    0 TCP Protocols
  17            0         1       0   0.00%   0.00%   0.00%    0 Probe Input
  18            4         1    4000   0.00%   0.00%   0.00%    0 RARP Input
  19            4         1    4000   0.00%   0.00%   0.00%    0 BOOTP Server
  20         1008        61   16524   0.00%   0.04%   0.00%    0 IP Background
  21           12        50     240   0.00%   0.00%   0.00%    0 IP Cache Ager
  22            0         1       0   0.00%   0.00%   0.00%    0 PAD InCall
  23            4         2    2000   0.00%   0.00%   0.00%    0 X.25 Background
  24            0         1       0   0.00%   0.00%   0.00%    0 Socket Timers
  25            0         1       0   0.00%   0.00%   0.00%    0 Inspect Timer
 --More--
```

FIGURE 3.9 The output of the show process CPU diagnostic command.

Bus Error: Occurs when the CPU attempts to use a memory location or device that does not exist. Usually, this situation occurs because of a software or hardware bug.

Parity Error: Occurs because of a hardware failure. In this case, the internal hardware checking fails.

Emulator Trap: Occurs when the processor performs an illegal operation. This type of error can be generated when there is a failure of hardware such as the CD-ROM.

Watchdog Timeout: Monitors certain types of system interruptions. If watchdog time fails or is reset, an error occurs.

Address Error: Occurs when any software tries to access data that is not placed in the correct memory blocks. This error type is also referred to as software forced crash.

The output of the show stack command is displayed in Figure 3.10.

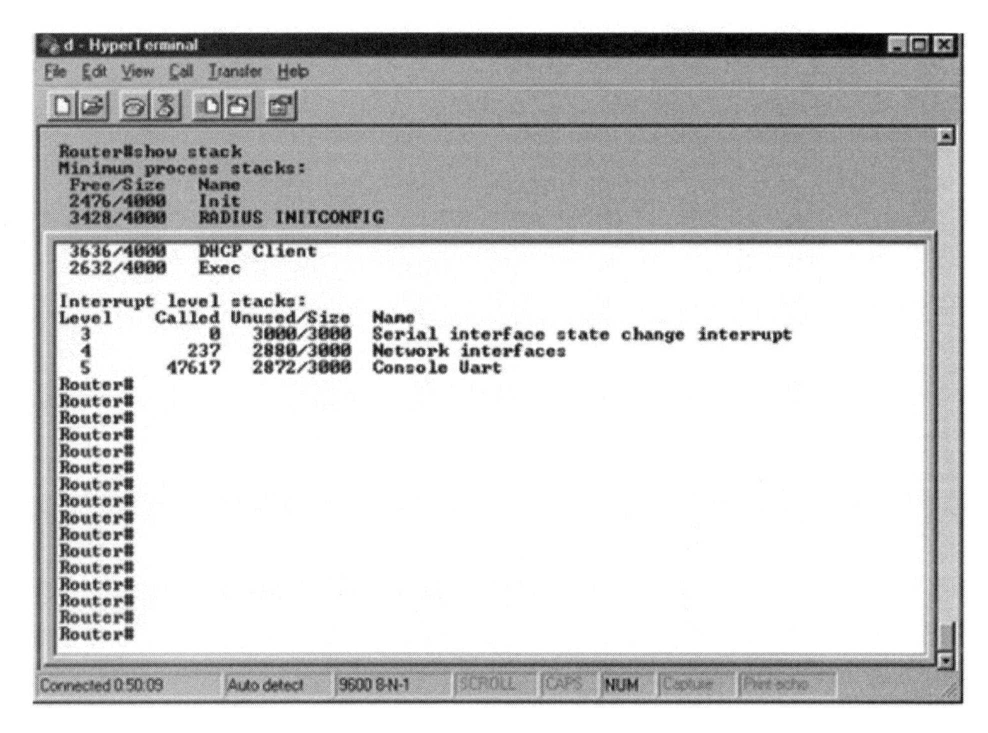

FIGURE 3.10 The output of the show stack diagnostic command.

The Show CDP Neighbors Command

The show cdp neighbors command displays information about the neighboring devices directly connected to the router. It provides reachability data, which helps determine the status of the devices at the Physical and Data-link layers. The output of the show cdp neighbors command is displayed in Figure 3.11.

Other Show Commands

A couple other show commands you might run into include:

Show debugging: Determines the type of debugging enabled on a particular router. This information is useful if you need to switch to a different type of debugging mode or if you need to disable a particular type of debugging when there are more than one debugs running.

Show logging: Displays the status of the syslog errors and event logging. It displays the host addresses, the type of logging being performed, and various logging statistics including the messages stored in log buffer. This command can run in the privilege exec mode only. The output of the show logging command is displayed in Figure 3.12.

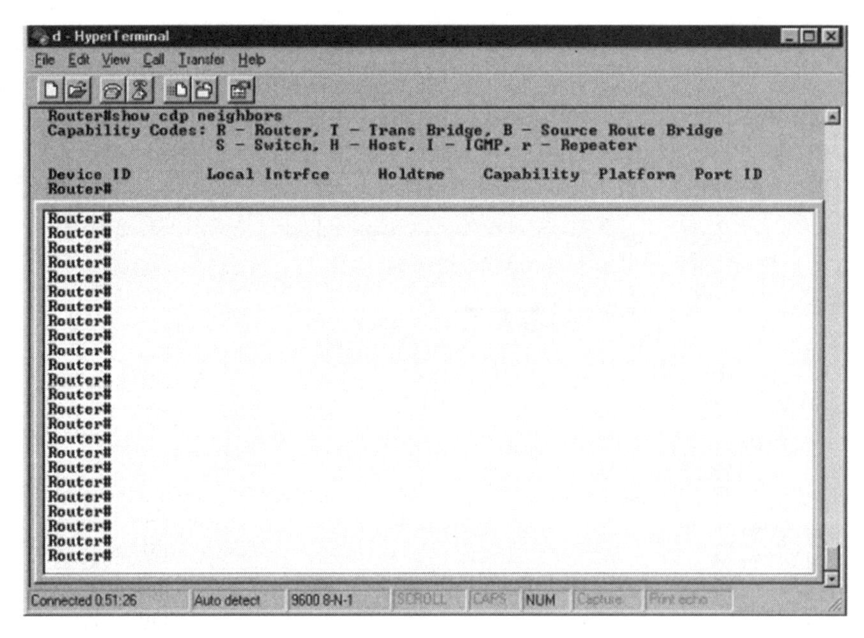

FIGURE 3.11 The output of the show cdp neighbors diagnostic command.

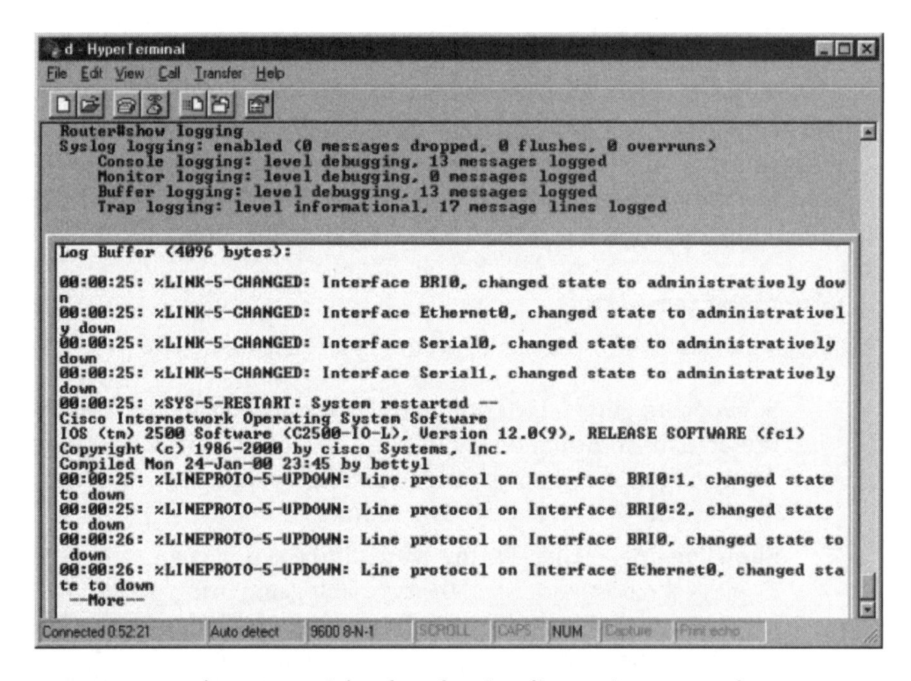

FIGURE 3.12 The output of the show logging diagnostic command.

Depending on the type of information that is required, you need to decide the most appropriate show command to use in a particular situation. Table 3.1 displays various show commands that can be used in various situations.

TABLE 3.1 The Show Commands with Respect to the Problem Area

Problem Area on the Network	Show Commands
Degradation of performance	show interfaces show buffers show memory show process cpu
Loss of functionality for protocols or connection	show protocol show [protocol] traffic show [protocol] interface show [protocol] access lists
General troubleshooting	show version show running-config show controllers show stack show interfaces show process mem show process cpu show buffers

A highly useful show command is show tech support, which displays the combined output of the show controllers, show stack, show interfaces, show process memory, show process cpu, and show buffers commands.

The Ping Command

The ping command checks the connectivity between the nodes on the network. It sends ICMP echo messages to the destination node and waits for a reply. If a reply to the echo message is not received, it confirms connectivity problems on the network. The ping command can be executed in both the user mode and the privileged exec mode.

In the user mode, default parameters are defined to check for connectivity. For example, five echo messages for 100 bytes each are sent by default to the destination host with a timeout interval of two seconds.

While troubleshooting connectivity problems, it is very important to identify the problem area. If the ping command does not receive a reply from the destination host, it should be used for destination hosts that are nearer to the source host. As soon as the ping command is able to receive a reply from a particular destination host, you can isolate the exact problem area on the network.

Connectivity problems might occur when the devices are extremely busy or have restricted access on the network. In the case of a connection timeout or devices with restricted access, you need to identify the busy or prohibited devices and perform remedial actions accordingly.

The debug ip icmp command may be used for troubleshooting connectivity issues.

The ping command can be used with AppleTalk and IPX. However, this is a Cisco-proprietary command, and as a result, non-Cisco devices such as NetWare servers do not receive responses for this command. To enable Cisco routers to generate pings that are compatible with the Novell environment, the IPX ping default novell command should be used. This command is a global configuration command for Novell. However, it is not applicable if the ping command requires compatibility with IPX and non-Cisco devices. To enable ping for IPX and non-Cisco devices, you need to access the privilege mode. In this mode, the option to use a Novell standard echo is available. This allows you to ping for IPX and generate echo messages for Novell standard ping. The test characters generated during a ping response for IPX have certain implications, listed in Table 3.2.

TABLE 3.2 Test Characters and Their Implications

Test Character	Implication
!	Echo message has received a reply
.	Router timeout waiting for a reply
U	Datagram unable to reach destination host
C	Received packet has experienced congestion
I	Test was interrupted by the end user
?	Type of the received packet is unknown
&	Lifetime of the packet has exceeded

For AppleTalk, ping sends AppleTalk Echo Protocol (AEP) packets to the destination node in the AppleTalk environment and receives replies. Table 3.3 displays the test characters and their implications, when ping for AppleTalk is used.

TABLE 3.3 Test Characters and Their Implications

Test Character	Implication
!	Echo message has received a reply
.	Router timeout waiting for a reply
B	Received datagram was bad or malformed
C	DDP checksum of the datagram was improper
E	Echo packet could not be transmitted to the destination host
R	Echo packet cannot find a route for transmission

The ping functionality can be extended in the privilege mode. In the extended mode, you can configure ping for the type of protocol, target IP address, repeat count, datagram size, timeout in seconds, source address, data pattern, and supported header options including Loose, Strict, Record, Timestamp or Verbose, along with a host of other options. To enter the extended ping mode, type "Yes" when prompted for entering the extended command prompt for ping. Table 3.4 displays the options that you need to specify while entering the privilege mode of the ping command.

TABLE 3.4 Ping Command Options and Their Descriptions

Command Option	Description
Source address	Specify any of the local IP addresses of the router or its interface.
Type of service [0]:	Specify the value for this bit as 1 to show the selection of Internet Service Quality.
Set DF bit in IP header? [no]	Specify the value as no. Specifying yes as the value will enable the Don't Fragment option and prevent the packet from being fragmented when it needs to pass through a segment with a smaller MTU. This would cause the operation to fail and generate an error.
Data Pattern [0xABCD]	Use this prompt to edit the 16-bit data pattern. This enables you to detect crosstalk and other cable problems.

(continued)

TABLE 3.4 *(continued)*

Command Option	Description
Sweep range of sizes [n]	Change the size of packets using this prompt.
Loose, Strict, Record, Timestamp, Verbose [none]	Select any one of these options to display the same prompt again, and allow the selection of more than one option. Selecting any of the options automatically includes Verbose. The preferred option is Record, because it displays the addresses of the hops to which the packets need to be transmitted.

For example, as a network administrator, you need to ping a host with IP address 201.202.203.254. Figure 3.13 displays the output of a successful ping command.

The Record option of the Loose, Strict, Record, Timestamp, Verbose [none] prompt is similar to the traceroute command, but it provides enhanced capabilities. The Record option provides data about the hops through which the echo request was transmitted and mentions the path used when the reply to the message was sent. This information is not provided with the traceroute command.

The Debug Command

The debug command is an advanced command that provides high-end options for monitoring and retrieving network data for troubleshooting. It can be executed only in the privilege exec mode. As it is highly resource-intensive, the debug command is used only for temporary or specific troubleshooting purposes.

Debug commands prevent high-speed switching of data packets and force the use of the route processor and process switching before data packets can be sent to the interfaces from which they are finally dispatched. This reduces the speed of operation of the router and increases the processing time, which in turn reduces network performance.

Process switching is performed when the router receives the first data packet for a particular destination host. During process switching, the data packet is transferred within the router using the switching technique. For example, when the data packet is transferred from the internal interface buffer to the main memory, it uses switching. This process is very slow and involves the use of the route processor to determine the interface of the destination host to which the data packet

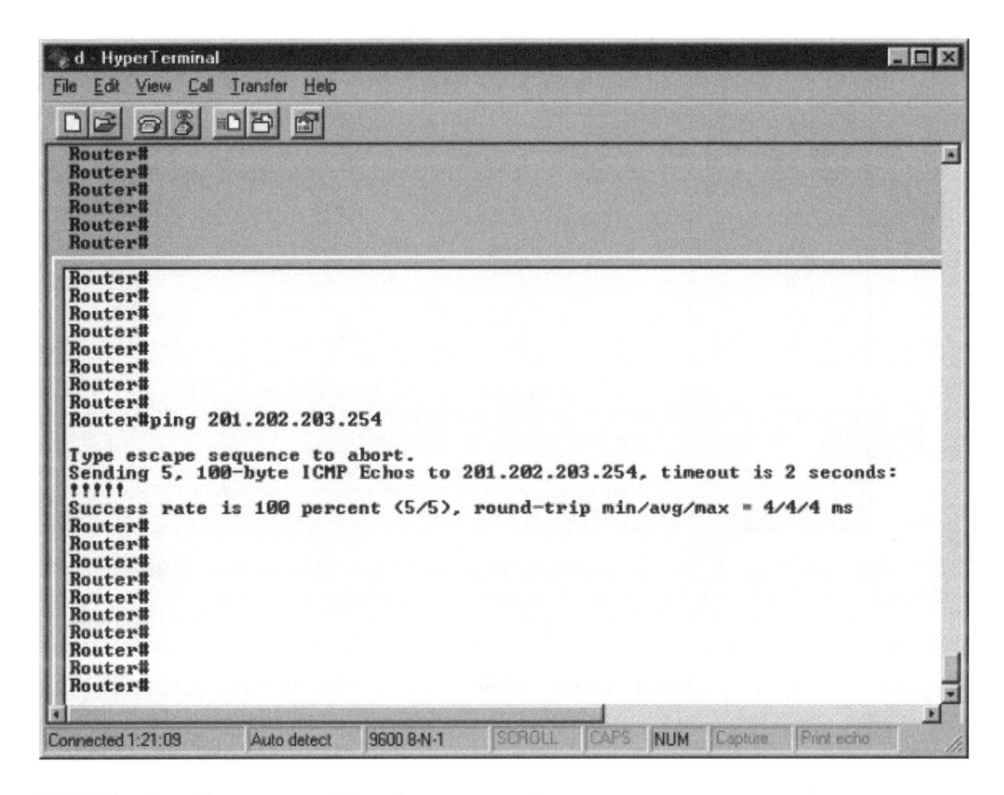

FIGURE 3.13 The output of the ping command.

needs to be sent. To increase the routing speed, the information collected during process switching is stored in the switching cache and is used to transmit subsequent data packets to the same destination host. This process is known as *fast switching*. However, during troubleshooting, the debug command needs to collect the data for each data packet that is transmitted. As a result, it forces all the data packets to be process switched, which heavily consumes router resources. As a result, the debug command should, therefore be used with caution while troubleshooting. To isolate problems and derive alternate solutions, you can use the debug cdp packets command to debug the packets transmitted from one host to another using the current router, as shown in Figure 3.14.

After the information received from the debug command has been viewed and analyzed, it should be disabled in order to revert to the normal mode in which high-speed switching techniques are used.

The debug all command should not be used at all, because it collects information about the entire process and all the data packets transmitted through the router, which may render the entire router unusable.

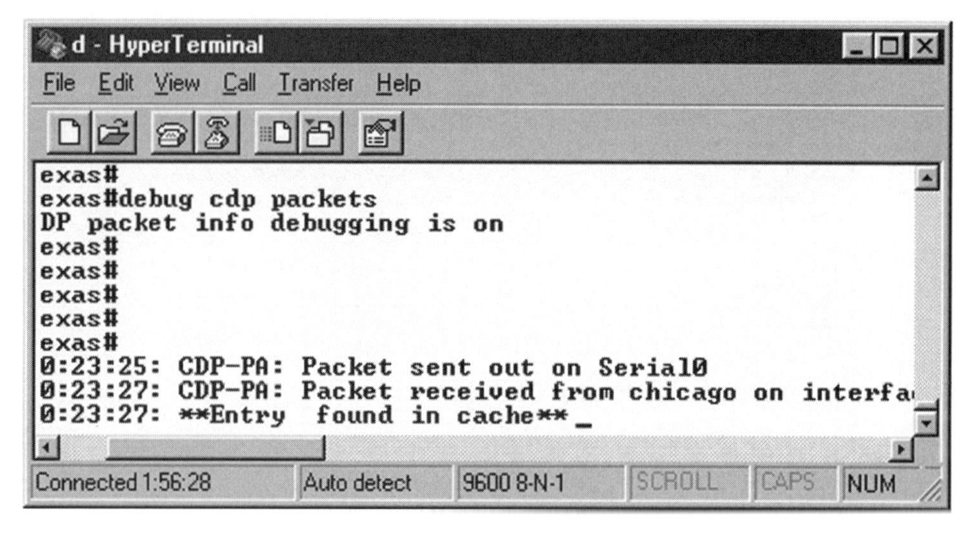

FIGURE 3.14 The output of the debug cdp packets diagnostic command.

Some guidelines to follow while executing the debug command are:

- Display the timestamp with each line of the debug command, using the command:

```
router(config)#service timestamps debug [ datetime | uptime]
```

- Use the debug command for specific purposes only, such as diagnosing a facility, task, or protocol. This helps in focusing on a particular area or problem and isolating the problem.
- If the debugging requirements are not exhaustive, use the event debugging technique, because it is less resource-intensive than packet debugging. Use this only if you need more information on the problem.
- Check the CPU utilization before running the debug command. If the CPU utilization is more than 50%, it is recommended not to use the debug command, because it may render the network nonfunctional.
- Study the usage pattern of the network and determine peak usage time intervals. Depending on the severity of the problem, determine whether it would be feasible to run the debug command during or after peak hours. This decision should be based on the fact that Cisco routers assign high priority to debug commands as compared to other processes running in the environment.
- Disable the debug command as soon as the required task has been accomplished. The no debug {argument} can be used to disable a particular type of debugging. To disable all types of debugging, use no debug all or undebug all.
- Use protocol analyzers or network management applications for troubleshooting, when feasible, instead of using debug.

- Use access lists with the debug command in order to narrow down the data that the debug command needs to collect. The debug IP packet detail command allows you to specify the name or number of the access list, which restricts the debug command to collect data only for packets pertaining to the specified access list. This reduces the network overhead and provides relevant information by narrowing down the debug criteria.

The output of the debug command can be logged and stored in four places: the console, the internal buffer of the router, the virtual terminal or Telnet, or the syslog server. Table 3.5 lists the various commands used for logging messages.

TABLE 3.5 Commands for Logging Messages

Logging Command Option	Description
logging console [level]	Enables console logging and sets the level of logging as console. To disable console logging, use the no logging console command.
logging buffered [level]	Directs the logging messages to the internal buffer and sets the level of logging as buffered. To disable this command use the no logging buffered command.
logging monitor [level]	Directs the logging messages to the virtual terminal or Telnet, specifying the level of logging to the virtual terminal lines. To disable this command use the no logging monitor command. Executing the command, terminal monitor from the virtual terminal, can also turn the logging on for the virtual terminal. To disable the command use the no monitor command.
logging trap [level]	Directs the logging messages to the syslog server and allows you to specify the level of messages. To disable this command use the no logging trap command. The default level of this type of logging is informational.
logging [ip-address]	Identifies the IP address of the syslog server to route the logging messages. If the logging messages need to be sent to more than one syslog server, this command needs to be executed with the IP address of each server. To disable the command for a server, use the no form command.

The console logging command generates the maximum overhead when executed, followed by the virtual terminal, syslog, and buffered logging commands.

The output of the debug command has eight levels of message logging, based on the severity of the message. Table 3.6 provides the level name, value, and description of the messages.

TABLE 3.6 Level Name, Value, and Description of Debug Log Messages

Level	Level Name	Syslog Definition	Description
0	Emergency	LOG_EMERG	The network is unusable.
1	Alerts	LOG_ALERT	The problem requires immediate action.
2	Critical	LOG_CRIT	The problem is critical.
3	Errors	LOG_ERR	An error has occurred.
4	Warning	LOG_WARNING	Warning against a preset criteria or a potential problem.
5	Notification	LOG_NOTICE	An important but normal event has occurred.
6	Informational	LOG_INFO	Information about the network status.
7	Debugging	LOG_DEBUG	Messages displayed while debugging.

The Trace Command

The trace command is used to determine the route followed by the data packets on the network. This command is available only for IP in the Cisco IOS. The traceroute command initiates a session by sending UDP or connectionless probes with a time-to-live (TTL) interval of 1. The TTL value increases until the probe reaches the destination host. The probe reaches the first hop when the TTL value is 1. The hop then responds with a time-exceeded message. When the TTL reaches the second hop, the TTL value increases to 2. This process goes on until the probe arrives at the destination host. The destination host then sends a port unreachable message to the source, because it is unable to deliver the packet to an application.

When a probe reaches a router, the TTL value decreases by 1.

Three types of probes are generated for each TTL value, and if a response is not received within a certain interval, the output is generated as an asterisk, indicating that the destination host was unreachable. The trace command ends when the probe reaches the destination port, the user interrupts the trace command using the escape sequence, or the maximum TTL value is exceeded.

Like ping, the trace command provides extended capabilities in the privilege exec mode. It allows you to change the operational parameters and specify the source address of the probes. In addition, it allows selection of the Loose, Strict, Record, and Timestamp options. For example, you may need to trace the host with IP address 1.1.1.1, as shown in Figure 3.15.

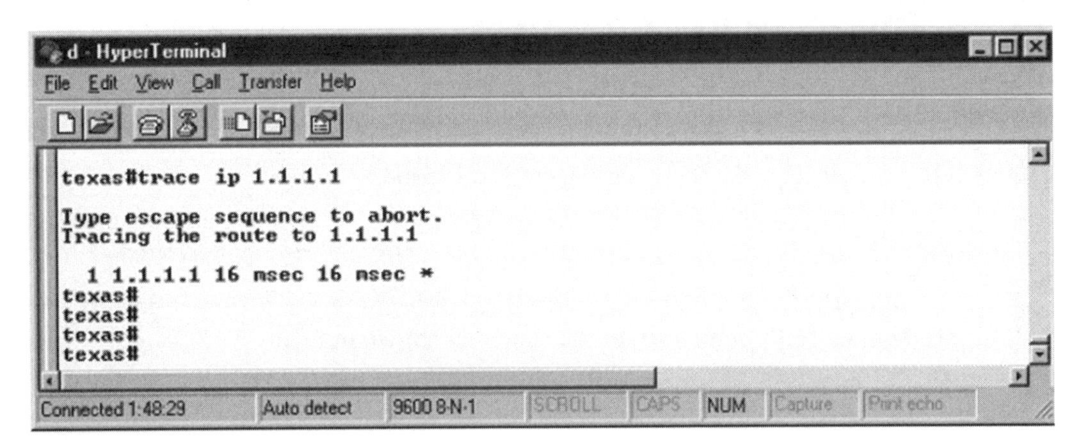

FIGURE 3.15 The output of the trace diagnostic command.

Cisco Discovery Protocol (CDP)

A proprietary protocol developed by Cisco, CDP functions at the Data-link layer and supports heterogeneous network layer protocols on different networks. It can run on Cisco devices such as routers, bridges, access servers, and switches. In addition, it can run on all types of media that support SNMP, Frame Relay, ATM, and LAN.

Each device that uses CDP advertises a minimum of one address at which it can receive SNMP messages, and a TTL limit that is set to 180 seconds, by default. It helps in fast detection of the interface states. When a CDP packet is transmitted, the value of TTL is nonzero once the interface is enabled; when the interface is in an idle state, the TTL value becomes zero.

The cdp command is used to determine the IP addresses of neighboring devices. This can be done by conducting a Telnet session with the neighboring device and executing the show cdp command. This command displays all the devices that are directly connected to the router, as shown in Figure 3.16.

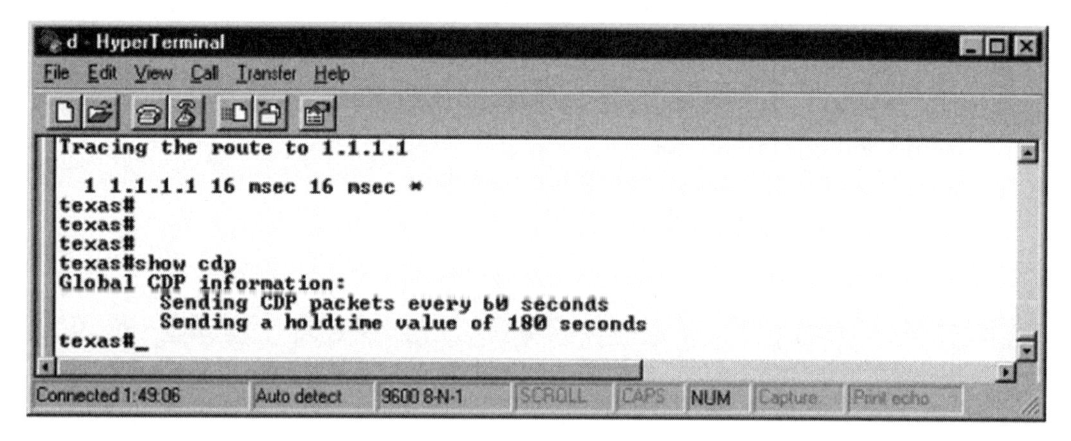

FIGURE 3.16 The output of the show cdp diagnostic command.

Similarly, the show cdp neighbors diagnostic command is used to identify the neighbors directly connected to the router, as shown in Figure 3.17.

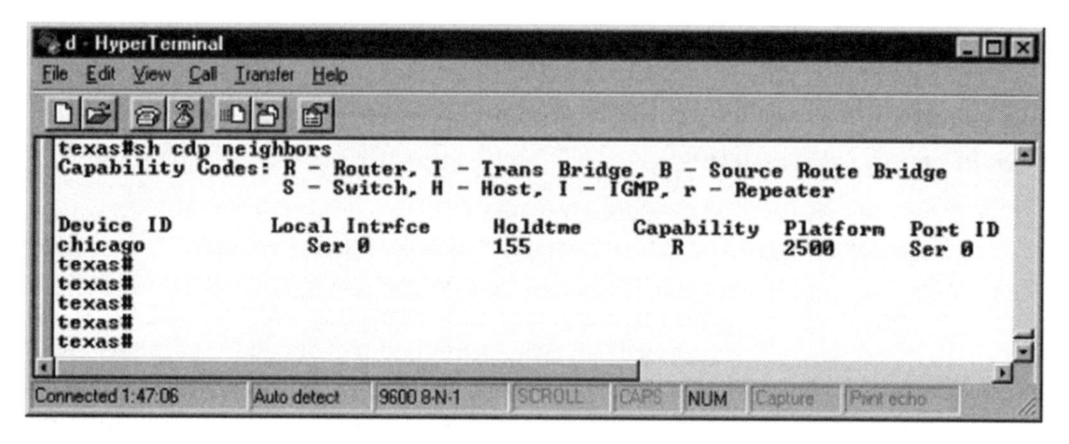

FIGURE 3.17 The output of the show cdp neighbors diagnostic command.

To configure the cdp command before it is used for troubleshooting purposes, you need to:

1. Configure the CDP transmission timer and hold time: Execute the cdp holdtime [seconds] and cdp timer [seconds] commands respectively. These commands need to be executed in the global configuration mode. The default value of the CDP timer is 60 seconds. This time limit is usually sufficient for running the command successfully. If the limit is reduced to an excessively low value, it may lead to high network traffic.
2. Disable and enable the protocol: Run the no cdp run command to disable it, and then run the cdp enable command to re-enable CDP globally. The cdp command is enabled by default. These commands need to be executed in the global configuration mode.
3. Disable and enable the protocol on the interface: Run the no cdp run command to disable CDP and then the cdp run command to enable it, in the interface configuration mode.
4. Monitor CDP: There are various commands that assist in managing CDP information:

   clear cdp counters: Resets the traffic counter value to zero.

   show cdp: This is one of the most important cdp commands. It displays global information about CDP including the frequency transmission rate, hold time of data packets, and other critical information.

   show cdp interface: Displays details about the interface where CDP is running.

   show cdp entry: Displays information about a particular neighbor. It can run with the protocol and version options to restrict the information to information about the protocol and version only.

In addition, the show cdp neighbors command is used to display information about networks that are connected to the router directly.

Core Dumps

Core dumps are used to retrieve a full copy of the memory image of the router in case of a crash. The memory image can then be used to identify the cause of the crash. Core dumps have the ability to transfer the binary image file using the FTP, TFTP, or RCP protocols. However, executing the core dump command disrupts the entire network operation, so make sure the command is used only when it is evident that the router will ultimately crash.

If you experience bottlenecks in the operations of the router and anticipate that a crash will occur, you can execute the write core command.

This command enables you to generate a core dump without reloading the router. However, there are certain prerequisites for executing this command, which include the following:

- The TFTP, FTP, or RCP server being used in the network must have sufficient storage capacity.
- The server must be accessible.
- You must know the file naming convention of the server in order to store the image file.
- You must know whether an empty file needs to be created on the server.

The write core command cannot be used if the server has already crashed. In that case, the exception dump ip-address global configuration command is used. This command tries to generate a core dump after the crash. The core dump is written to the file, [hostname]-core on the server, where hostname is the name of the router. The attempt to retrieve the core dump may not be successful always, depending on the severity of the crash. The data stored in the core dump file is used by the Cisco technical support group to decipher the cause of the crash and recover the system.

If you use TFTP for receiving the core dump file on the server, only the first 16 MB of the image file are transferred.

SUMMARY

In this chapter, you learned about the various network management tools such as CiscoView, CiscoWorks, Traffic Director RMON, VLAN Switch Management, NetSys Network Management Suite, WAN Manager, NetSys Baseliner 4.0, and NetSys SLM Suite. In addition, you learned the different Cisco diagnostic commands such as show, ping, debug, and trace. You also learned about the Cisco Discovery Protocol. In the next chapter, we will look at troubleshooting tools.

POINTS TO REMEMBER

- The Cisco network management tools have been developed on the Cisco AVVID guidelines to provide centralized monitoring and management capabilities to enterprises of network environments such as WAN, wireless and mobile networks, and VLAN setups.

- CiscoView is a device management tools that uses a Graphic User Interface (GUI) to display the status of Cisco devices on the network. It also provides real-time information about the status and configuration of Cisco devices on the network.
- CiscoWorks constitutes a family of products including LAN and WAN management solutions, IP telephony network management solutions, and mobile network management.
- The Traffic Director Remote Monitor is a traffic management tool that performs traffic analysis and performance management and determines traffic utilization, broadcast levels, error rates, and number of collisions on particular ports.
- The VLAN Director Switch Management application provides management capabilities for VLAN ports.
- Cisco WAN Manager (CWM) is a multiprotocol network management application based on SNMP. It provides a number of services such as fault, configuration, performance, account and security management, and planning, modeling, and analysis.
- NetSys is a simulation tool that provides network planners, analysts, and managers with planning and analytical capabilities.
- The NetSys Baseliner is a tool that diagnoses network problems by simulating the current configuration of the network.
- The NetSys SLM suite is a policy-based network management application, which establishes service levels while defining, monitoring, and assessing network connectivity, security, and performance policies.
- Diagnostic commands are built into Cisco IOS to provide information that is used for diagnosing network problems. The important diagnostic commands are show, debug, ping, and trace.
- Cisco IOS provides a range of show commands that display information about the rate of utilization of router resources, network interface status, and router configurations.
- The ping command is used to check the connectivity between the nodes on the network.
- The debug command is an advanced command that provides high-end options for monitoring and retrieving network data for troubleshooting.
- The trace command is used to determine the route followed by the data packets on the network.
- CDP is a Cisco-proprietary protocol that functions at the Data-link layer and supports heterogeneous network layer protocols on different networks.

4 Troubleshooting Tools

IN THIS CHAPTER

- Connectivity and Cable Testers
- Digital Interface Testers
- Network and Protocol Analyzers
- Network Monitoring and Management System
- Enterprise Management Systems
- CiscoWorks QoS Policy Manager
- Modeling and Simulation Tools
- Traffic Generators
- Cisco IOS Diagnostic Commands

Networks are sensitive and require constant monitoring during and after installation. As a network administrator, you should be able to diagnose network problems and take effective remedial actions using troubleshooting tools. This chapter surveys the tools for monitoring and troubleshooting network problems pertaining to Cisco internetworking devices, including routers and switches.

The troubleshooting tools discussed in the chapter include native software products developed by Cisco, Cisco IOS commands, and third-party tools that cater to specific troubleshooting needs. Each tool operates at a particular OSI layer and ascertains systematic diagnosis of network bottlenecks at that layer. However, some tools applied for troubleshooting Cisco devices such as routers and switches may impede their performance. As a result, you need to be judicious about applying a tool in a particular situation, especially in the case of Cisco IOS commands. Following the layered approach, troubleshooting tools have been discussed with respect to the seven layers of the OSI model, starting with the Physical layer.

CONNECTIVITY AND CABLE TESTERS

Cable testers are used at the Physical layer to check the network for transmission media-related problems such as high EDI rate, attenuation, and insufficient cable length. In addition, specialized equipment is available for testing cables at the lower and higher spectrum of cable testing. Let us discuss these in detail.

Low Spectrum Cable Testers

Low spectrum cable testing involves testing the transmission media for parameters pertaining to the physical connectivity of networks. These parameters are AC and DC voltage, current, capacitance, resistance, and cable continuity. Devices such as volt-ohm meters and digital multimeters check the connectivity at this level. Cable testers also check networks for attenuation, Near-End Crosstalk (NEXT), and noise. In addition, these devices test various types of cables, such as Shielded Twisted Pair (STP), Unshielded Twisted Pair (UTP), coaxial, twinax, and 10BaseT.

Some advanced cable testers provide Time Domain Reflectometers (TDR), traffic monitoring, and wire map functions. In addition, low spectrum cable testers provide limited protocol testing support using diagnostic commands, such as ping.

Low spectrum cable testers (scanners) provide information about the MAC layer, such as actual network utilization and packet error rates. This helps determine the rate of optimal network utilization and detect transit errors of data packets. In addition, cable testers help increase overall network performance by checking and troubleshooting physical connectivity.

Time Domain Reflectometers (TDRs) and Optical Time Domain Reflectometers (OTDRs) are devices that identify the cable break location, impedance mismatches, and other physical connectivity problems. Fiber optic cables require extra precaution before and after installation due to the high cost of setting up a fiber optic-based network. Therefore, it is recommended that the fiber optic cables are tested before installation. This process is called on-the-reel testing. When installation is complete, perform periodic checks on cables, using continuity testing tools such as a light source or a reflectometer. There are various types of reflectometers that provide light at standard wavelengths such as 850 nm, 1300 nm, and 1500 nm. These reflectometers work with optical power meters to measure the specified wavelengths and identify their attenuation and fiber loss. An optical power meter measures the power to and from an optical device. For optimal performance, it is a good idea to calibrate this meter every year.

Fiber optic cables are highly sensitive and should be kept clean. Many problems occur due to dirty cables. Use an optical cleaning kit to keep the optical cable connection clean.

OMNI is a commonly used cable scanner provided by Microtest that has the ability to test all the cables that support a wide range of 100 dB and a bandwidth of 300 MHz.

High Spectrum Cable Testers

High spectrum cable testing involves testing the transmission media to check for impedance mismatch, crimps, and other physical problems in metallic and fiber cables, respectively. Devices such as TDRs and OTDRs check connectivity at this level.

A TDR works by "bouncing" a signal at the end of a cable. Due to various problems in the cable, open or short circuits reflect the signal at different amplitudes. Using a TDR, you can determine the time taken by the signal to reflect after it reaches the end of the cable. The TDR calculates the time based on the following formula:

Distance (d) = Propagation Rate x Time to Measure the Distance to a Cable Fault

In addition to troubleshooting, TDRs measure the length of the cable. This is measured when the signal is returned at very low amplitude after it is reflected from the end of the cable.

OTDRs test fiber optic cables and detect the length of cable breaks, attenuations, and splice losses. In addition, they calculate the number of reflections to compute connector losses.

OTDRs are also used to measure fiber attenuation using pulse reflections, which occur at breaks or joints and uniformly scatter the reflections in a backward direction throughout the fiber optic cable.

Spectrum analyzers analyze the light with respect to the wavelength that helps detect channel crosstalk. It also conducts periodic laser tests on fiber optic cables for better performance and stability.

As a network administrator, you can specify the normal baseline or standard performance of fiber optic cables, based on the normal rate of attenuation and splicing. While monitoring the network, use these baseline statistics to detect cabling problems.

Fixed attenuators can be used to add fixed attenuation levels to connections, including five attenuators with 5 dB at a wavelength of 1310 nm and another 5 with 10 dB at the same wavelength.

DIGITAL INTERFACE TESTERS

Communication channels may encounter problems such as congestion due to heavy data traffic or bandwidth waste. To counter such problems, you can use different types of digital interface testers. Digital interface testers are third-party tools used for testing digital signals received by and transmitted from computers and other network devices such as modems, printers, and scanners. These testers are used for both serial and parallel interfaces pertaining to digital signals and data communication.

The digital interface testers identify problems of the data transmission line, capture corrupted data, and detect common problems related to communication channels. Examples of digital interface testers are:

- Breakout boxes
- Fox boxes
- Bit/Block Error Rate Testers (BERTs/BLERTs)

These digital interface–testing tools measure digital signals on peripheral devices such as PCs, printers, and Channel Service Unit/Digital Service Unit (CSU/DSU). They are also responsible for analyzing the data communication line, detecting corrupt data, and preventing corrupted data from reaching the destination node. In addition, these digital interface–testing tools monitor data that is transmitted from Data Terminal Equipment (DTE) through Data Communications Equipment (DCE). While monitoring the data transmission from a DTE through a DCE, these tools identify:

- Corrupted bit patterns
- Improper cabling within a network
- Problematic situations such as traffic congestion during data transmission

For example, if you need to identify a problem in the connection between a CSU/DSU and a router, a breakout box can be deployed between the unit and the router. This helps determine whether the CSU/DSU sets the CTS high before the router transmits the data and sets the DTR high. This test enables you to determine if the router is appropriately transmitting data on the network.

However, the main disadvantage of digital interface–testing tools is that they cannot be used to test media signals from Ethernet, Token Ring, or FDDI. Though these tools test the effectiveness of the communication channel and digital signals, they cannot be used to test protocol-related issues on a communication channel.

NETWORK AND PROTOCOL ANALYZERS

Network and protocol analyzers record, interpret, and analyze the operations of a protocol within a network. In addition, they filter traffic from a particular device and generate frames for transmission over the network. Protocol analyzers do not have any impact on network performance. They only provide details about network traffic and protocol paths, network configuration and operation, and offer potential solutions to critical network problems.

When a node transmits frames on the network, the protocol analyzer captures the frames and decodes the protocol layers in a recorded frame. This information is in the form of readable summary, which provides information about the protocol layer and the function of each byte in the frame. As the size of the network increases, the scope of the functions performed by the protocol analyzer increases. This is because the protocol analyzers detect and decode all the protocols used by the frames.

The protocol analyzer also generates and transmits frames for capacity planning and for performing load tests on the network. For example, if network performance regularly deteriorates at a particular time or in a particular region, one of the possible reasons can be heavy network traffic during that period or in a particular region. To detect and reduce such performance-related issues, the protocol analyzer should be able to send multiple captured frames.

A protocol analyzer works in two modes, capture and display. In the capture mode, it records the frames or the network traffic depending on certain performance criteria or a predefined threshold. For example, you may observe network downtime when data is transmitted to a particular network. To determine the exact cause of network failure, attach a threshold or filter to the protocol analyzer to capture the frames directed to that particular network. In such a situation, the protocol analyzer captures the frames directed to that particular network in the capture mode. The captured frames will have a timestamp attached to them that will determine the exact period during which the network performance deteriorates. This type of information is critical for organizations such as banks and stock exchanges that require seamless network connectivity.

In the display mode, a protocol analyzer decodes the captured frames and stores the information in a readable format for future interpretation. To view the captured frames, you can use thresholds. In addition, you can apply these thresholds to view only those frames that match a certain criteria.

Protocol analyzers are intelligent tools that use specialized techniques to diagnose problems based on the symptoms. The knowledgebase of this system includes:

- Theoretical databases, which store information about the standards
- Network-specific databases, which store information about the network topology

- End user experience, which includes records of the network problems that have occurred within the network

The expert system administrator uses this knowledgebase to generate a hypothesis that describes the most probable cause of the problem.

Based on the different characteristics and information captured, there are three types of protocol analyzers:

General purpose analyzers: Provide information about the network such as traffic monitoring, protocol capture, and network traffic modeling in the network design phase.

Software-based analyzers: Provide information restricted to a particular network or LAN. The software is usually installed on a PC on the network and performs internal network troubleshooting.

High-end analyzers: Provide relatively expansive protocol decoding and capture traffic at higher rates. A significant feature of the high-end analyzer is the "generate and capture" capability, which is used for network capacity planning and load testing.

One of the most commonly used protocol analyzers is Sniffer Pro. It diagnoses network problems based on the symptoms. This product can decode as many as 250 protocols.

Another effective network protocol analyzer tool is Ethereal, which examines data in an active network and has a GUI for viewing data. You can interactively view both the summary and detailed information about the data packets. The tool assembles the data packets transmitted in a TCP session and displays the ASCII data for the session. In addition, it provides powerful filters, which accommodate a large number of fields.

To demonstrate use of network analyzers, consider a situation in which a network has problems with workstations attached to a particular port of a switch. To resolve this problem, you need to analyze the packets that are transmitted by the workstation. Connect a network analyzer to the port and configure a span to monitor the incoming and outgoing packets to the port.

NETWORK MONITORING AND MANAGEMENT SYSTEM

To ensure consistent network performance, you need to constantly monitor and manage any network faults and configurations. As the size of the network increases, it becomes difficult to monitor resources and configuration. This in turn increases network vulnerabilities such as performance slowdown and failures. To monitor the performance of network resources when the size of the network increases, you need network management tools. To troubleshoot network problems, you need to perform:

Fault detection: Refers to the initial tasks to be performed when a network starts witnessing problems.

Configuration management: Refers to the set of tasks performed after diagnosing the problem area in the network and re-establishing the configuration of network components during and after problem resolution.

Performance management: Refers to the set of monitoring tasks for monitoring the prospective areas of network problems and taking remedial actions.

Resource management: Refers to the tasks that provide Quality of Service (QoS) to network users and applications. You need to prioritize the applications or end users that require more resources as compared to other end users or applications.

The currently available network management solutions provide detailed drill-down reports about the performance of network components. You can use these reports to detect the reason of network failure or performance slowdown.

Network monitoring is an effective tool to determine network capacity, establish a standard for network performance, and generate alerts when network problems occur. A performance standard is established by taking into consideration parameters such as network performance and traffic patterns on the network over a period. These statistics help derive an average rate of network performance. You can use this performance standard to compare the current network performance and check for any deterioration. Some network monitoring tools that offer remote and local network monitoring services are: CiscoWorks, RMON, and MIBs.

Some of the Cisco IOS commands used for collecting network-related information are debug diagnostic commands and show process cpu commands. The debug command runs in the privileged exec mode on the router console. It provides information about various router operations, such as traffic generated or received by the router on the network and errors on the router interface. The only limitation of this command is that it consumes a significant amount of network resources for the router. This leads to performance slowdowns for the routers. Hence, it is recommended that this tool be used only occasionally to determine specific network problems. The debug command can be used for a specific facility, task, or protocol that could be the cause of the problem. For example, if a particular protocol in the TCP/IP suite requires debugging, the debug command provides the facility to use the event or packet option. Event debugging consumes fewer resources as compared to packet debugging. A sample output of the debug command is displayed in Figure 4.1.

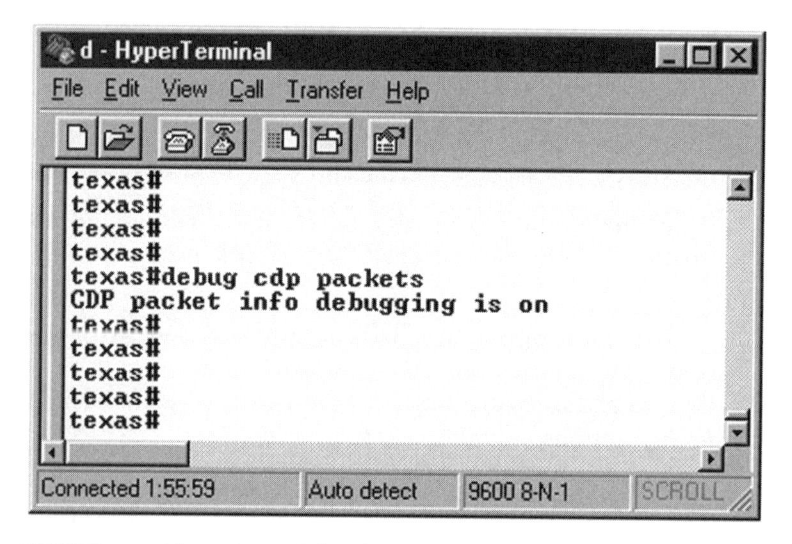

FIGURE 4.1 The output of the debug diagnostic command.

In addition to the debug command, there is a show process cpu command, which determines the amount of CPU utilization. This helps you identify router processes that are using the CPU inefficiently and, as a result, enable optimal CPU utilization. Figure 4.2 displays the output of the show process cpu command.

Configuration management tasks are usually performed manually. If you have made changes in the network configuration, ensure that the network components are concurrent with the latest configuration setup. For example, if a printer is removed from a network or if the IP address of the printer changes, update the static IP address of the new network printer on each system to ensure printer access over the network.

An Application layer protocol that facilitates exchange of management information between network devices is SNMP, which is a component of the TCP/IP suite. SNMP allows you to manage fault detection and recovery of your network. Its components include the SNMP agent and the SNMP management station. The SNMP agent is deployed on the router, and the SNMP manager queries this agent to collect network statistics. This way, you can determine how the network is functioning and anticipate any problems that may occur in the future. The SNMP framework is displayed in Figure 4.3.

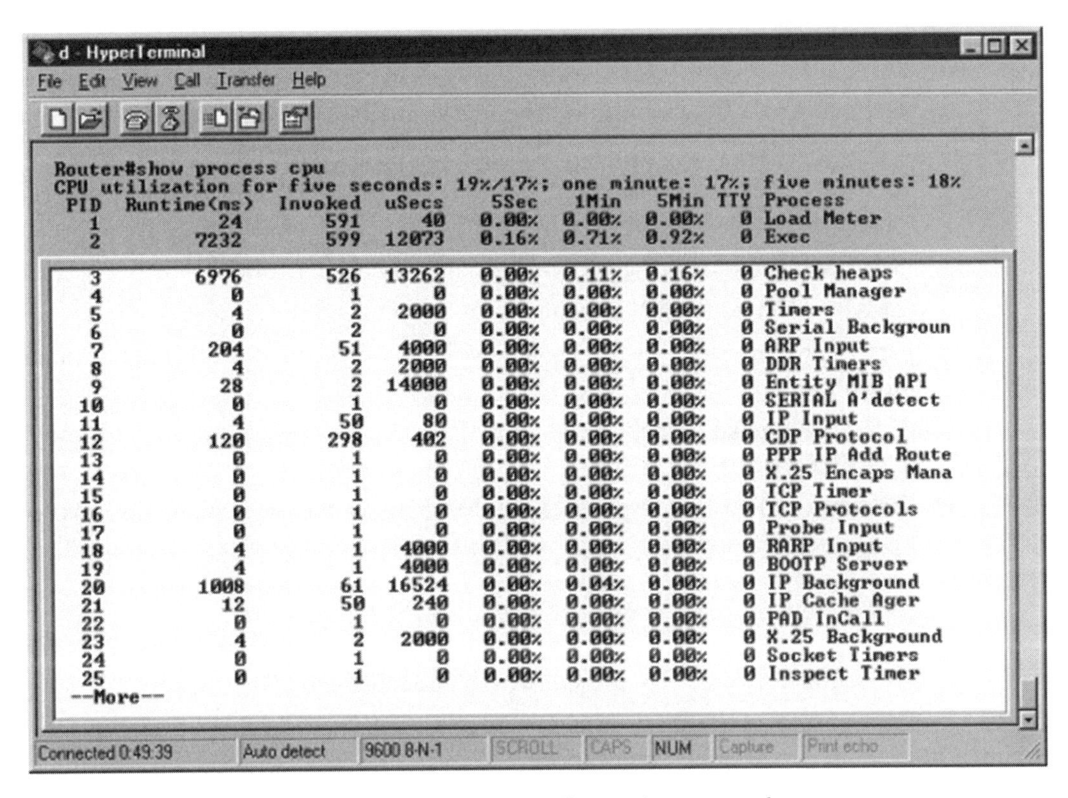

FIGURE 4.2 The output of the show process cpu diagnostic command.

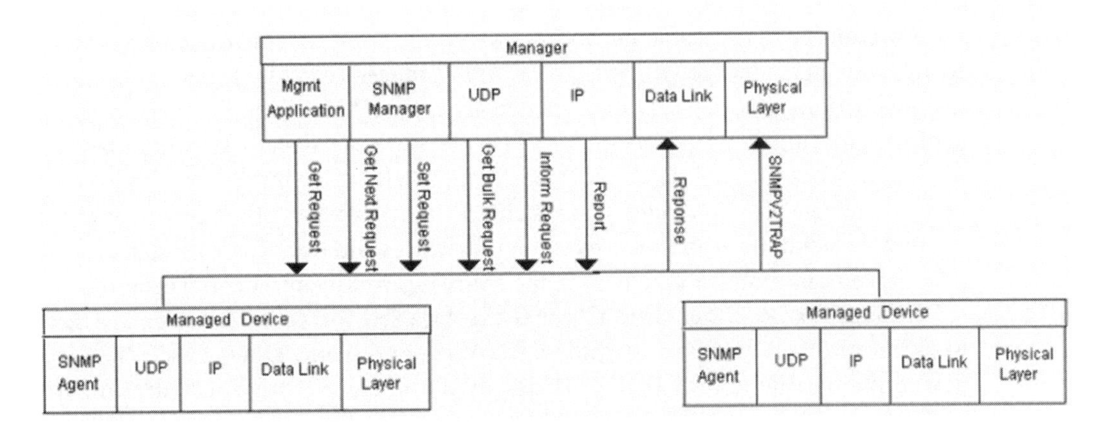

FIGURE 4.3 The SNMP framework has two managed devices.

As shown in Figure 4.3, SNMP commands perform different tasks:

Get Request: Gets information about the attributes of a managed object from the SNMP agent

Get Next: Performs the same task as the get request command for the next object in the managed device tree

Set: Sets the value for a particular managed object attribute

TRAP: Alerts the SNMP manager about events on the managed object

Get Bulk Request: Obtains a large volume of information through a single operation

Inform Request: Enables the SNMP manager to communicate with other SNMP managers

Response: Retrieves responses from the agents for the Get Request, GetNextRequest, GetBulkRequest, SetRequest, or InformRequest Protocol Data Unit (PDU)

There are four major constituents of the SNMP framework:

Managed devices: Are present within an embedded SNMP agent

Management server: Connects with managed devices, providing various services to management applications

Management protocol: Uses the SNMP protocol for receiving and sending encoded messages for the SNMP manager and agents

Management information model: Defines the managed resources within the SNMP framework using a pseudo object-oriented design in which all objects are stored virtually in MIB

An example of a scalable network management product developed by Cisco for service provider networks is the Cisco WAN Manager (CWM). It performs all the major functions of a network management system (NMS), including performance, configuration, and fault management. It provides statistical data for network management and stores it in the database. In addition, CWM easily integrates with existing network management and operating systems. While monitoring the network, you can use the CWM interface and tools to configure devices and track network statistics. As a result, you can monitor and manage the entire network from a central location.

Besides CWM, other products that can be used for performing partial network monitoring functions include CiscoView and Cisco DFM. CiscoView is a device management tool that provides dynamic status and configuration for Cisco devices. Cisco DFM proactively detects network faults and alerts network managers.

ENTERPRISE MANAGEMENT SYSTEMS

The complexity of networks has increased due to integration of heterogeneous devices including different types of media, protocols, and interconnectivity with networks that may have an entirely different setup. An enterprise with a large network setup that spans multiple LANs or a WAN requires constant monitoring and manageability to prevent network failure. Though network failures are inevitable, you can reduce network downtime if you are prepared. Proactive monitoring of enterprise networks ensures quick resolution of network faults, based on service level standards and performance baselines. This enables you to take precautionary measures before a network problem occurs. Today, enterprise management systems (EMSs) have become a necessity for large and small networks.

A conventional EMS provides an integrated set of tools to manage mission critical characterstics of a network such as network and resource availability, responsiveness, resilience, and security. It is characterized by Service Level Agreements (SLAs) and policies, drill-down views of the network infrastructure, and a "single pane of glass" view of the network, which enable you to centrally monitor and manage the network. The most common functions performed by an EMS are:

Fault management: Requires the system to process events generated within the network environment and produce alerts when a network fault occurs. In addition, it involves event filtering and trend monitoring by maintaining data.

Configuration management: Involves enabling the technical support staff to modify the configuration of managed devices if the configuration needs to be synchronized. Configuration management is critical for the network, providing facilities such as VPNs and service standards including QoS.

Accounting Management: Involves maintenance of user-based accounts for tracking payments. This type of service is most suitable for telecommunication networks.

Performance Management: Involves monitoring and maintaining capabilities to ensure consistent network performance with respect to the service standards. An EMS usually possesses tools that have a predefined matrix for measuring network performance. Various network resources are constantly monitored and alerts are generated depending on thresholds set by network operators.

Security Management: Ensures secure access through implementation of user authentication and authorization techniques along with data integrity and audit checks.

In addition to these functions, an NMS includes utilities such as MIB browsers. A conventional MIB browser allows you to view the MIB tree, using the point-and-click GUI of a particular device. The display shows the MIB variable, the values, and the structure of the MIB for a particular device.

Recent trends show that large and small networks usually converge to a common IP environment. Anticipating this trend as a major turnover, Cisco has developed a general network infrastructure called the Architecture for Voice, Video, and Interface Data (AVVID). While AVVID may not be exhaustive, it provides many network layouts in a common IP environment.

AVVID provides the base infrastructure needed to design a scalable and reliable network setup by providing resilience for networks, communication, application, and businesses. As a generalized set of standards and best practices for setting up a scalable network environment, AVVID covers various aspects of network installation. It provides standards for ensuring high availability, QoS security, and mobility. This architecture enables you to focus on business-specific network requirements, while following the guidelines provided by the AVVID norms to set up the network. This architecture can be used when designing or redesigning a network. If a network has performance issues, you can check the guidelines provided by AVVID and identify the relevant problem areas.

Managing an enterprise network requires products that can monitor and detect bottlenecks for all aspects of the network, including connectivity issues, interface problems, and configuration mismatches. One of the popular enterprise management applications provided by Cisco is CiscoWorks. It is a family of products that provides end-to-end management of networks and increases the flexibility of support for new services such as voice, wireless, and content management devices. CiscoWorks constitutes various applications that are used for managing LAN and WAN networks along with mobile and wireless networks. These applications provide a number of tools and techniques for efficiently managing network devices, configurations, users, and services.

NOTE

All the solutions available with CiscoWorks are based on the AVVID architecture guidelines.

Some of the products that assist in performing enterprise-wide management of network devices and resources within the CiscoWorks family include:

- Management Solutions for LANs
- Management Solutions for WANs
- Management Solutions for VPN/Security
- Policy-Based Management of Quality of Service Networks
- Management for Small Networks

Management Solutions for LANs

The CiscoWorks LAN Management Solution (LMS) provides a framework of techniques that assist you in managing and monitoring LANs. The applications included in LMS are:

Cisco nGenius Real-Time Monitor: Provides multiuser Web access to information about RMON. It is a real-time monitor that uses RMON-enabled catalyst switches, internal network analysis modules, and LAN switch probes.

CiscoWorks Device Fault Manager (DFM): Follows an unconventional top-down approach for analyzing network problems. It identifies the fault condition using a problem signature, which represents a set of symptoms that occur due to the fault. This tool creates a causality mapping between the fault condition and the symptom, which in turn, determines the problem. This information is then coded into the DFM analysis model, which diagnoses the fault conditions based on events that already exist in the analysis model. Depending on the event that defines the fault condition, remedial measures can be taken. DFM works as an independent product or combines with various network and/or enterprise management systems for proactively detecting faults.

CiscoWorks Campus Manage: Manages and analyzes the complex structure of physical and logical network layouts. It provides powerful Layer 2 tools for configuring and managing physical and logical networks.

CiscoWorks Resource Manager Essential (RME): Covers inventory and configuration management aspects of the network. It is a powerful Web-based management application that manages inventory, configuration, and software updates for Cisco routers and switches, providing facilities to schedule periodic updates and generating alerts in case of new updates.

CiscoView: Provides updated status, statistics, and configuration information about Cisco products including routers, switches, hubs, and access servers. CiscoView is one of the most commonly used network management products by Cisco. It is a GUI-based management application, which displays the status of Cisco devices using color-coding schemes that define various severity levels for devices. This allows you to understand the status of network devices at a glance, and you do not have to go through all the statistics. This allows centralized management of network devices even on a remote network setup. In addition, it provides real-time monitoring and tracking of data about network device performance, traffic, and usage based on matrices such as percentage utilization, frames transmitted and received, and errors generated. CiscoView also provides the capability to modify configuration for trap, IP route, VLANs, and bridges.

Management Solutions for WANs

The CiscoWorks Routed WAN Management Solution allows easy fault detection and recovery, which reduces network downtime and performance downturns. CiscoWorks provides Routed WAN Management Solution for enterprise-wide management of WANs. It provides applications for configuring, administering, and troubleshooting routed WANs. In addition, it provides various tools for configuring and optimizing bandwidth use across WAN links. The available applications are:

CiscoWorks Internetworks Performance Monitor (IPM): Monitors network congestion and latency. As a part of the routed WAN management solution, this tool allows you to monitor the performance of multiple protocols across heterogeneous networks. In addition, it measures the response rate and availability of IP networks on a hop-by-hop basis.

IPM is used with SNA networks for mainframes. It measures the response time between the routers and the mainframe system.

CiscoWorks ACL Manager: Is an add-on application used with Resource Manager Essentials (RME) for managing information about Access Control Lists (ACLs). In addition, CiscoWorks RME and CiscoWorks View are used with routed WAN Management Solution.

Management Solutions for VPN/Security

Managing enterprise-wide networks is incomplete without taking into account the management of VPNs and security issues. The VPN/Security Management Solutions (VMS) provides services such as configuring, monitoring, and managing VPNs, firewalls, networks, and host-based intrusion detection. Constituents of VMS include:

Management Center for PIX Firewalls: Provides centralized management of Cisco PIX firewalls displaying information about a PIX device manager.

Management Center for IDS Sensors: Configures and deploys switch and network IDS sensors and detects intrusion and other security violations.

Management Center for VPN Routers: Manages security while configuring and deploying VPN connections for each network site.

Monitoring Center for Security: Provides a central console for monitoring events related to network, switch, and host IDS along with Cisco PIX and Cisco IOS devices.

Auto Update Server: Deploys updates and the latest configurations for the PIX firewall while working with the PIX firewall management center.

Cisco Secure Policy Manager: Provides predefined policies that serve as service standards for Cisco PIX- and IOX-based firewalls along with IPSec VPN routers.

Cisco IDS Host Sensor and Console: Prevents attacks from NIMDA and Code Red worm viruses. As a security and intrusion detection tool, it can identify an attack and prevent unauthorized access to network resources.

CISCOWORKS QoS POLICY MANAGER

Based on the AVVID infrastructure, the CiscoWorks QoS Policy Manager (QPM) provides centralized monitoring of QoS, policy control, and automated policy deployment. While QoS solutions are associated with 24X7 network availability, consistent performance, and reduced network downtime, QPM provides predefined policies for measuring system performance based on these policies. It enables you to set thresholds that generate alerts each time there is violation of a policy.

The product provides intelligent QoS monitoring that automatically checks for reasons of failure, enhancing the voice, video, and Internet application performance at reduced costs of management.

MODELING AND SIMULATION TOOLS

Modeling and simulation tools perform stress testing, network reconfiguration, and network setup redesign. Consider the implications of experimenting with expensive and mission critical network systems. The risk factor is quite high in attempting to experiment on a live network. However, it is important to check network traffic modulations, network performance, required configuration, and other network statistics in various mission-critical conditions. This would enable you to set up customized networks that suit the requirements of the organization. To allow this facility, network modeling and simulation techniques are used. In case there are problems in the network design that cause performance bottlenecks, simulation and modeling enable you to redesign networks.

The products used for modeling and simulation employ object-oriented techniques to check a particular network setup with a certain set of protocols and network devices, for example, traffic and routing algorithms for a LAN or WAN setup. For example, to check the network performance and response rate of a LAN using a particular routing algorithm, the simulator would allow you to import the required configuration and test the network performance for a particular scenario. The output generated from these tools provides information on response rates; node, link, and LAN utilization; packets lost; and network throughput.

The Cisco Service Level Management Suite provides "What If" simulation techniques for testing various permutations and combinations of network devices, protocols, routing algorithms, and network traffic. This is done to determine the network performance baseline in different situations. In addition, these techniques can determine the impact of configuration changes on the network, specifically if you need to deploy new network devices or increase the size of the network. This helps increase the rate of network fault tolerance while designing a real network setup.

Another tool that provides modeling and simulation capabilities is the NetSys Connectivity Service Manager, which allows you to monitor the actual network configuration data and use it to establish service level standards. It applies the View, Isolate, Solve, Test, and Apply (VISTA) methodology for diagnosing and troubleshooting network problems.

The NetSys Performance Manager is another tool that complements the Service Manager. It provides the ability to establish models for routing and flow transport on Cisco devices. This would enable you to analyze results of different traffic flows, topologies, routing configurations, and IOS.

The NetSys WAN Service Manager is another tool that provides modeling and simulation capabilities that help analyze the behavior of Cisco devices. Using this tool, you can gauge the impact on network performance for various traffic, topology, and network devices.

The Cisco NetSys Service Level Management suite enables you to import real-time data and configurations from the network, using configuration files. This allows you to view real Layer 2 and 3 multi-protocol network topologies with traffic flows and network device configurations. As a result, you prevent the risk of experimenting on a real network to analyze various situations of network downtime and their causes.

TRAFFIC GENERATORS

Traffic generators are network management functions that check for the correct functioning of network components and their associated links. The traffic generator function results in a continuous generation of data packets, which can be followed by an echo indicating that the data packet is sent or received. However, association of an echo message with data packets is optional. The source node generates the call for the traffic generator function. In addition, the traffic generator function monitors absorption of data packets transmitted by the source node until an acknowledgment is received from the destination node. The distribution period of the data packets is decided by the account number assigned by the traffic generator function to each data packet. In addition, the traffic generator sends a message that contains the name of the network component that is sending the data packets and the number of data packets received. The functions of the traffic generator are:

- Testing the inter-packet arrival gap. For this, Network Intrusion Detection Systems (NIDS) use traffic generators within high-speed networks.
- Generate multi-protocol traffic streams and raw packet rates. For this, Network Verification Services (NVS) use traffic generators.

CISCO IOS DIAGNOSTIC COMMANDS

Cisco IOS commands are built in with the software and provide both basic and high-level troubleshooting. Using these commands, you can gather information about the state of the network and devices. This data helps identify potential network performance bottlenecks and decide a corrective action to remove these bottlenecks. However, a major drawback of the Cisco IOS commands is that they consume a significant amount of resources and lead to deterioration in the performance of routers. As a result, these commands need to be properly planned and should be used only on a temporary basis.

There is a wide range of show and debug commands, which can be used for viewing status information and debugging at the event or process level. In addition, the ping and trace commands are available for checking connectivity and routing status over heterogeneous networks.

The Show Command

Cisco provides various IOS commands that display the status of routers, their interface, and utilization of router resources. In addition, the show commands enable isolation of network faults and monitor network behavior during installation and redesign. The functions performed by show commands are:

- Isolate faults related to applications, interface, nodes, and media
- Display router behavior during initial installation
- Monitor the network for congestion
- Monitor day-to-day operations of the network
- Provide status updates for servers and clients

Some of the commonly used show commands for troubleshooting are:

Show version: Displays details such as the system hardware configuration, software version, names and sources of configuration files, and boot images.

Show running-config: Displays the current router configuration.

Show startup-config: Shows the router configuration stored in NVRAM.

Show interfaces: Displays statistical data about all the interfaces configured on the router or access server.

Show controllers: Views the statistics of the interface card controller.

Show flash: Provides the layout and content of flash memory.

Show buffers: Shows data for buffer pools on the router.

Show memory summary: Displays the memory pool statistics and summary information about the system memory allocator and provides a block-by-block listing of the memory use.

Show process cpu: Provides information about processes that are active on the router.

Show stack: Shows information about the utilization of processes, and interrupts routines and provides information about the reason for the last system reboot.

Show cdp neighbors: Displays information about the degree of reachability of directly connected Cisco devices. This is a Cisco-proprietary command that assists in determining the operational status of the Physical and Data-link layers.

Show debugging: Displays information about the type of debugging that is enabled for a router.

Like the debug commands, some of the show commands listed are accessible only at the privilege exec mode (enable mode) of the router. Table 4.1 displays the show commands and the services of each command.

TABLE 4.1 The Show Commands and Their Services

Command	Service
access-lists	Lists the access lists of the routers
arp	Displays the ARP table
cdp	Displays the CDP information
clock	Displays the system clock

TABLE 4.1 *(continued)*

Command	Service
configuration	Displays the contents of the nonvolatile memory
controllers	Displays the interface controller status
flash	Displays information about the flash filesystem
frame-relay	Displays information about the Frame Relay
history	Displays information about the session command history
hosts	Displays the domain name, nameservers, and the host table
interfaces	Displays the interface status and configuration
ip	Displays information about IP protocol
ipx	Displays information about the Novell IPX protocol
isdn	Displays information about ISDN
ntp	Displays information about the Network Time Protocol (NTP)
protocols	Displays information about the currently active network routing protocols
running-config	Displays the current operating configuration
sessions	Displays information about the TELNET connections
startup-config	Displays the startup configuration
terminal	Displays the terminal configuration parameters
users	Displays information about the terminal lines
version	Displays information about the hardware and the software of the system

The Debug Command

The debug command provides detailed information about network traffic on an interface. It also provides error messages generated by nodes along with information about protocol-specific diagnostic packets. Debug commands run in the privilege exec mode of the CLI. However, as stated earlier, this tool needs to be used with caution to minimize the impact on the network. This means that you should use debug commands only for detecting and isolating network problems and not for day-to-day monitoring of network statistics. Debug commands are usually processor intensive, which increases the CPU utilization tremendously and has a negative impact on network performance. This occurs when the network router is preloaded with other tasks.

The adverse impact of debug commands can be avoided by following certain guidelines:

- Before using the debug command, use the no logging console global configuration to prevent other users from logging on to the console.
- Executing the debug command at the console port generates character-by-character interrupts, which increases the processor load. To prevent this, it is recommended that you use Telnet to access the privilege exec mode and the terminal monitor command to copy the output of the debug command to the current terminal. This enables remote viewing of the debug command output without connecting to the console port.
- Spool the output to a file to save the debug command output.
- Before using the debug command, check if network traffic is at its peak. Execute the command when the network traffic is not at its peak so that the data packet transmission is not hampered.
- Disable the command after the debugging process is over. To disable a particular debug command, execute the no debug command. You can also disable all types of debugging commands by using the no debug all command.

The output of the debug command differs depending on the type of debug command that is executed. While some debug commands generate a single line of output, others provide multiple lines of output per packet. Using the debug command, you can debug almost all networking features. Table 4.2 displays some the services that use the debug command:

TABLE 4.2 The Debug Command and Its Services

Command	Service
nbf	NetBIOS
arp	ARP and HP probe transactions
lapb	LAPB transactions
cls	CLS
apollo	Apollo
domain	Domain Name System (DNS)
broadcast	MAC broadcast packets

TABLE 4.2 *(continued)*

Command	Service
aps	Automatic Protection
cbus	CiscoBus events
ipx	IPX
atm	ATM signaling
llc2	LLC2 type II
nhrp	NHRP
callback	Callback activity
sna	SNA
modem	Modem activation
clns	CLNS
serial	Serial interface
compress	COMPRESS traffic
dhcp	DHCP client activity
custom-queue	Custom output queueing
decnet	DECnet
ipc	Interprocess communications
dialer	Dial on Demand
token	Token Ring
dnsix	Dnsix
lnx	qllc/llc2 conversion
cdp	CDP information
dxi	ATM-DXI
eigrp	EIGRP
x25	X.25
ethernet-interface	Ethernet network interface
translate	Protocol translation
filesys	Filesystem
sscop	SSCOP
fras	FRAS Debug

(continued)

TABLE 4.2 *(continued)*

Command	Service
ip	IP
stun	STUN
adjacency	Adjacency
isis	IS-IS
kerberos	KERBEROS
lane	LAN emulation
access-expression	Boolean access expression
lat	LAT
ppp	PPP
sdllc	SDLLC
local-ack	Local acknowledgment
async	Async interface
mop	DECnet MOP server
chat	Chat scripts activity
ntp	NTP
tacacs	TACACS
pad	X.25 PAD
lex	LAN extender
source	Source bridging
probe	Probe proxy requests
radius	RADIUS
rif	RIF cache
sdlc	SDLC
packet	Log unknown packets
confmodem	Modem configuration database
telnet	Telnet connections
smrp	SMRP
channel	Channel interface
xremote	XREMOTE
snmp	SNMP

TABLE 4.2 *(continued)*

Command	Service
frame-relay	Frame Relay
spanning	Spanning tree
dspu	DSPU
standby	Hot standby
fastethernet	Fast Ethernet interface
vlan	VLAN
tarp	TARP
tbridge	Transparent bridging
v120	V120
tftp	TFTP
apple	AppleTalk
entry	Incoming queue entries
tunnel	Tunnel information
priority	Priority output queueing
vg-anylan	VG-AnyLAN interface
vines	VINES
snapshot	Snapshot activity
dlsw	Data Link Switching events
xns	XNS
smf	Software MAC filter

Regardless of the precautionary conditions applied while debugging, it is advisable to use third-party tools for troubleshooting, especially when you need to gather information about protocols. A more feasible alternate for gathering information about protocols is a third-party protocol analyzer.

The Ping Commands

The ping command is used to monitor host reachability and connectivity over a network. The command can be executed from the user and privilege exec modes. In the user exec mode, ping allows connection with remote devices, basic level testing, and temporary modification of terminal settings.

The ping command checks basic network connectivity for AppleTalk, IP, Novell, Apollo, ISO Connectionless Network Service (CLNS), VINES, DECnet, or XNS network.

In the case of IP, the ICMP echo messages are sent by the ping command to check the interconnectivity among hosts in a network. If a problem occurs while echoing packets, ICMP reports errors and provides relevant information pertaining to IP packet addressing. For example, Figure 4.4 shows the output of a successful ping command, when ICMP successfully echoes the data packets.

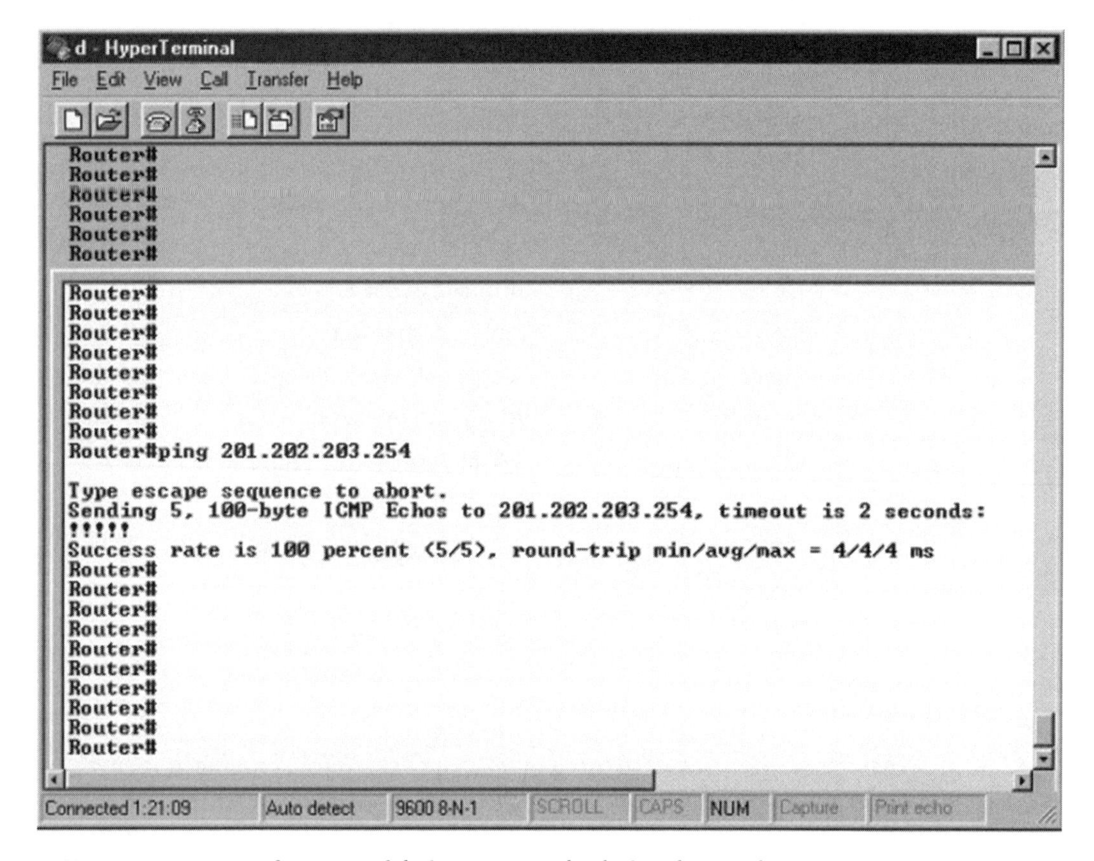

FIGURE 4.4 Output of a successful ping command echoing data packets.

The privilege exec mode allows you to specify the supported IP header options, which enable routers to perform detailed testing.

In the initial stages, some echo messages may timeout due to the need for setting the Address Resolution Protocol (ARP). The ping command works on a round trip path, in which the destination node sends a reply to the echo message, sent from the source node. While troubleshooting, do remember that the ping command usually fails because the destination host does not find a path to send a reply to the echo message.

The Trace Command

The trace command determines the routes taken by data packets that are transmitted from the source to the destination host. This command receives the error messages generated by routers when the datagram exceeds its TTL and provides the appropriate remedial action.

For example, you can use the trace diagnostic command to identify the route from the host to a specified IP address. Figure 4.5 shows the output of a successful trace command.

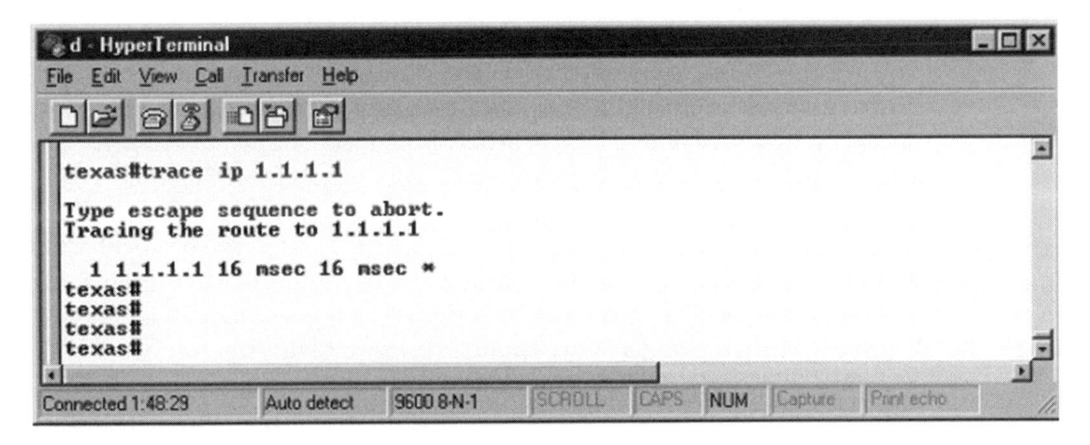

FIGURE 4.5 The trace diagnostic command showing its output.

The initial probe datagrams are transmitted with a TTL value of 1. The first router discards the probe datagrams and sends back a Time Exceeded error message. After this, the trace command sends a number of probe datagrams and displays the time of the round trip, increasing the TTL value by 1 after every third probe.

An outgoing datagram can receive two types of error message:

Time exceeded: Is generated when an intermediate router has seen and discarded the probe datagram.

Port unreachable: Is generated when the destination host receives the probe datagram and discards it because it could not be passed to any application.

The trace command is terminated when the destination host responds or the maximum TTL limit is exceeded. In addition, the trace command is terminated when it is interrupted using the escape sequence.

SUMMARY

In this chapter, you learned about troubleshooting tools developed by Cisco that operate at the different layers of the OSI model. We also reminded you about the judicious use of these tools and Cisco IOS commands, because they can hamper the performance of devices such as routers and switches. In the next chapter, we move on to troubleshooting tools specific to TCP/IP.

POINTS TO REMEMBER

- Cable testers are used at the Physical layer and are of two types, low spectrum and high spectrum.
- Low spectrum cable testers (scanners) provide information about the MAC layer such as the actual amount of network utilization and packet error rates.
- High spectrum cable testing involves checking for impedance mismatch, crimps, and various other physical problems in metallic and fiber cables.
- Digital interface testing tools cannot test media signals from Ethernet, Token Ring, or FDDI or protocol-related issues on a communication channel.
- Protocol analyzers provide details about the network traffic and protocol paths and offer potential solutions to critical network problems without affecting network performance.
- A protocol analyzer works in two modes: capture and display.
- Examples of protocol analyzers are Sniffer Pro and Ethereal.
- Different management tasks include fault detection and configuration, performance, and resource management.
- Examples of Cisco IOS commands are debug diagnostic and show process cpu.
- SNMP works on the Application layer and allows management of fault detection and network recovery.
- The four major components of an SNMP framework are managed devices, management server, management protocol, and management information model.
- Cisco WAN Manager (CWM) performs all the major functions of a network management system while CiscoView and Cisco DFM perform partial network monitoring functions.
- An EMS manages fault, configuration, accounting, performance, and security within a network.
- The Architecture for Voice, Video, and Interface Data (AVVID) network infrastructure provides many network layouts in a common IP environment.
- All solutions available with CiscoWorks, an EMS application, are based on the AVVID architecture guidelines.

- Applications to optimize and configure WAN links include CiscoWorks Internetworks Performance Monitor (IPM) and CiscoWorks ACL Manager.
- Modeling and simulation techniques perform stress testing and network reconfiguration and are used to redesign a network setup.
- The Cisco Service Level Management Suite provides the complete "What If" simulation techniques for determining the network performance baselines in different situations.
- The NetSys Connectivity Service Manager uses the View, Isolate, Solve, Test, and Apply (VISTA) methodology for diagnosing and troubleshooting network problems.
- The NetSys WAN Service Manager provides simulations to analyze the behavior of Cisco devices and their effect on network performance.
- A traffic generator utilizes Network Intrusion Detection Systems (NIDS) and Network Verification Services (NVS).
- The four types of Cisco IOS commands are show, debug, ping, and trace.
- Cisco IOS commands are used to view status information, debug at different levels, and check connectivity and routing status.

5 Troubleshooting TCP/IP

IN THIS CHAPTER

- TCP/IP Router Diagnostic Tools
- Troubleshooting Techniques

Transmission Control Protocol/Internet Protocol (TCP/IP) is the basic data transfer protocol of the Internet. TCP/IP is also used for data transmission over private networks, such as intranets and extranets. When a network connection is established using TCP/IP, TCP divides a stream of data into fixed size data packets while IP provides IP addresses to each data packet. When the data packets are received at the destination, IP arranges the data packets as per their IP addresses, and TCP converts the fixed size data packets into a stream of data. There can be instances when there are connectivity problems in a network connection. For example, the source computer is unable to establish connection or data packets are not received at the destination computer due to hardware failure. You may use troubleshooting tips and techniques pertaining to TCP/IP networks in these situations.

This chapter discusses the various TCP/IP router diagnostic tools, such as the ping, trace, show, and debug commands. These tools are also referred to as diagnostic commands. We will also discuss the troubleshooting techniques that are followed in TCP/IP networks.

TCP/IP ROUTER DIAGNOSTIC TOOLS

There are four major diagnostic commands that are executed on Cisco routers for troubleshooting TCP/IP network problems. These are:

- ping
- trace
- show
- debug

Ping

The ping command is a diagnostic tool to test the connectivity of hosts in the TCP/IP network. This command also isolates TCP/IP network problems. For example, if there is a connectivity related problem in a TCP/IP network, you can execute a ping command. The hosts that do not respond to the command indicate that there is a connectivity problem. The administrator can now concentrate only on the hosts that are not responding. The process of executing a ping command is depicted in Figure 5.1.

In Figure 5.1, the ping packet from PC1 with IP address 127.0.0.1 is sent to PC2 with IP address 202.0.0.2. However, the ping packet is not able to reach PC2 due to a hardware fault in the switch of PC2. As a result, a request timed out message is sent back to PC1. A ping packet is also sent from PC1 having IP address 127.0.0.1 to PC1 having IP address 202.0.0.1, which in turn sends a reply to the sender PC1.

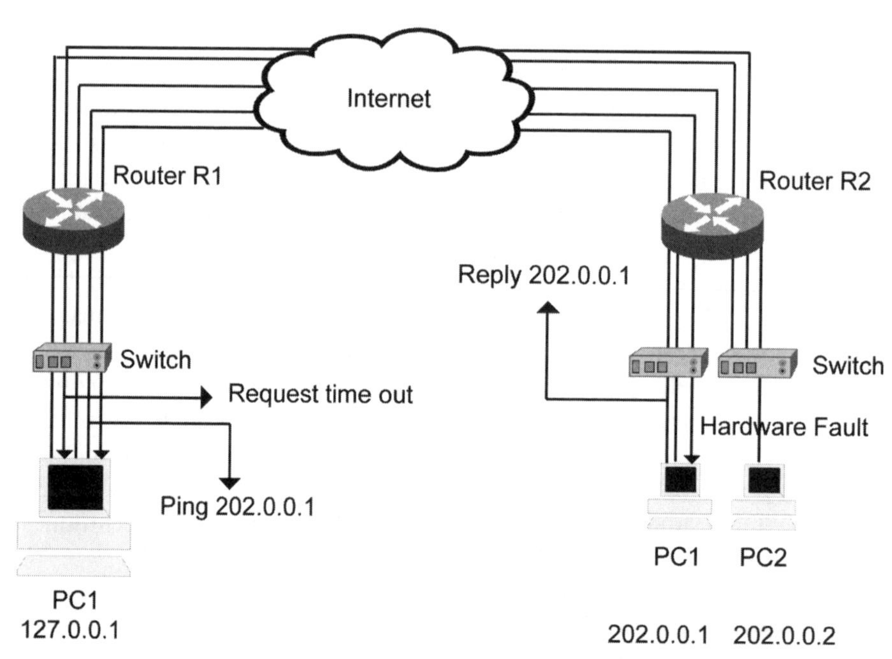

FIGURE 5.1 Testing TCP/IP network connectivity using the ping command.

The ping command sends a fixed 32-byte packet to the host for testing the network connectivity. This packet is an Internet Control Message Protocol (ICMP) echo packet. When the host receives an ICMP echo packet, it replies by echoing the packet back to the source. The connectivity of the host is tested in this manner. Cisco provides two modes to execute the ping command: the user exec mode and privilege exec mode.

Ping Command in the User Exec Mode

The user exec mode is used to perform a simple TCP/IP network connectivity test. This mode is for users who do not have administrator rights. The syntax of the ping command in user exec mode is:

```
Ping [Protocol] (Host name | Host address)
```

Different parameters of the ping command in the user exec mode are:

Protocol: Identifies the protocol for executing a ping command. IP is the default protocol. However, other protocols, such as IPX and AppleTalk, can be specified.

Host Name: Identifies the destination computer. The IP address of the destination computer is found using the Domain Name System (DNS).

Address: Identifies the IP address of the system to ping.

Listing 5.1 shows a sample output of the ping command in the user exec mode.

LISTING 5.1 Sample Output of ping Command in the User Exec Mode

```
texas#
texas#
texas#ping 192.12.1.10
Type escape sequence to abort.
Sending 5, 100-byte ICMP Echoes to 192.12.1.10, timeout is
  2 seconds:
!!!!!
Success rate is 100 percent (5/5), round-trip min/avg/max
  = 1/2/4 ms
texas#
texas#
texas#
```

In Listing 5.1, output of the ping command displays some characters, such as ! and ?. Each character has a unique description, as listed in Table 5.1.

TABLE 5.1 ping Command Characters and Descriptions

Character	Description
!	Indicates receipt of echo reply packet.
.	Indicates that the network server timed out while expecting the echo reply packet.
U	Indicates that the destination host is unreachable.

(continued)

TABLE 5.1 *(continued)*

Character	Description
N	Indicates that network is unreachable.
P	Indicates that protocol is unreachable.
Q	Indicates a source quench message sent by the destination computer or an Internet gateway as a request to the source computer to diminish the rate of sending data packets. This message is sent when the destination computer or an Internet gateway receives data packets at a rate greater than their processing speed.
M	Indicates the inability to fragment.
?	Indicates an unknown packet type.

ping Command in the Privilege Exec Mode

The privilege exec mode is also known as an extended ping and is used to perform a detailed check of TCP/IP network connectivity. Only administrators can enter into the privilege exec mode. To enter this mode, type the enable command at the user prompt. The syntax is

```
hostname> enable
hostname#
```

in which hostname> shows the user exec mode. Different parameters of the ping command in the privilege exec mode are:

Protocol [ip]: Indicates that the default protocol is IP.

Target IP address: Prompts to enter the IP address or the hostname of destination system.

Repeat count [5]: Prompts to enter the number of packets to be sent to the destination system. By default, five packets are sent.

Datagram size [100]: Prompts to enter the size of a packet in bytes to ping. The default packet size is 100 bytes.

Timeout in seconds [2]: Prompts to enter the timeout interval. The default timeout interval is 2 seconds.

Extended commands [n]: Prompts to specify any extended command, such as the data pattern and header options. To specify extended commands, enter "Yes." However, the default value is "No."

Listing 5.2 shows a sample output of the ping command in the privilege exec mode.

LISTING 5.2 Sample Output of the ping Command in the Privilege Exec Mode

```
texas#
texas#
texas#
texas#ping
Protocol [ip]:
Target IP address: 201.16.1.2
Repeat count [5]:
Datagram size [100]:
Timeout in seconds [2]:
Extended commands [n]:

Type escape sequence to abort.
Sending 5, 100-byte ICMP Echos to 201.16.1.2, timeout is
   2 seconds:
!!!!!
Success rate is 100 percent (5/5), round-trip min/avg/max
   = 1/2/4 ms
texas#
texas#
texas#
texas#
texas#
```

The sample output of the ping command when "Yes" is entered in the Extended commands [n] field is as shown in Listing 5.3.

LISTING 5.3 Output of the ping Command

```
texas#
texas#
texas#ping
Protocol [ip]:
Target IP address: 192.12.1.93
Repeat count [5]:
Datagram size [100]:
Timeout in seconds [2]:
Extended commands [n]: y
Source address or interface: 192.12.1.2
Type of service [0]:
Set DF bit in IP header? [no]:
```

```
Validate reply data? [no]:
Data pattern [0xABCD]:
Loose, Strict, Record, Timestamp, Verbose[none]: r
Number of hops [ 9 ]:
Loose, Strict, Record, Timestamp, Verbose[RV]:
Sweep range of sizes [n]:
Type escape sequence to abort.
Sending 5, 100-byte ICMP Echoes to 192.12.1.93, timeout is
  2 seconds:
Packet has IP options:  Total option bytes= 39, padded
    length=40
Record route: 0.0.0.0 0.0.0.0 0.0.0.0 0.0.0.0
    0.0.0.0 0.0.0.0 0.0.0.0 0.0.0.0 0.0.0.0
Reply to request 0 (1 ms). Received packet has options
Total option bytes= 40, padded length=40
Record route: 192.12.1.2 192.12.0.13 192.12.12.1
    192.12.1.93 192.12.0.14 192.12.0.21 192.12.1.2 *
    0.0.0.0 0.0.0.0
End of list
Reply to request 1 (4 ms). Received packet has options
Total option bytes= 40, padded length=40
Record route: 192.12.1.2 192.12.0.13 192.12.12.1
    192.12.1.93 192.12.0.14 192.12.0.21 192.12.1.2 *
    0.0.0.0 0.0.0.0
 End of list
Reply to request 2 (4 ms). Received packet has options
Total option bytes= 40, padded length=40
Record route: 192.12.1.2 192.12.0.13 192.12.12.1
    192.12.1.93 192.12.0.14 192.12.0.21 192.12.1.2 *
    0.0.0.0 0.0.0.0
End of list
Reply to request 3 (1 ms). Received packet has options
Total option bytes= 40, padded length=40
Record route: 192.12.1.2 192.12.0.13 192.12.12.1
    192.12.1.93 192.12.0.14 192.12.0.21 192.12.1.2 *
    0.0.0.0 0.0.0.0
 End of list
Reply to request 4 (1 ms). Received packet has options
Total option bytes= 40, padded length=40
Record route: 192.12.1.2 192.12.0.13 192.12.12.1
    192.12.1.93 192.12.0.14 192.12.0.21 192.12.1.2
    0.0.0.0 0.0.0.0
End of list
Success rate is 100 percent (5/5), round-trip min/avg/max
```

```
    = 1/2/4 ms
texas#
texas#
texas#
```

The ping command is the most useful command for testing the connectivity between hosts and routers in TCP/IP networks.

Trace

The trace command is a diagnostic tool that traces the path between two hosts to locate any connectivity problems. The syntax of the trace command is

```
trace [Host name | Host address]
```

The parameters of the trace command are:

Host name: Specifies host name to trace.

Host address: Specifies host address to trace.

When a trace command is executed, the following taks are performed:

1. The source host sends a packet to the local router with time-to-live (TTL) value set to 1.
2. The local router discards the packet and sends an ICMP packet with time exceeded back to the source host.
3. The source host resends additional packets with increased TTL values.

Three steps are repeated to generate a list of routers, as well as the path and time taken to reach those routers. Three probes are generated for every TTL value, and in case a response is not received within a certain time interval, the output is an asterisk, indicating that the destination host was unreachable. The trace command ends when:

- The destination host replies on receipt of the packet.
- The maximum TTL value of 30 is exceeded.
- A user interrupts the trace command by using the escape sequence.

Listing 5.4 shows a sample output of the trace command.

LISTING 5.4 Sample Output of the trace Command

```
texas#
texas#
```

```
texas#
texas#
texas#
texas#
texas#trace ip 201.16.1.2

"Type escape sequence to abort."
Tracing the route to 201.16.1.2
1 1.1.1.2 0 msec 16 msec 0 msec
2 5.5.5.6 20 msec 16 msec 16 msec
3 201.16.1.2 20 msec 16 msec *

texas#
texas#
texas#
texas#
texas#
```

*The * character indicates that the source host has reached timeout.*

Show Commands

The show command is a diagnostic tool to determine the exact cause of a performance slowdown in a TCP/IP network. Cisco IOS provides a range of show commands that display information about the rate of utilization of router resources, network interface status, and router configurations. The show commands can be executed in both the user and privilege modes.

Table 5.2 lists several show commands supported by Cisco.

TABLE 5.2 Cisco-supported Show Commands

Command	Description
access-lists	List access lists
arp	ARP table
cdp	CDP information
clock	Display the system clock

TABLE 5.2 *(continued)*

Command	Description
configuration	Contents of nonvolatile memory
controllers	Interface controller status
flash	Display information about the flash filesystem
frame-relay	Frame-Relay information
history	Display the session command history
hosts	IP domain name, nameservers, and host table
interfaces	Interface status and configuration
ip	IP information
ipx	Novell IPX information
isdn	ISDN information
ntp	Network time protocol
protocols	Active network routing protocols
running-config	Current operating configuration
sessions	Information about Telnet connections
startup-config	Contents of startup configuration
terminal	Display terminal configuration parameters
users	Display information about terminal lines
version	System hardware and software status

These commands are used to obtain essential information, such as IP traffic and IP neighbor reachability. The major show commands are:

- show ip route
- show ip ospf database
- show ip ospf interface
- show ip ospf neighbor
- show ip access-list
- show ip arp
- show ip protocols
- show ip traffic

Show ip Route

This command displays records in the IP routing table. An IP routing table stores all the IP routes and their corresponding information.

Listing 5.5 shows a sample output of the show ip route command without specifying any IP address.

LISTING 5.5 Sample Output of the show ip route Command

```
texas#
texas#
texas#show ip route
Codes: C - connected, S - static, I - IGRP, R - RIP,
       M - mobile, B - BGP
       D - EIGRP, EX - EIGRP external, O - OSPF,
       IA - OSPF inter area
       E1 - OSPF external type 1,
       E2 - OSPF external type 2, E - EGP
       i - IS-IS, L1 - IS-IS level-1, L2 - IS-IS level-2,
       - candidate default
       U - per-user static route

Gateway of last resort is not set
  C  172.16.0.0/16 is directly connected, 172.16.1.1
  C  1.0.0.0/8 is directly connected, 1.1.1.1
  R  192.16.1.0/24 [120/1] via 1.1.1.2, 00:06:19, Serial0

texas#
texas#
texas#
```

This output shows the routing table database. Networks connected with C are directly connected to the router, and R shows the networks learned by the routing protocol RIP.

You can also specify an IP address in the show ip route command to display detailed information, such as the network source interface.

Listing 5.6 shows a sample output of the show ip route command with a specified IP address.

LISTING 5.6 Sample Output of the show ip route Command

```
texas#
texas#
texas#show ip route 192.12.50.0
```

```
Routing entry for 192.12.50.0/24
  Known via "eigrp 100", distance 90, metric 2195456, type
    Internal
  Redistributing via eigrp 100
  Last update from 192.12.30.6 on Serial1, 00:02:03 ago
  Routing Descriptor Blocks:
  *192.12.30.6, from 192.12.30.6, 00:02:03 ago, via
  Serial1
    Route metric is 2195456, traffic share count is 1
    Total delay is 21000 microseconds, minimum bandwidth
    is 1544 Kbit
    Reliability 128/255, minimum MTU 1500 bytes
    Loading 1/255, Hops 1
texas#
```

Show ip ospf Database

This command displays detailed information regarding Open Shortest Path First (OSPF) ports and interfaces. The output of the show ip ospf command depends on the specified keyword. For example, when a router keyword is specified, information regarding router link states is displayed. Similarly, when the network keyword is specified, information regarding network link states is displayed.

Listing 5.7 shows a sample output of the show ip ospf command.

LISTING 5.7 Sample Output of the show ip ospf Command

```
texas#
texas#
texas#
texas#
texas#
texas#
texas#
texas#show ip ospf database
     OSPF Router with ID (172.16.1.1) (Process ID 10)

Router Link States (Area 0)

Link ID       ADV Router       Age       Seq#        Checksum Link count
1.1.1.2       1.1.1.2          146       0x80000003  0x9084    1
Net Link States (Area 0)
Link ID       ADV Router       Age       Seq#        Checksum
1.1.1.2       1.1.1.2          146       0x80000003  0x9084
```

Show ip ospf Interface

This command displays information regarding the interfaces on which OSPF is running. Listing 5.8 shows a sample output of the show ip ospf interface command.

LISTING 5.8 Sample Output of the show ip ospf interface Command

```
texas#
texas#
texas#
texas#show ip ospf interface
Serial0 is up, line protocol is up
Internet Address 1.1.1.1/8 , Area 0
  Process ID 10, Router ID 172.16.1.1, Network Type
BROADCAST, Cost: 64
  Transmit Delay is 1 sec, State DR, Priority 1
  Designated Router (ID) 172.16.1.1, Interface address
1.1.1.1
  Backup Designated router (ID) 1.1.1.2, Interface address
1.1.1.1
  Timer intervals configured, Hello 10, Dead 40, Wait 40, Retransmit 5
    Hello due in 00:00:02
  Neighbor Count is 1, Adjacent neighbor count is 1
    Adjacent with neighbor 1.1.1.2 (Backup Designated Router)
  Suppress hello for 0 neighbor (s)
Ethernet0 is up, line protocol is up
  Internet Address 172.16.1.1/16, Area 0
  Process ID 10, Router ID 172.16.1.1, Network Type
BROADCAST, Cost: 10
  Transmit Delay is 1 sec, State DR, Priority 1
  Designated Router (ID) 172.16.1.1, Interface address 172.16.1.1
  Backup Designated router (ID) 1.1.1.2, Interface address 172.16.1.1
  Timer intervals configured, Hello 10, Dead 40, Wait 40, Retransmit 5
    Hello due in 00:00:02
  Neighbor Count is 1, Adjacent neighbor count is 1
    Adjacent with neighbor 1.1.1.2 (Backup Designated Router)
  Suppress hello for 0 neighbor(s)
texas#
```

Show ip ospf Neighbor

This command is used to display information regarding the neighboring routers. Listing 5.9 shows a sample output of the show ip ospf neighbor command.

LISTING 5.9 Sample Output of the show ip ospf neighbor Command

```
texas#
texas#
texas#
texas#
texas#
texas#
texas#
texas#
texas#
texas#show ip ospf neighbor
Neighbor ID     Pri   State        Dead Time   Address      Interface
192.16.16.1     1     FULL/        00:00:02    1.1.1.2      Serial0
texas#
texas#
texas#
texas#
texas#
texas#
```

Show ip Access-list

This command is used to display information regarding specific or all current IP access lists in the 1 to 99 range. This information enables you to troubleshoot problems due to access or security settings.

Listing 5.10 shows a sample output of the show ip access-list command.

LISTING 5.10 Sample Output of the show ip access-list Command

```
texas#
texas#
texas#
texas#
texas#
texas#
texas#show ip access-list
Extended IP access list 105
    105 deny tcp any any eq 23 (0 matches)
    105 permit ip any any 21 (0 matches)
Extended IP access list 101
    105 deny tcp any any eq 23 (0 matches)
    101 permit tcp any any eq 20 (0 matches)
    101 permit tcp any any eq 21 (0 matches)
```

```
texas#
texas#
texas#
texas#
```

Show ip arp

This command displays information, such as the IP address and encapsulation type, present in the router's Address Resolution Protocol (ARP) cache.

Listing 5.11 shows a sample output of the show ip arp command.

LISTING 5.11 Sample Output of the show ip arp Command

```
texas#
texas#
texas#
texas#
texas#
texas#show ip arp
Protocol Address      Age(min)  Hardware Addr   Type  Interface
Internet 172.16.1.1          -  000C.4684.6106 ARPA  Ethernet0
Internet 172.16.1.2          1  000C.8730.4419 ARPA  Ethernet0
texas#
texas#
texas#
texas#
texas#
texas#
texas#
texas#
texas#
```

In Listing 5.11, a hyphen in the Age field indicates that the address is local. In addition, ARPA in the Type field indicates that the encapsulation type is Ethernet. SNAP in this field indicates that the encapsulation type is RFC 1042, and SAP indicates that the encapsulation type is IEEE 802.3.

Show ip Protocols

This command displays information, such as route redistribution, regarding IP routing protocols that are executed via the router.

Listing 5.12 shows a sample output of the show ip protocols command.

LISTING 5.12 Sample Output of the show ip protocols Command

```
texas#show ip protocols
Routing Protocol is "rip"
  Sending updates every 30 seconds, next due in 26 seconds
  Invalid after 180 seconds, hold down 180, flushed after 240
  Outgoing update filter list for all interfaces is
  Incoming update filter list for all interfaces is
  Redistributing:   rip
  Default version control: send version 1, receive any version
    Interface         Send Recv   Key-chain
    Serial0           1    1 2
    Ethernet0         1    1 2
  Routing for Networks:
    172.16.0.0
    1.0.0.0
  Routing Information Sources:
    1.1.1.1               120       00:00:00
  Distance: (default is 120)
```

Show ip Traffic

This command displays information collected by a router regarding the IP traffic.

The output of this command is organized as per the IP protocol and is shown in Listing 5.13.

LISTING 5.13 Sample Output of the show ip traffic Command texas#

```
texas#
texas#show ip traffic
IP statistics:
  Rcvd:  500 total, 500 local destination
         0 format errors, 0 checksum errors,
         0 bad hop count
         0 unknown protocol, 0 not a gateway
         0 security failures, 0 bad options,
         0 with options
  Opts:  0 end, 0 nop, 0 basic security,
         0 loose source route
         0 timestamp, 0 extended security, 0 record route
         0 stream ID, 0 strict source route, 0 alert,
         0 cipso
         0 other
  Frags: 0 reassembled, 0 timeouts, 0 couldn't reassemble
         0 fragmented, 0 couldn't fragment
```

```
    Bcast: 38 received, 52 sent
    Mcast: 390 received, 393 sent
    Sent:  396 generated, 0 forwarded
           0 encapsulation failed, 0 no route
  ICMP statistics:
    Rcvd: 0 format errors, 0 checksum errors, 0 redirects, 0
    unreachable
          0 echo, 0 echo reply, 0 mask requests, 0 mask
          replies, 0 quench
          0 parameter, 0 timestamp, 0 info request, 0 other
          0 irdp solicitations, 0 irdp advertisements
    Sent: 0 redirects, 0 unreachable, 0 echo, 0 echo reply
          0 mask requests, 0 mask replies, 0 quench, 0
          timestamp
          0 info reply, 0 time exceeded, 0 parameter problem
          0 irdp solicitations, 0 irdp advertisements
  UDP statistics:
    Rcvd: 0 total, 0 checksum errors, 0 no port
    Sent: 0 total, 0 forwarded broadcasts
  TCP statistics:
    Rcvd: 0 total, 0 checksum errors, 0 no port
    Sent: 0 total
  Probe statistics:
    Rcvd: 0 address requests, 0 address replies
          0 proxy name requests, 0 where-is requests, 0
          other
    Sent: 0 address requests, 0 address replies (0 proxy)
          0 proxy name replies, 0 where-is replies
  EGP statistics:
    Rcvd: 0 total, 0 format errors, 0 checksum errors, 0 no
    listener
    Sent: 0 total
  IGRP statistics:
    Rcvd: 0 total, 0 checksum errors
    Sent: 0 total
  OSPF statistics:
    Rcvd: 0 total, 0 checksum errors
          0 Hello, 0 database desc, 0 link state req
          0 link state updates, 0 link state acks
    Sent: 0 total
  IP-IGRP2 statistics:
    Rcvd: 402 total
    Sent: 406 total
  PIMv2 statistics: Sent/Received
```

```
  Total: 0/0, 0 checksum errors, 0 format errors
  Registers: 0/0, Register Stops: 0/0
IGMP statistics: Sent/Received
  Total: 0/0, Format errors: 0/0, Checksum errors: 0/0
  Host Queries: 0/0, Host Reports: 0/0, Host Leaves: 00
  DVMRP: 0/0, PIM: 0/0
ARP statistics:
  Rcvd: 0 requests, 0 replies, 0 reverse, 0 other
  Sent: 1 requests, 5 replies (0 proxy), 0 reverse
texas#
texas#
texas#
```

Debug Commands

The debug commands are diagnostic tools used to collect detailed information while diagnosing a facility, task, or protocol. However, debugging commands should be used only if needed, because most of the debugging commands require every packet to be scanned by the route processor. This causes additional overhead on the router.

The significant debug command used for debugging TCP/IP networks is debug ip rip.

Debug ip rip

This command displays information regarding Routing Information Protocol (RIP) transactions, such as updates of a routing table, that are sent by a serial interface, as shown in Listing 5.14.

LISTING 5.14 Sample Output of the debug ip rip Command

```
texas#debug ip rip
RIP protocol debugging is on
RIP: received update from 1.1.1.2 on Serial0
     192.16.1.0 in 1 hops
RIP: sending update to 255.255.255.255 via Serial0
(1.1.1.1)
subnet 172.16.0.0, metric 1

RIP: sending update to 255.255.255.255 via Ethernet0
(172.16.1.1)
 subnet 1.0.0.0, metric 1
 subnet 192.16.1.0, metric 2
RIP: received update from 1.1.1.2 on Serial0
     192.16.1.0 in 1 hops
```

```
RIP: sending update to 255.255.255.255 via Serial0
 (1.1.1.1)
 subnet 172.16.0.0, metric 1
RIP: sending update to 255.255.255.255 via Ethernet0
 (172.16.1.1)
 subnet 1.0.0.0, metric 1
 subnet 192.16.1.0, metric 2
RIP: received update from 1.1.1.2 on Serial0
     192.16.1.0 in 1 hops
RIP: sending update to 255.255.255.255 via Serial0
 (1.1.1.1)
 subnet 172.16.0.0, metric 1
```

Debug ip icmp

The debug ip icmp command enables you to determine if the router is sending and receiving ICMP messages. ICMP messages can include redirect or network unreachable messages. This command is used to troubleshoot end-to-end connectivity problems.

Debug ip packet

The debug ip packet command displays all the IP debugging information and IP security options. The IP debugging information includes packets received, generated, and forwarded. In addition, use this command to analyze messages sent from the local host to the remote host or vice versa while troubleshooting an end-to-end connection problem within a network.

Debug arp

The debug arp command checks the flow of data at the ARP. This command is used in a TCP/IP network when some nodes respond and some nodes do not. For this, the debug arp command checks if the router is sending or receiving ARP requests and replies.

TROUBLESHOOTING TECHNIQUES

TCP/IP networks, like OSI networks, require systematic troubleshooting techniques to be applied to identify and rectify network problems. A systematic treatment ensures that problems are timely detected and resolved, and the network is up and working. Network problems can range from simple connectivity problems between two hosts to incorrectly configured routers within a network.

To solve a specific network problem that is hampering functioning of the network, apply the problem-resolution model developed by Cisco. This model provides a systematic approach toward detecting the problem, identifying probable causes of the problem, selecting a probable solution to the problem, and finally, isolating the problem to solve it.

Problem Isolation in TCP/IP Networks

Problem isolation is a continuous phase of the process of identifying and solving a problem in TCP/IP networks.

Consider a scenario in which the local host is unable to access the remote host connected within a TCP/IP network. There can be several reasons, such as an incorrect address, subnet mask, or incorrect value specified in the default gateway entry of the local host. In addition, there can be connectivity problems between a specific local host and the remote host.

To isolate the actual problem of this scenario, follow a systematic path, as shown in Figure 5.2.

Figure 5.2 displays specific steps to identify the cause of the connectivity problem between the local host and the remote host. These steps can be performed in a sequence to isolate the problem between the local host and the remote host, as listed here:

1. Check the configuration of the local host and check for any incorrectly specified IP address, subnet mask address, and the default gateway. Specify correct values for an incorrectly specified address.
2. Check the connectivity between the local host and the router nearest to the local host. For this, use the ping and trace commands to test the connectivity. Use the ping command to test the connectivity to the nearest router and subsequently test the connectivity to the next routers. In addition, use the IP addresses to detect any problems with the DNS server.
3. Use the show command to determine the router configuration and its state if the router does not respond to the ping command.
4. Check the routing table by using the show ip route command.
5. Check the router's fast-switching cache by using the show ip cache command.
6. After checking the state and configuration of the routers, check the configuration of the remote host. For the remote host, check the correct values of IP address, subnet mask address, and default gateway address.

The execution of these steps is explained with this example.

Define Solutions, a U.S.-based company, deals with the manufacturing and marketing of computer hardware, such as monitors, peripherals, and CPUs. The headquarters of Define Solutions is located in Denver, where it coordinates with its suppliers across the world.

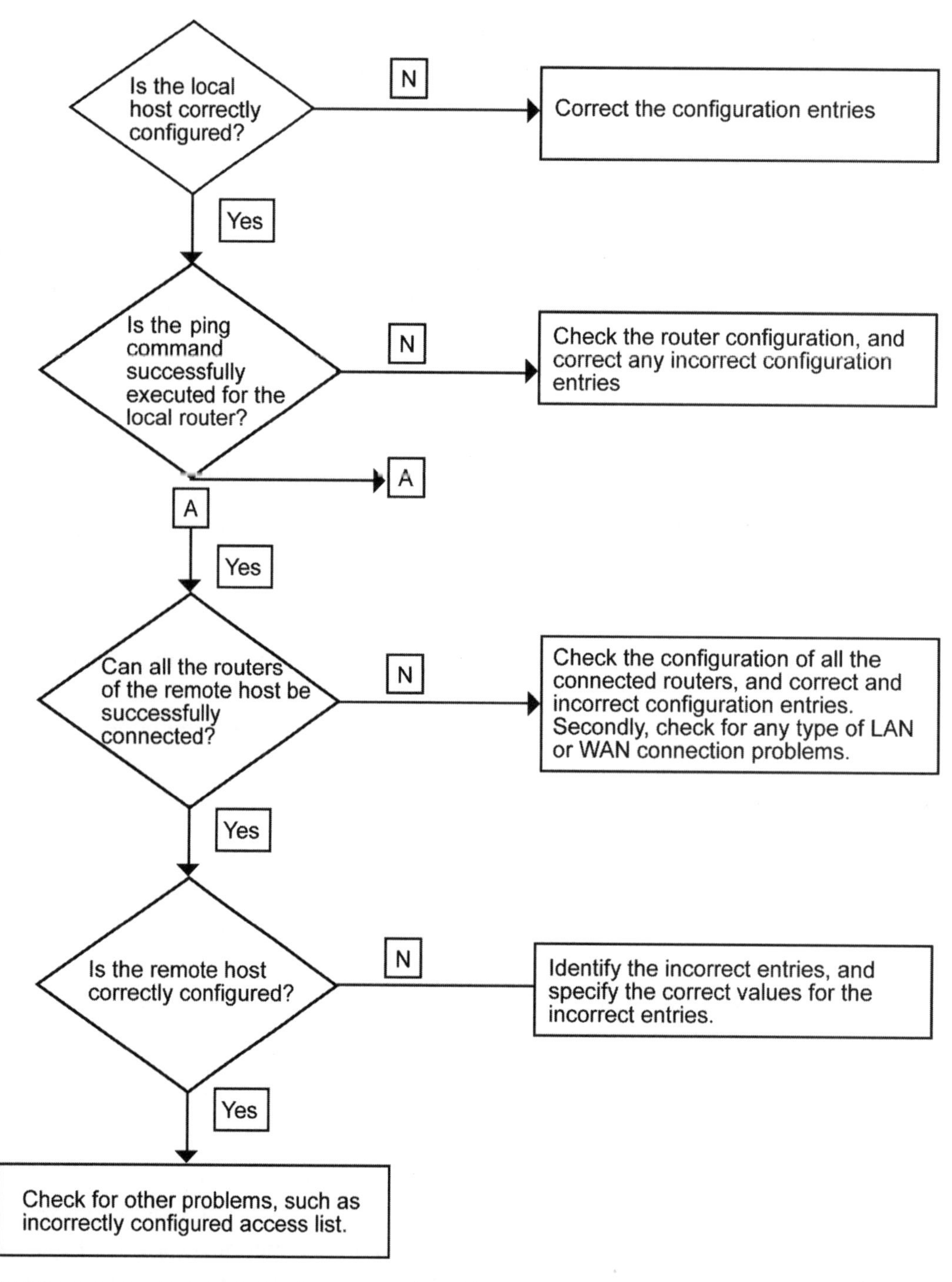

FIGURE 5.2 Problem isolation for a network problem.

Define Solutions has a TCP/IP-based network, which is divided into two subnets, subnet 1 and subnet 2. The network architecture of Define Solutions is shown in Figure 5.3.

In Figure 5.3, hosts A, B, and C are connected to subnet 1, and host X is connected to subnet 2. Host X has been recently added to the network, and the network administrator has incorrectly configured its subnet mask. The subnet mask of host B is also incorrectly configured. As a result, some hosts can access host X, but others cannot.

Such connectivity problems are common in TCP/IP-based networks. These errors occur due to several reasons, such as an incorrectly configured default gateway address or subnet mask address. For example, connectivity between host B and host X exists, but host A cannot connect to host X. This is because host X has an incorrectly configured subnet mask address.

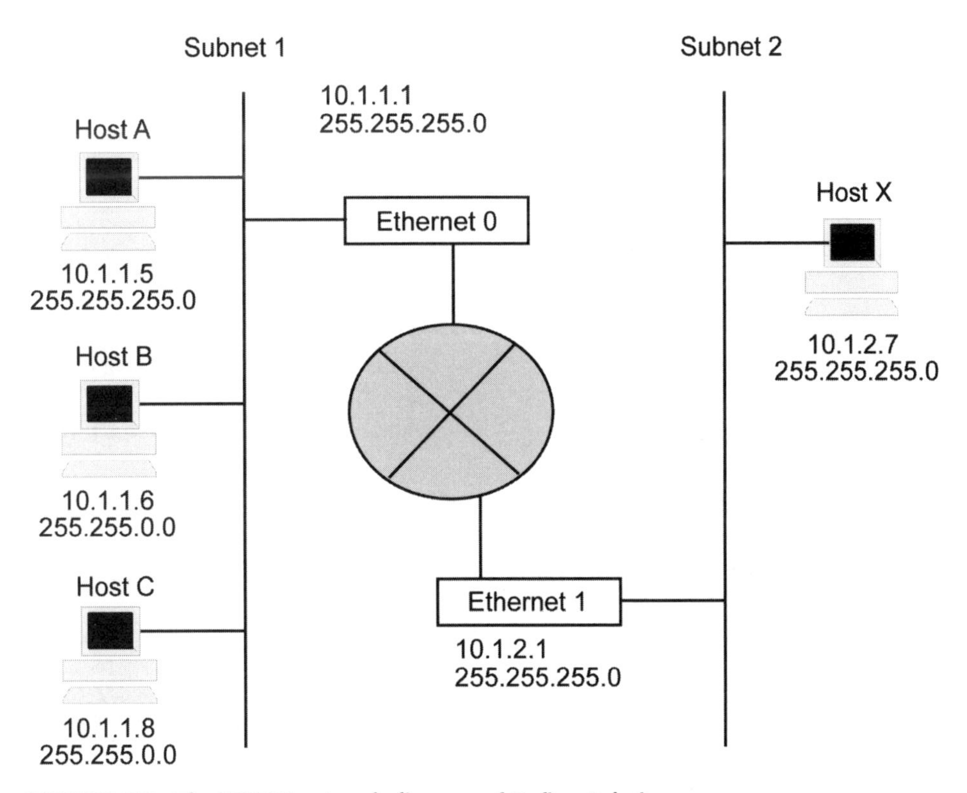

FIGURE 5.3 The TCP/IP network diagram of Define Solutions.

TCP/IP Problems and Symptoms

Now that you have learned the diagnostic commands used for troubleshooting TCP/IP network problems, let us discuss various connectivity related problems that are encountered in TCP/IP networks. Consider a scenario in a TCP/IP network in which hosts A and B are able to access host X, but hosts D and E cannot connect to host X. There can be several reasons for this problem, such as an incorrectly configured access list.

Table 5.3 displays various problems and their probable symptoms that can create connectivity problems in TCP/IP-based networks.

Based on the symptoms of the problems discussed in Table 5.3, specific action plans are devised and implemented to counter them.

Based on the symptoms listed in Table 5.3, problems of the TCP/IP-based networks are identified and isolated. After the problem is isolated, network administrators devise an action plan, which is implemented to resolve the isolated problem.

TABLE 5.3 TCP/IP Problems and Symptoms

Problem	Symptom
Host cannot access other networks through the router.	Misconfigured access list. Default gateway not specified on the local host. Discontiguous network due to link failure or poor design.
Host cannot access offnet hosts through the router.	Misconfigured subnet mask on the local host. Router between the connecting hosts is nonfunctional. Default gateway not specified on the local host.
Some network services are available while some network services are not accessible.	Extended access list is incorrectly configured.
Router receives duplicate routing updates and packets.	Bridge or repeater is working in parallel with the router.
Some protocols are being routed while some protocols are not being routed.	The access list is incorrectly configured.
Router becomes nonfunctional when redistribution is used.	Problem with the default administrative distance. Missing redistribute or default-metric command.

TABLE 5.3 *(continued)*

Problem	Symptom
Router as well as the host cannot connect to certain parts of the network.	Mismatch between the subnet mask configuration between the host and the router. Default gateway is not specified. Access list is incorrectly configured.
Users can access some hosts in a network but cannot access other hosts.	Default gateway is not specified on the remote hosts. The access list is incorrectly configured. Subnet mask or the address on the host or the router is incorrectly configured.
Users cannot connect to the network when one redundant path is nonfunctional.	The routing process has not converged. The access list or the routing filters are incorrectly configured. Discontiguous network due to link failure.

Table 5.4 displays the various problems you might encounter in a TCP/IP-based network and suggests possible action plans.

Sometimes, the network problems can be temporary and arise due to slow router convergence. This happens because some of the routing protocols may take a long time to converge, which results in network problems. For example, RIP takes a longer time to converge than IGRP.

TABLE 5.4 TCP/IP Problems and Action Plans

Problem	Action Plan
Incorrectly configured access list or filters.	Use the ping or trace command to isolate the router with a misconfigured access list. Use the show IP route command to check the routing table. Check for protocol exchanges by using commands, such as debug ip rip. Disable the access list temporarily, by using the no ip access-group command. Debug access lists.
Router between the connecting hosts is nonfunctional or down.	Use the ping command to identify and isolate the problem area in the entire

(continued)

TABLE 5.4 *(continued)*

Problem	Action Plan
	network. This application of the ping command is called *pinging outward.* Check for incorrect router configurations and correct them. Check for intermediate LAN or WAN problems.
Default gateway is not specified on the local or remote host.	Check the routing table of the host by using the netstat –rn command. If no default gateway is specified, use the route add default address 1 command. In this command, address 1 is the IP address of the default gateway. If the default gateway is already specified, boot with the default gateway, specify the IP address of the default gateway in the host at the location, files/etc/defaultrouter.
The subnet mask is incorrectly configured on the local host, remote host, or the router.	Check the hosts files, files/etc/netmasks and /etc/re.local. Check the configuration of the router using the show ip interface command. Check the configuration of the host using the ifconfig –a command.
Discontiguous network due to poor design or link failure.	Check the routes by using the show ip route command. Use the ping or trace command to determine where the network traffic gets congested or stopped. Assign a secondary address if a backup path is available. Fix the nonfunctional link in the network. Redo the topology or reassign the addresses.
Router has not converged.	Check the problem using the show ip route command.

Another major problem with TCP/IP networks is that the router starts receiving duplicate routing updates on different interfaces, which results in loss of connectivity and poor network performance. This happens because a bridge or a repeater may have been placed parallel to the router, which enables the router to view other routers on multiple interfaces. To counter such problems, use the show ip route command to identify routes for each interface. In addition, you can use the debug ip command to examine the routing updates received by the router.

You may also encounter problems while redistributing routes in TCP/IP networks. When routes are redistributed from a routing protocol in a domain to a routing protocol being used in another domain, translation problems may hamper connectivity. These translation problems occur because metrics of a protocol may not get translated into metrics of another protocol. This can lead to unsecured exchange of dynamic routing information between different routing protocols, such as RIP and IGRP. Such unsecured exchange of routing information may create routing loops, which can degrade network performance and operations. To counter such problems, use the default-metric or the redistribute protocol metric command.

Pinging Loopback and Local IP Addresses

To check if the TCP/IP network is working properly, execute the loopback ping command.

At the prompt of your computer, specify:

```
C:\ping TCP/IP ip address
```

An error generated by this command indicates that TCP/IP is not properly installed. However, if this ping command is successfully executed, test the functioning of your host computer by using the local IP address with the ping command. To test the local IP address, type this command at the prompt:

```
C:\ping local IP address
```

If this command displays an error, it indicates a problem with the network adapter.

Pinging the Router

Use the ping command to test the connectivity between your computer and the router. For testing the connectivity between the router and the host computer, use the console connection or Telnet to access the IP interface of the router.

Pinging DNS, Default Gateway, and WINS Servers

Use the ping command to test the connectivity with the DNS server and the WINS server. In addition, use the ping command to check the connectivity of your computer with the default gateway of your network.

To test the connectivity of your host with the DNS server, use the IP address of the DNS server with the ping command, as shown here:

```
Ping <IP address of the DNS Server>
```

You can get the IP address of the DNS server by using the ipconfig command. After executing the ipconfig command, view the IP addresses of the default gateway and the WINS server.

Use the ping command to test the connectivity of the host with the default gateway of the host and the WINS server. For this, use the IP addresses of the default gateway and the WINS server with the ping command. If the ping command is not executed successfully, check the IP address specified for the default gateway and the WINS server.

The WINS server is used for NetBIOS name resolution.

Pinging a Remote IP Address

Use the ping command to test the connectivity between the local host and a remote host. For this, specify the IP address of the remote host with the ping command. If the ping command displays an error message, it implies that either the subnet mask address of the remote host is incorrectly specified or the IP address of the remote address is incorrect.

Using the Tracert Tool

Use the Tracert tool to trace a TCP/IP packet crossing its way to the destination. The syntax of the Tracert tool is:

```
Tracert [-d] [-h maximum hops] [-j host-lists]
       [-w timeout] target-name
```

in which:

-d: Indicates not to resolve the addresses to the hostnames.

-h: Specifies the maximum number of hops that need to be taken to search for the target host.

-j: Specifies a loose source route along the host list.

-w: Specifies the number of milliseconds to wait before the next reply.

Target-name: Specifies the IP address of the target host.

SUMMARY

In this chapter, we learned about TCP/IP router diagnostic tools, such as the ping, trace, show, and debug commands. We also reminded you about troubleshooting techniques for different problems. In the next chapter, we move on to troubleshooting RIP environments.

POINTS TO REMEMBER

- The ping command is a diagnostic tool to test connectivity of hosts in a TCP/IP network and isolate TCP/IP network problems.
- The user exec mode is used to perform a simple TCP/IP network connectivity test.
- Parameters of the ping command in the user exec mode are protocol, host name, and address.
- The privilege exec mode, also known as an extended ping, is used to perform a detailed check of TCP/IP network connectivity.
- Different parameters of the ping command in the privilege exec mode are protocol, target IP address, repeat count [5], datagram size [100], timeout in seconds [2], and extended commands [n].
- The trace command is a diagnostic tool to trace the path between two hosts to locate any connectivity problems.
- The parameters of the trace command are host name and host address.
- The trace command ends when the destination host replies on receipt of a packet, the maximum TTL value of 30 is exceeded, or a user interrupts the trace command by using the escape sequence.
- The show command is a diagnostic tool to determine the exact cause of a performance slowdown in a TCP/IP network.
- The show ip route command displays records in the IP routing table.
- The show ip ospf command displays detailed information regarding OSPF ports and interfaces.
- The show ip ospf interface command displays information regarding the interfaces on which OSPF is running .
- The show ip ospf neighbor command is used to display information regarding the neighboring routers connected with OSPF.
- The show ip access-list command is used to display information regarding specific or all current IP access lists in the 1 to 99 range.
- The show ip arp command displays information present in the router's ARP cache.
- The show ip protocols command displays information regarding IP routing protocols that are executed on the router.
- The show ip traffic command displays information collected by the router regarding IP traffic.
- The debug ip rip command displays information regarding RIP routing transactions that are sent by a serial interface.
- The debug ip icmp command enables you to determine if the router is sending and receiving ICMP messages.
- The debug ip packet command displays all the IP debugging information and IP security options.

- The debug arp command checks the flow of data at the ARP.
- RIP takes a longer time to converge than IGRP.
- To counter unsecured exchange of routing information and routing loops, use the default-metric or redistribute protocol metric command.
- To check if a TCP/IP network is working properly, execute the loopback ping command.
- Use the Tracert tool to trace a TCP/IP packet crossing its way to the destination.

6 Troubleshooting RIP Environments

IN THIS CHAPTER

- Features of RIP
- Problem Isolation in RIP Environments
- Misconfiguration
- Configuration Problems
- Classless Routing
- Timer Problem
- Looping

Routing Information Protocol (RIP) is one of the first distance vector routing protocols to use hop count as a metric to calculate the best routing path to a destination network. Hop count indicates the number of routers crossed to reach a destination network. RIP uses the Bellman-Ford algorithm to compute the metric used for routing path decision.

With time, networks became more complex, and the RIPv1 protocol was found to be inefficient in handling most of the functions. This incompetence was resolved with the release of RIPv2. Both RIPv1 and RIPv2 are simple in comparison to the new generation of routing protocols that includes EIGRP, OSPF, and IS-IS.

FEATURES OF RIP

Before getting into the problems associated with RIP, let us review some features specific to distance vector routing protocols and RIP.

The characteristics of RIP include:

- Classful routing protocol; that is, it will not send the subnet mask information about the network routes mentioned in the routing updates.

- Networks that implemented RIPv1 often encountered classful routing loops, because RIPv1 does not carry subnet information. RIPv2 addressed this shortcoming. It was designed to carry the subnet mask information by supporting VLSM.
- Deals with several timers—such as update timer, invalid timer, hold-down timer, and flush timer—that define the operation behavior of the RIP environment.
- Uses a UCP datagram to transmit routing updates. Each routing update can carry about 25 route entries.
- Enables routers to advertise their routing updates to their directly connected neighbors and can receive routing advertisements about other remote networks from its directly connected neighbors. The frequency of such updates is 30 seconds.
- Being a classful routing protocol, it is incapable of supporting Variable Length Subnet Masking (VLSM).
- Does not support discontiguous network; that is, the subnets of a major network cannot be separated by another major network.
- Runs into the issues of routing loops while in the process of building up the routing table due to its tendency to converge. This results in late convergence, because RIP-enabled routers learn routing information from their directly connected neighbor. To counter the problems related to routing loops, RIP uses solutions, such as maximum hop count, split horizon, route poisoning, poison reverse, hold-down timer, and triggered updates.

The characteristics of RIPv1 and RIPv2 are listed in Table 6.1.

TABLE 6.1 Characteristics of RIPv1 and RIPv2

Feature	RIPv1	RIPv2
Category	Distance vector	Distance vector
Class type	Classful	Classless
VLSM	Not Supported	Supported
Authentication	Not Supported	Supported
Advertising address	Broadcast; 255.255.255.255	Multicast; 224.0.0.9
Autosummarization	Supported	Supported
Metric	Hop count	Hop count
Max hop count	15	15
Periodic interval	30 seconds	30 seconds

Though RIP is the one of the simplest routing protocols to deploy, there are numerous issues that lead to its unpredictable behavior. By understanding the basic functionalities and following a few guidelines, you will be successful in eliminating most of the RIP problems. The common problems encountered in RIP environments are:

- Misconfiguration
- Classless routing
- Timers
- Looping
- Version problems; incompatibility issues

PROBLEM ISOLATION IN RIP ENVIRONMENTS

There are a number of commands available in the Cisco IOS with which to isolate and troubleshoot problems in RIP networks. Table 6.2 lists the various show command options for RIP.

TABLE 6.2 Show Commands and Explanations

Command	Explanation
show ip rip	Displays all RIP routes. This command also displays this information for routes redistributed into RIP.
show ip protocols	Displays current RIP status. It includes RIP timer, filtering, version, RIP enabled interface, and RIP peer information.
show ip interface	Displays the details of all the interfaces of the router.

We will now discuss sample output of the various show commands. Listing 6.1 displays the output of the show ip route rip command.

LISTING 6.1 Output for show ip route rip

```
Router# show ip route rip
Router#sh ip route
Codes: C - connected, S - static, I - IGRP, R - RIP,
    M - mobile, B - BGP
    D - EIGRP, EX - EIGRP external, O - OSPF,
    IA - OSPF inter area
```

```
    N1 - OSPF NSSA external type 1,
    N2 - OSPF NSSA external type 2
    E1 - OSPF external type 1,
    E2 - OSPF external type 2,
    E - EGP
    i - IS-IS, L1 - IS-IS level-1, L2 - IS-IS level-2,
    ia - IS-IS inter area
    * - candidate default, U - per-user static route,
    o - ODR
    P - periodic downloaded static route

Gateway of last resort is not set.

  R  150.150.1.0/24 [120/1] via 150.150.2.1, 03:20:04, Serial1
  R  150.150.4.0/24 [120/1] via 150.150.3.3, 03:20:04, Serial0
```

Listing 6.2 displays the output of the show ip protocols command.

LISTING 6.2 Output for show ip protocols

```
    router> show ip protocols
    Routing Protocol is "rip"
    Sending updates every 30 seconds with +/-50%,
next due in 35 seconds
    Timeout after 180 seconds, garbage collect after 120 seconds
    Outgoing update filter list for all interface is not set
    Incoming update filter list for all interface is not set
    Default redistribution metric is 1
    Redistributing: kernel connected
    Default version control: send version 2, receive version 2
    Interface        Send  Recv
    Routing for Networks:
    eth0
    eth1
    1.1.1.1
    203.181.89.241
    Routing Information Sources:
    Gateway BadPackets BadRoutes  Distance Last Update
```

Listing 6.3 displays the output of the show ip interface command.

LISTING 6.3 Output for show ip interface

```
Router# show ip interface
Ethernet0 is up, line protocol is up
```

```
Internet address is 10.0.1.10, subnet mask is 255.255.255.0
Broadcast address is 255.255.255.255
Address determined by setup command
MTU is 1500 bytes
Helper address is not set
Directed broadcast forwarding is enabled
Multicast groups joined: 224.0.0.1 224.0.0.2
Outgoing access list is not set
Inbound access list is not set
Proxy ARP is enabled
Security level is default
Split horizon is enabled
ICMP redirects are always sent
ICMP unreachables are always sent
ICMP mask replies are never sent
IP fast switching is enabled
IP fast switching on the same interface is disabled
IP SSE switching is disabled
Router Discovery is disabled
IP accounting is disabled
TCP/IP header compression is disabled
Probe proxy name replies are disabled
Gateway Discovery is disabled
Serial0 is up, line protocol is up
Internet address is 200.2.2.1, subnet mask is 255.255.255.0
Broadcast address is 255.255.255.255
Address determined by setup command
MTU is 1500 bytes
Helper address is not set
Directed broadcast forwarding is enabled
Multicast groups joined: 224.0.0.1 224.0.0.2
Outgoing access list is not set
Inbound  access list is not set
Proxy ARP is enabled
Security level is default
Split horizon is enabled
ICMP redirects are always sent
ICMP unreachables are always sent
ICMP mask replies are never sent
IP fast switching is enabled
IP fast switching on the same interface is disabled
IP SSE switching is disabled
Router Discovery is disabled
IP accounting is disabled
```

```
TCP/IP header compression is disabled
Probe proxy name replies are disabled
Gateway Discovery is disabled
```

At times, monitoring is not enough to trace a problem. In such circumstances you need to debug the different events of RIP. Table 6.3 lists the different debug commands that can be used for thorough packet level troubleshooting for problems in RIP networks.

TABLE 6.3 Debug Commands and Explanations

Command	Explanation
debug rip events {}	This command displays RIP routing transactions. Sending and receiving packets, timers, and changes in interfaces are events shown.
debug rip packet {}	This command displays detailed information about the RIP packets. The origin and port number of the packet as well as a packet dump are displayed.
debug rip zebra {}	This command displays communication between rip and zebra. The main information will include addition and deletion of paths to the kernel and the sending and receiving of interface information.
show debugging rip {}	This command displays all information currently set for debug in RIP.

MISCONFIGURATION

Misconfiguration leads to disruption of routing communication between RIP-enabled routers. In addition, it impedes the advertising and receiving of valid routing updates between the routers. Following are the various possible areas pertaining to misconfiguration that may require troubleshooting in RIP environments:

- Network command misconfiguration
- Active VLSM subnets in the network topology
- Missing/incorrect network statement
- Inactive interfaces

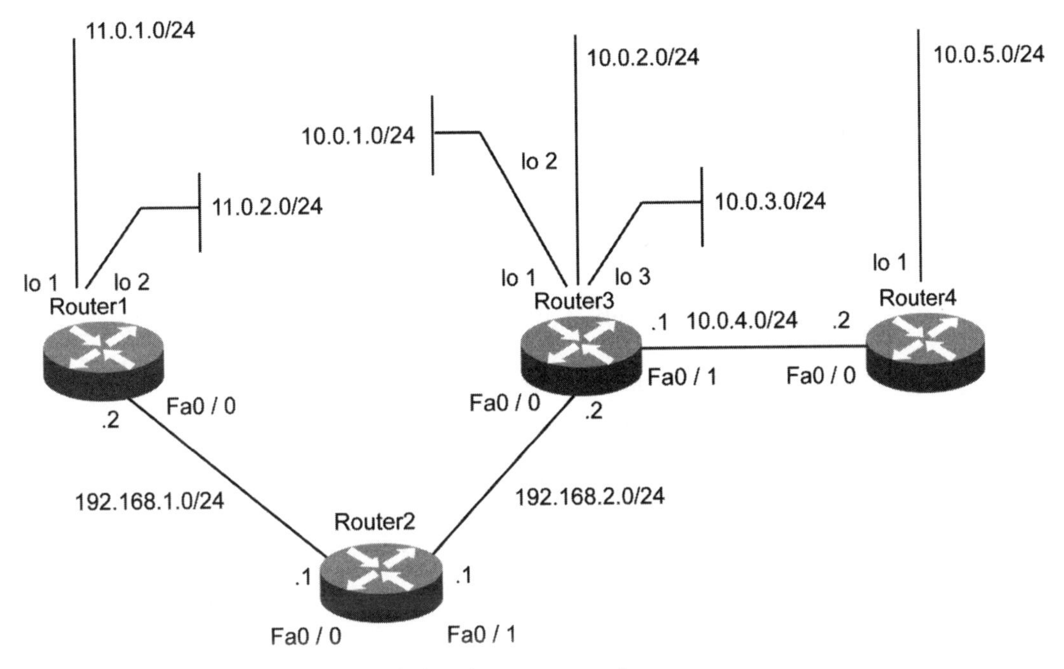

FIGURE 6.1 A routing scenario with four interconnected routers.

- Distribute list out is blocking the routes
- Advertised network interface is down
- Outgoing interface is defined as passive
- Neighbor statement misconfiguration
- Split-horizon is enabled

Consider a scenario in which four routers Router1, Router2, Router3, and Router4 are connected, as shown in Figure 6.1.

The scenario depicted in Figure 6.1 will be used to troubleshoot different problems as discussed in the following sections.

The output of the command router rip from all the four routers used in Figure 6.1 is displayed in the following code listings. Listing 6.4 shows the output of the command router rip from Router1.

LISTING 6.4 Output of the Command router rip from Router1

```
Router1#conf t
Router1(global)#interface fa0/0
Router1(global-if)#ip address 192.168.1.2 255.255.255.0
Router1(global-if)#no shut
```

```
Router1(global)#interface loopback 1
Router1(global-if)#ip address 11.0.1.1 255.255.255.0
Router1(global)#interface loopback 2
Router1(global-if)#ip address 11.0.2.1 255.255.255.0
Router1(global-if)#exit
Router1(global)# router rip
Router1(global-router)#network 192.168.1.0
Router1(global-router)#network 11.0.0.0
```

Listing 6.5 shows the output of the command router rip from Router2.

LISTING 6.5 Output of the Command router rip from Router2

```
Router2#conf t
Router2(global)#interface fa0/0
Router2(global-if)#ip address 192.168.1.1 255.255.255.0
Router2(global-if)#no shut
Router2(global)#interface fa0/1
Router2(global-if)#ip address 192.168.2.1 255.255.255.0
Router2(global-if)#no shut
Router2(global-if)#exit
Router2(global)# router rip
Router2(global-router)#network 192.168.1.0
Router2(global-router)#network 192.168.2.0
```

Listing 6.6 shows the output of the command router rip from Router3.

LISTING 6.6 Output of the Command router rip from Router3

```
Router3#conf t
Router3(global)#interface fa0/0
Router3(global-if)#ip address 192.168.2.2 255.255.255.0
Router3(global-if)#no shut
Router3(global)#interface loopback 1
Router3(global-if)#ip address 10.0.1.1 255.255.255.0
Router3(global)#interface loopback 2
Router3(global-if)#ip address 10.0.2.1 255.255.255.0
Router3(global)#interface loopback 3
Router3(global-if)#ip address 10.0.3.1 255.255.255.0
Router3(global)#interface fa0/1
Router3(global-if)#ip address 10.0.4.1 255.255.255.0
Router3(global-if)#no shut
Router3(global-if)#exit
Router3(global)# router rip
```

```
Router3(global-router)#network 192.168.2.0
Router3(global-router)#network 10.0.0.0
```

Listing 6.7 shows the output of the command router rip from Router4.

LISTING 6.7 Output of the Command router rip from Router4

```
Router4#conf t
Router4(global)#interface fa0/0
Router4(global-if)#ip address 10.0.4.2 255.255.255.0
Router4(global-if)#no shut
Router4(global)#interface loopback 1
Router4(global-if)#ip address 10.0.5.1 255.255.255.0
Router4(global-if)#exit
Router4(global)# router rip
Router4(global-router)#network 10.0.0.0
```

The output of the command show ip protocols from all the four routers used in Figure 6.1 is listed in the following listings. Listing 6.8 shows the output of the command show ip protocols from Router1.

LISTING 6.8 Output of the Command show ip protocols from Router1

```
Router1#sh ip protocols
Routing Protocol is "rip"
  Sending updates every 30 seconds, next due in 21 seconds
  Invalid after 180 seconds, hold down 180, flushed after 240
  Outgoing update filter list for all interfaces is
  Incoming update filter list for all interfaces is
  Redistributing:   rip
  Default version control: send version 1, receive any version
    Interface        Send  Recv   Key-chain
    FastEthernet0      1    1 2
  Routing for Networks:
    192.168.1.0
    11.0.0.0
  Routing Information Sources:
    192.168.1.1            120       00:00:0
  Distance: (default is 120)
```

Listing 6.9 shows the output of the command show ip protocols from Router2.

LISTING 6.9 Output of the Command show ip protocols from Router2

```
Router2#sh ip protocols
Routing Protocol is "rip"
  Sending updates every 30 seconds, next due in 31 seconds
  Invalid after 180 seconds, hold down 180, flushed after 240
  Outgoing update filter list for all interfaces is
  Incoming update filter list for all interfaces is
  Redistributing:   rip
  Default version control: send version 1, receive any version
    Interface         Send  Recv   Key-chain
    FastEthernet0      1     1 2
    FastEthernet0      1     1 2
  Routing for Networks:
    192.168.1.0
    192.168.2.0
  Routing Information Sources:
    192.168.1.2             120        00:00:03
    192.168.2.2             120        00:00:06
  Distance: (default is 120)
```

Listing 6.10 shows the output of the command show ip protocols from Router3.

LISTING 6.10 Output of the Command show ip protocols from Router3

```
Router3#sh ip protocols
Routing Protocol is "rip"
  Sending updates every 30 seconds, next due in 19 seconds
  Invalid after 180 seconds, hold down 180, flushed after 240
  Outgoing update filter list for all interfaces is
  Incoming update filter list for all interfaces is
  Redistributing:   rip
  Default version control: send version 1, receive any version
    Interface         Send  Recv   Key-chain
    FastEthernet0      1     1 2
    FastEthernet0      1     1 2
  Routing for Networks:
    10.0.0.0
    192.168.2.0
  Routing Information Sources:
    192.168.2.1             120        00:00:09
    10.0.4.2                120        00:00:09
  Distance: (default is 120)
```

Listing 6.11 shows the output of the command show ip protocols from Router4.

LISTING 6.11 Output of the Command show ip protocols from Router4

```
Router4#sh ip protocols
Routing Protocol is "rip"
  Sending updates every 30 seconds, next due in 22 seconds
  Invalid after 180 seconds, hold down 180, flushed after 240
  Outgoing update filter list for all interfaces is
  Incoming update filter list for all interfaces is
  Redistributing:   rip
  Default version control: send version 1, receive any version
    Interface         Send  Recv   Key-chain
    FastEthernet0      1     1 2
  Routing for Networks:
    10.0.0.0
  Routing Information Sources:
    10.0.4.1              120       00:00:09
Distance: (default is 120)
```

The command to view the routing table of a router is show ip route.

The output of the command show ip route for all the four routers as given in Figure 6.1 is listed in the following listings. Listing 6.12 shows the output of the command show ip route from Router1.

LISTING 6.12 Output of the Command show ip route from Router1

```
Router1#sh ip route
Codes: C - connected, S - static, I - IGRP, R - RIP,
       M - mobile, B - BGP
       D - EIGRP, EX - EIGRP external, O - OSPF,
       IA - OSPF inter area
       E1 - OSPF external type 1,
       E2 - OSPF external type 2, E - EGP
       i  - IS-IS, L1 - IS-IS level-1, L2 - IS-IS level-2,
       *  - candidate default
       U  - per-user static route

Gateway of last resort is not set
  C  11.0.1.0/24 is directly connected, Loopback1
  C  192.168.1.0/24 is directly connected, FastEthernet0/0
  C  11.0.2.0/24 is directly connected, Loopback2
  R  192.168.2.0/24 [120/1] via 192.168.1.1, 00:06:29, FastEthernet0/0
```

```
R  10.0.1.0/24 [120/2] via 192.168.1.1, 00:08:29, FastEthernet0/0
R  10.0.2.0/24 [120/2] via 192.168.1.1, 00:03:36, FastEthernet0/0
R  10.0.3.0/24 [120/2] via 192.168.1.1, 00:02:15, FastEthernet0/0
R  10.0.4.0/24 [120/2] via 192.168.1.1, 00:06:23, FastEthernet0/0
R  10.0.5.0/24 [120/3] via 192.168.1.1, 00:02:43, FastEthernet0/0
```

Listing 6.13 shows the output of the command show ip route from Router2.

LISTING 6.13 Output of the Command show ip route from Router2

```
Router2#sh ip route
Codes: C - connected, S - static, I - IGRP, R - RIP,
       M - mobile, B - BGP
       D - EIGRP, EX - EIGRP external, O - OSPF,
       IA - OSPF inter area
       E1 - OSPF external type 1,
       E2 - OSPF external type 2, E - EGP
       i - IS-IS, L1 - IS-IS level-1, L2 - IS-IS level-2,
       - candidate default
       U - per-user static route

Gateway of last resort is not set
  C  192.168.1.0/24 is directly connected, FastEthernet0/0
  C  192.168.2.0/24 is directly connected, FastEthernet0/1
  R  11.0.1.0/24 [120/1] via 192.168.1.2, 00:08:20, FastEthernet0/0
  R  11.0.2.0/24 [120/1] via 192.168.1.2, 00:01:42, FastEthernet0/0
  R  10.0.1.0/24 [120/1] via 192.168.2.2, 00:02:22, FastEthernet0/1
  R  10.0.2.0/24 [120/1] via 192.168.2.2, 00:01:24, FastEthernet0/1
  R  10.0.3.0/24 [120/1] via 192.168.2.2, 00:07:38, FastEthernet0/1
  R  10.0.4.0/24 [120/1] via 192.168.2.2, 00:03:43, FastEthernet0/1
  R  10.0.5.0/24 [120/2] via 192.168.2.2, 00:07:25, FastEthernet0/1
```

Listing 6.14 shows the output of the command show ip route from Router3.

LISTING 6.14 Output of the Command show ip route from Router3

```
Router3#sh ip route
Codes: C - connected, S - static, I - IGRP, R - RIP,
       M - mobile, B - BGP
       D - EIGRP, EX - EIGRP external, O - OSPF,
       IA - OSPF inter area
       E1 - OSPF external type 1,
       E2 - OSPF external type 2, E - EGP
       i - IS-IS, L1 - IS-IS level-1, L2 - IS-IS level-2,
       * - candidate default
       U - per-user static route
```

```
Gateway of last resort is not set
  C  10.0.1.0/24 is directly connected, Loopback1
  C  10.0.2.0/24 is directly connected, Loopback2
  C  10.0.3.0/24 is directly connected, Loopback3
  C  192.168.2.0/24 is directly connected, FastEthernet0/0
  C  10.0.4.0/24 is directly connected, FastEthernet0/1
  R  192.168.1.0/24 [120/1] via 192.168.2.1, 00:07:44, FastEthernet0/0
  R  10.0.5.0/24 [120/1] via 10.0.4.2, 00:01:14, FastEthernet0/1
  R  11.0.1.0/24 [120/2] via 192.168.2.1, 00:07:44, FastEthernet0/0
  R  11.0.2.0/24 [120/2] via 192.168.2.1, 00:09:26, FastEthernet0/0
```

Listing 6.15 shows the output of the command show ip route from Router4.

LISTING 6.15 Output of the Command show ip route from Router4

```
Router4#sh ip route
Codes: C - connected, S - static, I - IGRP, R - RIP,
       M - mobile, B - BGP
       D - EIGRP, EX - EIGRP external, O - OSPF,
       IA - OSPF inter area
       E1 - OSPF external type 1,
       E2 - OSPF external type 2, E - EGP
       i - IS-IS, L1 - IS-IS level-1, L2 - IS-IS level-2,
       * - candidate default
       U - per-user static route

Gateway of last resort is not set
  C  10.0.5.0/24 is directly connected, Loopback1
  C  10.0.4.0/24 is directly connected, FastEthernet0/0
  R  10.0.1.0/24 [120/1] via 10.0.4.1, 00:01:32, FastEthernet0/0
  R  10.0.2.0/24 [120/1] via 10.0.4.1, 00:02:27, FastEthernet0/0
  R  10.0.3.0/24 [120/1] via 10.0.4.1, 00:07:35, FastEthernet0/0
  R  192.168.2.0/24 [120/1] via 10.0.4.1, 00:05:40, FastEthernet0/0
  R  192.168.1.0/24 [120/2] via 10.0.4.1, 00:07:17, FastEthernet0/0
  R  11.0.1.0/24 [120/3] via 10.0.4.1, 00:01:24, FastEthernet0/0
  R  11.0.2.0/24 [120/3] via 10.0.4.1, 00:09:35, FastEthernet0/0
```

CONFIGURATION PROBLEMS

The problems encountered while configuring and installing RIP in a network are:

- RIP routes are missing from the routing table.
- RIP is not installing all possible equal-cost paths.

RIP Routes Missing from Routing Table

This condition arises due to a missing or misconfigured network statement. The most obvious area to check is the router configuration specific to routing, which also includes the network statement configuration.

When a network command is added to the routing configuration of the router, it performs these functions:

- Activates the RIP routing engine and sends and receives the RIP routing updates.
- Advertises in the routing update about the information regarding the connected interfaces and the IP address assigned to the network configured using the network command.

As a result, if the network command is missing or not configured properly, it will lead to inconsistency or missing routes in the routing table.

Consider a scenario in which Router4 (as shown in Figure 6.1) is unable to ping any IP address that belongs to networks 11.0.1.0 and 11.0.2.0. Any host (including router interfaces) belonging to networks 11.0.1.0 and 11.0.2.0 is unreachable from the Router4. This problem can arise due to two reasons:

- Missing network command
- Configuration of passive interface

Missing Network Command

As discussed earlier, when a network command is configured, it activates the router to send the routing updates to its neighboring routers. This can happen only if one or more interfaces belonging to the router are assigned the IP address that represents the configured network. The current state of the IP routing table of Router4 is shown in Listing 6.16.

LISTING 6.16 Current State of the IP Routing Table of Router4

```
Router4#sh ip route
Codes: C - connected, S - static, I - IGRP, R - RIP,
       M - mobile, B - BGP
       D - EIGRP, EX - EIGRP external, O - OSPF,
       IA - OSPF inter area
       E1 - OSPF external type 1,
       E2 - OSPF external type 2, E - EGP
       i - IS-IS, L1 - IS-IS level-1, L2 - IS-IS level-2,
       * - candidate default
       U - per-user static route
```

```
Gateway of last resort is not set
  C  10.0.5.0/24 is directly connected, Loopback1
  C  10.0.4.0/24 is directly connected, FastEthernet0/0
  R  10.0.1.0/24 [120/1] via 10.0.4.1, 00:01:32, FastEthernet0/0
  R  10.0.2.0/24 [120/1] via 10.0.4.1, 00:02:27, FastEthernet0/0
  R  10.0.3.0/24 [120/1] via 10.0.4.1, 00:07:35, FastEthernet0/0
  R  192.168.2.0/24 [120/1] via 10.0.4.1, 00:05:40, FastEthernet0/0
  R  192.168.1.0/24 [120/2] via 10.0.4.1, 00:07:17, FastEthernet0/0
```

From the routing table in Listing 6.16, it is evident that Router4 does not have any knowledge about networks 11.0.1.0 and 11.0.2.0. This may happen when Router4 is not receiving routing updates from the mentioned networks from its connected neighbors. To check this, confirm whether or not Router1 is advertising the 11.0.0.0 networks in its routing update. Use the command show ip protocols to confirm the networks that are being advertised. Listing 6.17 shows the output of the command show ip protocols by Router1 (of Figure 6.1) in its routing update.

LISTING 6.17 Output of the Command show ip protocols from Router1

```
Router1#sh ip protocols
Routing Protocol is "rip"
  Sending updates every 30 seconds, next due in 21 seconds
  Invalid after 180 seconds, hold down 180, flushed after 240
  Outgoing update filter list for all interfaces is
  Incoming update filter list for all interfaces is
  Redistributing:   rip
  Default version control: send version 1, receive any version
    Interface        Send  Recv   Key-chain
    FastEthernet0      1    1 2
  Routing for Networks:
    192.168.1.0
  Routing Information Sources:
    192.168.1.1            120      00:00:0
  Distance: (default is 120)
```

From the Listing 6.17, it is clear that the RIP in Router1 is active only for network 192.168.1.0 and not for network 11.0.0.0. This problem occurs due to a missing network command. To resolve this, configure the router with the network command to start sending routing updates about the 11.0.0.0 network to the directly connected neighboring routers. The mentioned configuration for Router1 is shown in Listing 6.18.

LISTING 6.18 Configuration for Router1

```
Router1#conf t
Router1(global)#router rip
Router1(global-router)#network 11.0.0.0
```

Once the router accepts the changes, recheck the configuration by using the commands show ip protocols in Router1 and show ip route in Router4.

The recheck of output of the command show ip protocols for Router1 is shown in Listing 6.19.

LISTING 6.19 Recheck of Output of the Command show ip protocols for Router1

```
Router1#show ip protocols
Routing Protocol is "rip"
  Sending updates every 30 seconds, next due in 21 seconds
  Invalid after 180 seconds, hold down 180, flushed after 240
  Outgoing update filter list for all interfaces is
  Incoming update filter list for all interfaces is
  Redistributing:   rip
  Default version control: send version 1, receive any version
    Interface          Send  Recv   Key-chain
    FastEthernet0       1     1 2
  Routing for Networks:
    192.168.1.0
    11.0.0.0
  Routing Information Sources:
    192.168.1.1            120       00:00:0
   Distance: (default is 120)
```

As per the output of the command in Listing 6.19, it is clear that Router1 is now routing for networks 192.168.1.0 and 11.0.0.0. This output confirms that at least Router1 is now sending updates about network 11.0.0.0 to its connected routers. The next step would be to check whether Router4 has updated its routing table with a new entry about 11.0.0.0, as required. Listing 6.20 shows the output of the command show ip route for Router4.

LISTING 6.20 Recheck of Output of the Command show ip route for Router4

```
Router4#sh ip route
Codes: C - connected, S - static, I - IGRP, R - RIP,
       M - mobile, B - BGP
       D - EIGRP, EX - EIGRP external, O - OSPF,
       IA - OSPF inter area
       E1 - OSPF external type 1,
       E2 - OSPF external type 2, E - EGP
```

```
       i - IS-IS, L1 - IS-IS level-1, L2 - IS-IS level-2,
       * - candidate default
       U - per-user static route

Gateway of last resort is not set
  C   10.0.5.0/24 is directly connected, Loopback1
  C   10.0.4.0/24 is directly connected, FastEthernet0/0
  R   10.0.1.0/24 [120/1] via 10.0.4.1, 00:01:32, FastEthernet0/0
  R   10.0.2.0/24 [120/1] via 10.0.4.1, 00:02:27, FastEthernet0/0
  R   10.0.3.0/24 [120/1] via 10.0.4.1, 00:07:35, FastEthernet0/0
  R   192.168.2.0/24 [120/1] via 10.0.4.1, 00:05:40, FastEthernet0/0
  R   192.168.1.0/24 [120/2] via 10.0.4.1, 00:07:17, FastEthernet0/0
  R   11.0.1.0/24 [120/3] via 10.0.4.1, 00:01:24, FastEthernet0/0
  R   11.0.2.0/24 [120/3] via 10.0.4.1, 00:09:35, FastEthernet0/0
```

The output of command show ip route in Router4 now displays the route entries for the 11.0.1.0 and 11.0.2.0 networks learned via the RIP routing protocol. Therefore, the problem of missing routes due to the absence of the network 11.0.0.0 command in Router1 is resolved.

Sometimes, after making the configuration changes, they are not reflected in the routing tables of the router. This happens due to delay in convergence. This means that the routers will send updates only when it is their time to advertise (default is 30 seconds). To change the routing table of a router, use the command clear ip route *, which refreshes the current routing table.

After observing how a missing network command restricts Router4 from receiving updates from networks 11.0.1.0 and 11.0.2.0, let us look at the other option: configuration of a passive interface.

Configuration of Passive Interface

As seen in Figure 6.1, Router4 is unable to ping any IP address that belongs to networks 11.0.1.0 and 11.0.2.0. The same problem can occur due to configuration of the passive interface. The passive interface command converts a routing interface into a silent interface. When configured for an interface, this command holds the interface from sending any routing updates, but like any other router interface, it can track the RIP routing update broadcasts on the link. The command also enables the router to receive responses/routing packets from any other router and modify the routing table based on that. This configuration is not RIP-specific, and therefore, will be the same for all routing protocols, such as IGRP, BGP, and OSPF. Passive interface configuration helps prevent the router from sending its routing table to any other router via the configured interface.

To disable the passive interface command, you can use the command no passive-interface IFNAME. Passive interface sets the specified interface to passive mode. On a passive mode interface, all receiving packets are processed as normal, and RIP neither multicasts nor unicasts RIP packets except with the RIP neighbor specified neighbor command.

Apart from the passive interface command, the other commands used are:

version VERSION: An RIP command used to set the RIP process's version (version can be 1 or 2).

ip rip send version VERSION: An interface command used to override the router's RIP version setting. This command enables the selected interface to send packets with RIP Version 1, RIP Version 2, or both. In the case of Version 1 or 2, packets are both broadcast and multicast.

ip rip receive version VERSION: An interface command used to set the version for incoming RIP packets. This command enables the selected interface to receive packets in RIP Version 1, RIP Version 2, or both.

Figure 6.2 depicts a passive interface configuration in Router1.

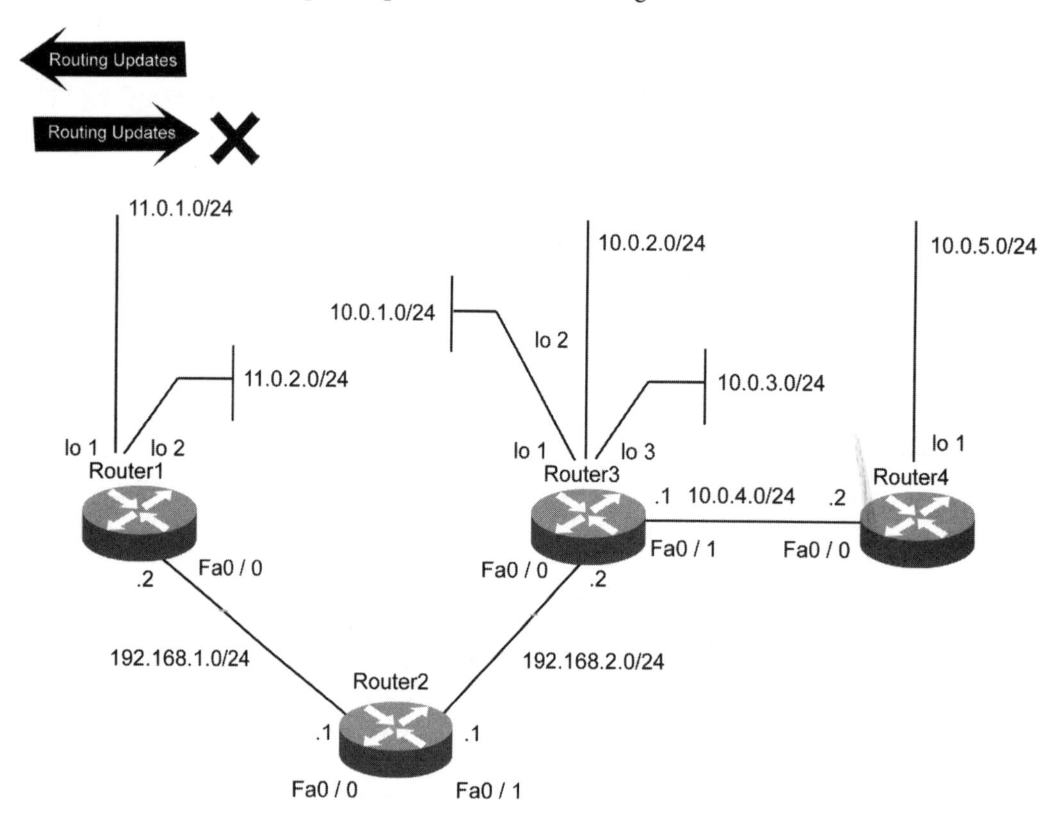

FIGURE 6.2 Passive interface configuration in Router1.

To verify if Router1 has been configured with a passive interface, we will look into its configuration using the command show running-config, shown in Listing 6.21.

LISTING 6.21 Configuration of Router1 Using the show running-config Command

```
Router1#show running-config
router rip
 network 192.168.1.0
 network 11.0.0.0
passive-interface FastEthernet0/0
```

From the configuration of Router1, it is clear that the interface FastEthernet0/0 of Router1 has been configured as passive (as per the command passive-interface FastEthernet0/0). Therefore, this interface of Router1 configured as passive now will only be able to track the RIP routing updates and hold all its outgoing RIP routing updates via the configured interface, as depicted in Listing 6.21. When the interface is removed from the passive interface mode, it will immediately start sending its routing table to neighboring routers and will reach Router4. Router4 will modify its routing table as per the received updates and will be able to reach any host in networks 11.0.1.0 and 11.0.2.0. To remove an interface from passive mode, use the command no passive-interface fa0/0, as shown in Listing 6.22.

LISTING 6.22 Removing an Interface Using the no passive-interface fa0/0 Command for Router4

```
Router1#conf t
Router1(config)#router rip
Router1(config-router)#no passive-interface fa0/0
```

RIP Is Not Installing All Possible Equal-Cost Paths

The RIP routing protocol supports equal-cost path load balancing; that is, when multiple equal cost paths exist, a router will use them to transmit data. Equal cost path refers to multiple routes to a common destination with the same metric values. By default, RIP can use up to four equal cost paths and can be configured to use a maximum of six equal cost paths. If the command is not configured accurately, it can lead to missing equal cost routes in the routing table. Figure 6.3 displays the redundant link between Router1 and Router4.

Consider a scenario in which there is a redundant link between Router1 and Router4 as shown in Figure 6.3. To access networks 10.0.1.0, 10.0.2.0 and 10.0.3.0 from Router1, there are two options:

- Router1 to Router4 and finally to Router3
- Router1 to Router2 and finally to Router3

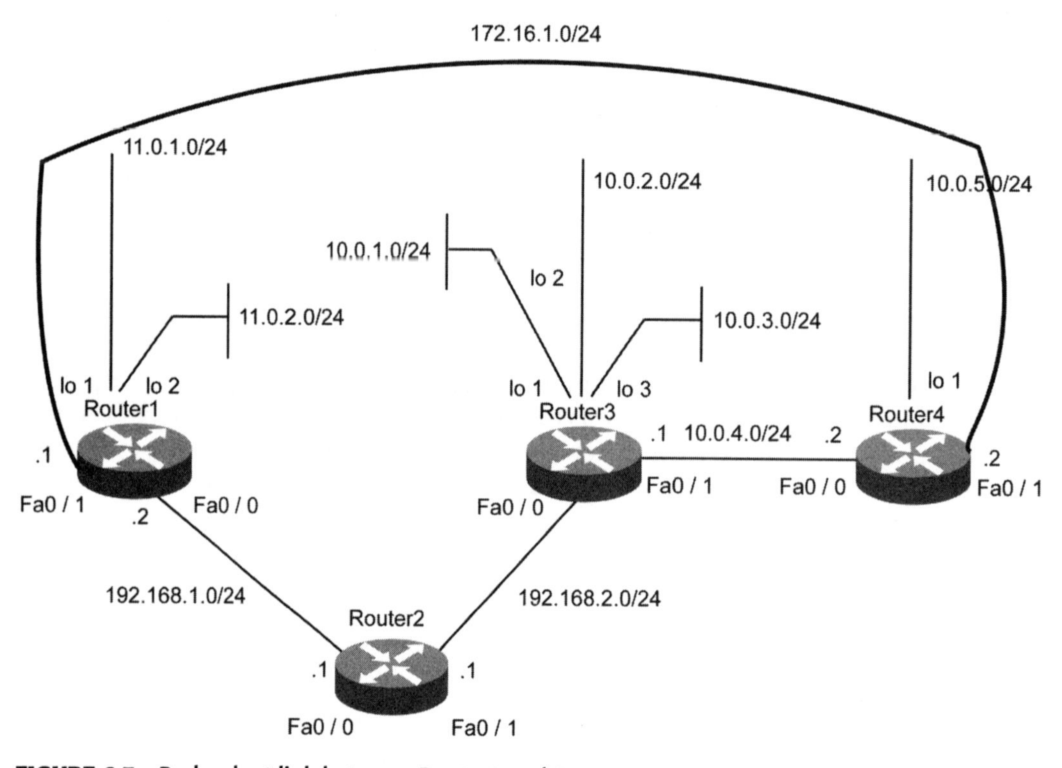

FIGURE 6.3 Redundant link between Router1 and Router4.

A detailed understanding of both possible paths reveals the common path cost. Therefore, both these paths should exist in the routing table of Router1.

The routing table of Router1 shows only one path for destination networks 10.0.1.0, 10.0.2.0, and 10.0.3.0, as shown in Listing 6.23.

LISTING 6.23 Routing Table of Router1

```
Router1#sh ip route
Codes: C - connected, S - static, I - IGRP, R - RIP,
       M - mobile, B - BGP
       D - EIGRP, EX - EIGRP external, O - OSPF,
```

```
       IA - OSPF inter area
       E1 - OSPF external type 1,
       E2 - OSPF external type 2, E - EGP
       i - IS-IS, L1 - IS-IS level-1, L2 - IS-IS level-2,
       * - candidate default
       U - per-user static route

Gateway of last resort is not set
  C  11.0.1.0/24 is directly connected, Loopback1
  C  192.168.1.0/24 is directly connected, FastEthernet0/0
  C  11.0.2.0/24 is directly connected, Loopback2
  R  192.168.2.0/24 [120/1] via 192.168.1.1, 00:06:29, FastEthernet0/0
  R  10.0.1.0/24 [120/2] via 192.168.1.1, 00:08:29, FastEthernet0/0
  R  10.0.2.0/24 [120/2] via 192.168.1.1, 00:03:36, FastEthernet0/0
  R  10.0.3.0/24 [120/2] via 192.168.1.1, 00:02:15, FastEthernet0/0
  R  10.0.4.0/24 [120/2] via 192.168.1.1, 00:06:23, FastEthernet0/0
  R  10.0.5.0/24 [120/3] via 192.168.1.1, 00:02:43, FastEthernet0/0
```

To identify the possible solution to the problem, it is important to understand the RIP updates that are being received by the Router1. To observe the routing updates, use the command debug ip rip. Listing 6.24 shows the routing updates in Router1.

LISTING 6.24 Routing Updates in Router1

```
Router1#debug ip rip
RIP:   received v1 update from 172.16.1.2 on FastEthernet0/1
10.0.5.0 in 1 hops
10.0.4.0 in 1 hops
10.0.3.0 in 2 hops
10.0.2.0 in 2 hops
10.0.1.0 in 2 hops
RIP:   received v1 update from 192.168.1.1 on FastEthernet0/0
192.168.2.0 in 1 hops
10.0.3.0 in 2 hops
10.0.2.0 in 2 hops
10.0.1.0 in 2 hops
RIP:   sending v1 update to 255.255.255.255 via
FastEthernet0/1(172.16.1.1)
subnet 11.0.1.0, metric 1
subnet 11.0.2.3, metric 1
RIP:       sending v1 update to 255.255.255.255 via
FastEthernet0/0(192.168.1.2)
subnet 11.0.1.0, metric 1
subnet 11.0.2.3, metric 1
```

From the debugging messages in Listing 6.24, it is evident that Router1 is receiving updates about the networks 10.0.1.0, 10.0.2.0, and 10.0.3.0 from two different sources but with the same metric value. Even after receiving updates from two sources, Router1 registers once, therefore, the origin of error is at Router1. To troubleshoot this, let us look into the configuration of Router1 using the command show run given in Listing 6.25.

LISTING 6.25 Configuration of Router1 Using the Command show run

```
Router1#show run
router rip
  network 192.168.1.0
  network 11.0.0.0
  maximum-path 1
```

By default, RIP-enabled routers can install four equal cost paths to a common destination, whereas in this scenario, the administrator has configured the router to install only one route to the destination. When only one path is desired for routing, the routers are configured using the command maximum-path 1. Therefore, the solution is to execute the command maximum-path 6 for the router to maintain a maximum of six equal cost paths for a destination. After this, Router1 will show two equal cost path routes each for networks 10.0.1.0, 10.0.2.0, and 10.0.3.0. This can be verified by looking at the routing table of Router1, as given in Listing 6.26.

LISTING 6.26 Creation of Equal Cost Path Routes Using the Command maximum-path 6 for Router1

```
Router1#sh ip route
Codes: C - connected, S - static, I - IGRP, R - RIP,
       M - mobile, B - BGP
       D - EIGRP, EX - EIGRP external, O - OSPF,
       IA - OSPF inter area
       E1 - OSPF external type 1,
       E2 - OSPF external type 2, E - EGP
       i - IS-IS, L1   IS-IS level-1, L2 - IS-IS level-2,
       - candidate default
       U - per-user static route

Gateway of last resort is not set
  C  11.0.1.0/24 is directly connected, Loopback1
  C  192.168.1.0/24 is directly connected, FastEthernet0/0
  C  11.0.2.0/24 is directly connected, Loopback2
```

```
R  192.168.2.0/24 [120/1] via 192.168.1.1, 00:06:29, FastEthernet0/0
R  10.0.1.0/24 [120/2] via 192.168.1.1, 00:08:29, FastEthernet0/0
               [120/2] via 172.16.1.2,  00:08:29, FastEthernet0/1
R  10.0.2.0/24 [120/2] via 192.168.1.1, 00:03:36, FastEthernet0/0
               [120/2] via 172.16.1.2,  00:03:36, FastEthernet0/1
R  10.0.3.0/24 [120/2] via 192.168.1.1, 00:02:15, FastEthernet0/0
               [120/2] via 172.16.1.2,  00:02:15, FastEthernet0/1
R  10.0.4.0/24 [120/2] via 192.168.1.1, 00:06:23, FastEthernet0/0
R  10.0.5.0/24 [120/3] via 192.168.1.1, 00:02:43, FastEthernet0/0
```

CLASSLESS ROUTING

Currently, the most acceptable method of routing is classless routing, which is deployed on the Internet. Classless routing differs from classful routing in that the IP addresses consist of a network number that is a combination of the network number and the subnet mask.

In the classless routing protocol, the subnet mask being advertised with the network information is not restricted to those defined by the address classes but can contain a variable number of bits to represent a network number. This further leads to aggregation of several networks into a single entry in a routing table and significantly reduces the routing overheads.

Classless routing protocols do not abide by the bit boundaries of the IP address class, which is different from classful routing protocols. Also, classless routing protocols have a field specifically to carry the subnet mask of the routes being advertised. Consider the example shown in Figure 6.4.

In Figure 6.4, three routers—Router1, Router2, and Router3—are configured in a RIPv2 configuration. The configuration of Router1 is shown in Listing 6.27.

LISTING 6.27 Configuration of Router1

```
Router1#conf t
Router1(global)#interface fa0/0
Router1(global-if)#ip address 192.168.1.2 255.255.255.0
Router1(global-if)#no shut
Router1(global)#interface loopback 1
Router1(global-if)#ip address 11.0.1.1 255.255.255.0
Router1(global)#interface loopback 2
Router1(global-if)#ip address 11.0.2.1 255.255.255.0
Router1(global-if)#exit
Router1(global)#router rip
```

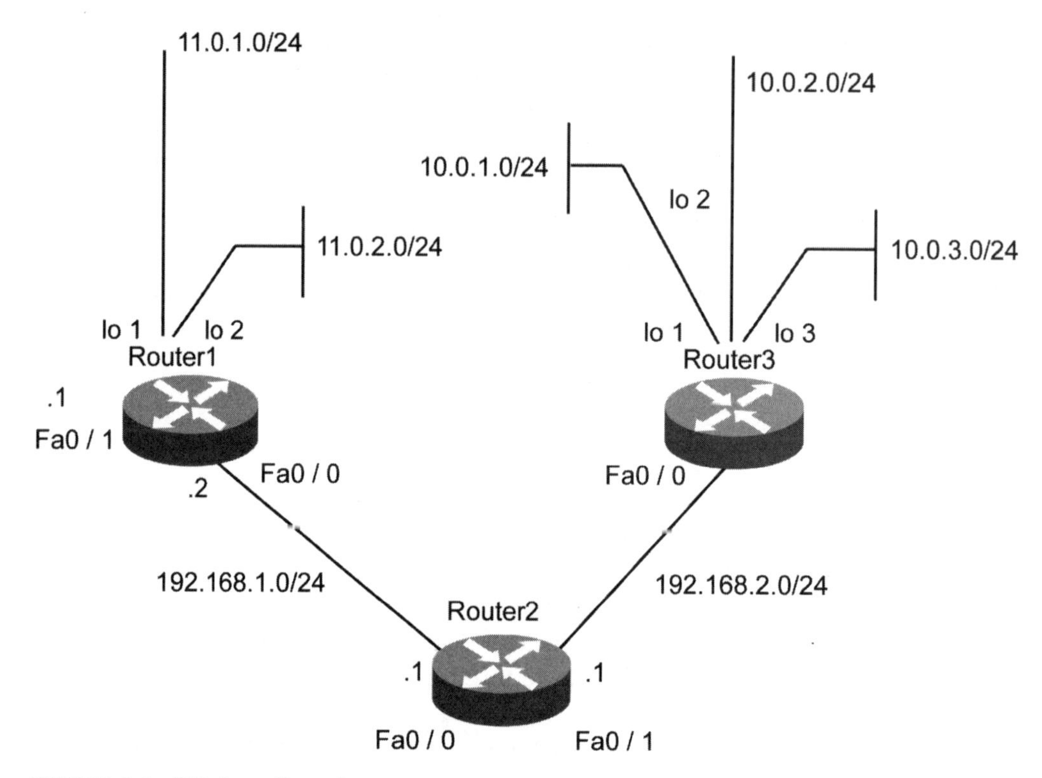

FIGURE 6.4 RIPv2 configuration.

```
Router1(global-router)#version 2
Router1(global-router)#network 192.168.1.0
Router1(global-router)#network 11.0.0.0
```

The configuration of Router2 is shown in Listing 6.28.

LISTING 6.28 Configuration of Router2

```
Router2#conf t
Router2(global)#interface fa0/0
Router2(global-if)#ip address 192.168.1.1 255.255.255.0
Router2(global-if)#no shut
Router2(global)#interface fa0/1
Router2(global-if)#ip address 192.168.2.1 255.255.255.0
Router2(global-if)#no shut
Router2(global-if)#exit
Router2(global)# router rip
Router2(global-router)#version 2
```

```
Router2(global-router)#network 192.168.1.0
Router2(global-router)#network 192.168.2.0
```

The configuration of Router3 is shown in Listing 6.29.

LISTING 6.29 Configuration of Router3

```
Router3#conf t
Router3(global)#interface fa0/0
Router3(global-if)#ip address 192.168.2.2 255.255.255.0
Router3(global-if)#no shut
Router3(global)#interface loopback 1
Router3(global-if)#ip address 10.0.1.1 255.255.255.0
Router3(global)#interface loopback 2
Router3(global-if)#ip address 10.0.2.1 255.255.255.0
Router3(global)#interface loopback 3
Router3(global-if)#ip address 10.0.3.1 255.255.255.0
Router3(global)#interface fa0/1
Router3(global-if)#ip address 10.0.4.1 255.255.255.0
Router3(global-if)#no shut
Router3(global-if)#exit
Router3(global)# router rip
Router3(global-router)#version 2
Router3(global-router)#network 192.168.2.0
Router3(global-router)#network 10.0.0.0
```

To verify the RIPv2 protocol configuration, use the command show ip protocols. It also displays the RIP version packets that are being advertised and received by the routers, as shown in Listing 6.30.

LISTING 6.30 Verifying RIPv2 Protocol Configuration

```
Router1#sh ip protocols
Routing Protocol is "rip"
Sending updates every 30 seconds, next due in 21 seconds
Invalid after 180 seconds, hold down 180, flushed after 240
Outgoing update filter list for all interfaces is
Incoming update filter list for all interfaces is
Redistributing:   rip
Default version control: send version 2, receive version 2
    Interface          Send  Recv   Key-chain
    FastEthernet0       2     2
Routing for Networks:
```

```
    192.168.1.0
    11.0.0.0
Routing Information Sources:
    192.168.1.1          120      00:00:0
Distance: (default is 120)
```

Listing 6.31 shows the verifying of RIPv2 protocol configuration using the command show ip protocols for Router2.

LISTING 6.31 Verifying RIPv2 Protocol Configuration Using the Command show ip protocols for Router2

```
Router2#sh ip protocols
Routing Protocol is "rip"
  Sending updates every 30 seconds, next due in 31 seconds
  Invalid after 180 seconds, hold down 180, flushed after 240
  Outgoing update filter list for all interfaces is
  Incoming update filter list for all interfaces is
  Redistributing:   rip
  Default version control: send version 2, receive version 2
    Interface        Send Recv   Key-chain
    FastEthernet0      2     2
    FastEthernet0      2     2
  Routing for Networks:
    192.168.1.0
    192.168.2.0
  Routing Information Sources:
    192.168.1.2          120      00:00:03
    192.168.2.2          120      00:00:06
  Distance: (default is 120)
```

Listing 6.32 shows the verifying of RIPv2 protocol configuration using the command show ip protocols for Router3.

LISTING 6.32 Verifying RIPv2 Protocol Configuration for Router3

```
Router3#sh ip protocols
Routing Protocol is "rip"
Sending updates every 30 seconds, next due in 19 seconds
Invalid after 180 seconds, hold down 180, flushed after 240
Outgoing update filter list for all interfaces is
Incoming update filter list for all interfaces is
Redistributing:   rip
Default version control: send version 2, receive version 2
```

```
Interface       Send  Recv   Key-chain
FastEthernet0  2     2
FastEthernet0  2     2
Routing for Networks:
10.0.0.0
192.168.2.0
Routing Information Sources:
192.168.2.1   120    00:00:09
Distance: (default is 120)
```

Listing 6.33 shows debugging messages from Router1, Router2, and Router3 that display the way these routers multicast their routing updates to the multicast address 224.0.0.9. The command to debug a router RIP message is debug ip rip.

LISTING 6.33 Debugging Using the debug ip rip Command on Router1

```
Router1#debug ip rip
RIP protocol debugging is on
Router1#
00:23:43: RIP: received v2 update from 192.168.1.1 on fa0/0
00:23:44:    192.168.2.0/24 via 0.0.0.0 in 1 hops
00:23:44:    10.0.0.0/24 via 0.0.0.0 in 2 hops
00:23:50: RIP: sending v2 update to 224.0.0.9 via fa0/0  (192.168.1.2)
00:23:50: RIP: build update entries
00:23:50:    11.0.0.0/24 via 0.0.0.0, metric 1, tag 0
```

Listing 6.34 shows the output of debug ip rip command for Router2.

LISTING 6.34 Debugging Using debug ip rip Command on Router2

```
Router2#debug ip rip
RIP protocol debugging is on
Router2#
00:23:43: RIP: received v2 update from 192.168.1.1 on fa0/0
00:23:44:    11.0.0.0/24 via 0.0.0.0 in 1 hops
00:23:43: RIP: received v2 update from 192.168.2.2 on fa0/0
00:23:44:    10.0.0.0/24 via 0.0.0.0 in 1 hops
00:23:50: RIP: sending v2 update to 224.0.0.9 via fa0/0  (192.168.1.1)
00:23:50: RIP: build update entries
00:23:50:    192.168.2.0/24 via 0.0.0.0, metric 1, tag 0
00:23:50: RIP: sending v2 update to 224.0.0.9 via fa0/0  (192.168.2.1)
00:23:50: RIP: build update entries
00:23:50:    192.168.1.0/24 via 0.0.0.0, metric 1, tag 0
```

Listing 6.35 shows the output of debug ip rip command for Router3.

LISTING 6.35 Debugging Using debug ip rip Command on Router3

```
Router3#debug ip rip
RIP protocol debugging is on
Router3#
00:23:43: RIP: received v2 update from 192.168.2.1 on fa0/0
00:23:44:    192.168.1.0/24 via 0.0.0.0 in 1 hops
00:23:44:    11.0.0.0/24 via 0.0.0.0 in 2 hops
00:23:50: RIP: sending v2 update to 224.0.0.9 via fa0/0  (192.168.2.2)
00:23:50: RIP: build update entries
00:23:50:    10.0.0.0/24 via 0.0.0.0, metric 1, tag 0
```

In these debug messages, via 0.0.0.0 shows that the advertising router interface address is the best address through which to reach the advertise network.

To verify whether these routers were able to learn the topology using the RIPv2 protocol, check their routing tables. To check the routing tables of these routers, use the command show ip route, as shown in Listing 6.36.

LISTING 6.36 Using the show ip route Command

```
Router1#sh ip route
Codes: C - connected, S - static, I - IGRP, R - RIP,
       M - mobile, B - BGP
       D - EIGRP, EX - EIGRP external, O - OSPF,
       IA - OSPF inter area
       E1 - OSPF external type 1,
       E2 - OSPF external type 2, E - EGP
       i - IS-IS, L1 - IS-IS level-1, L2 - IS-IS level-2,
       - candidate default
       U - per-user static route

Gateway of last resort is not set
  C  11.0.1.0/24 is directly connected, Loopback1
  C  192.168.1.0/24 is directly connected, FastEthernet0/0
  C  11.0.2.0/24 is directly connected, Loopback2
  R  192.168.2.0/24 [120/1] via 192.168.1.1, 00:06:29, FastEthernet0/0
  R  10.0.1.0/24 [120/2] via 192.168.1.1, 00:08:29, FastEthernet0/0
  R  10.0.2.0/24 [120/2] via 192.168.1.1, 00:03:36, FastEthernet0/0
  R  10.0.3.0/24 [120/2] via 192.168.1.1, 00:02:15, FastEthernet0/0
  R  10.0.4.0/24 [120/2] via 192.168.1.1, 00:06:23, FastEthernet0/0
  R  10.0.5.0/24 [120/3] via 192.168.1.1, 00:02:43, FastEthernet0/0
```

Listing 6.36 shows the routing table calculated by Router1 for the topology being discussed. You cannot identify whether the router tracked the routes using an RIPv1 or RIPv2 routing protocol. All the routes tracked by the router will be identified using R.

RIPv2 Cannot Reach Discontiguous Networks

This section discusses the default state of autosummarization, which is active in the case of RIPv2. The scenario shows that RIPv2 using its default configuration cannot reach a discontiguous network. To facilitate communication between these discontiguous networks, you should disable the autosummary (by default, autosummary is active in RIPv1 and RIPv2). To understand this, take, for example, the scenario presented in Figure 6.5.

In this scenario, Router1 is connected to networks belonging to a major network 10.0.0.0, and similarly, Router3 is connected to some subnets of the same network. The problem faced here is that Router3 is unable to ping 10.0.1.0 or 10.0.2.0 networks.

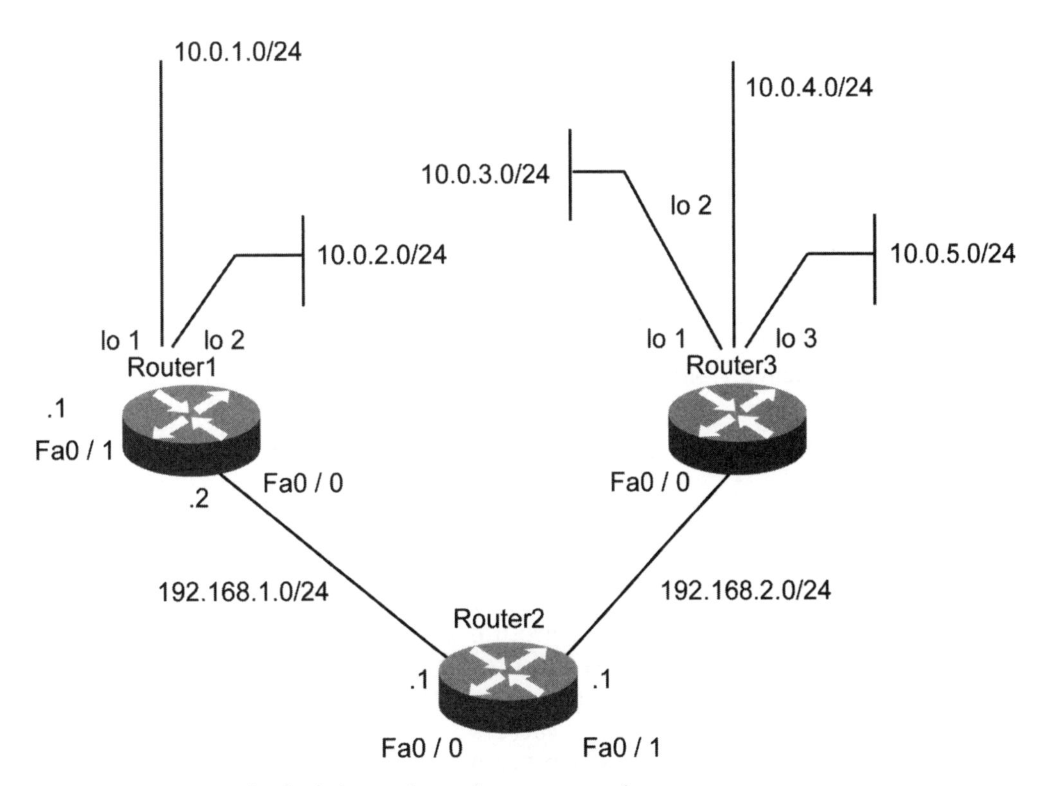

FIGURE 6.5 Scenario depicting a discontiguous network.

The configurations of the routers according to Figure 6.5 are discussed, followed by debug messages between them with their routing table. Listing 6.37 shows the configuration for Router1.

LISTING 6.37 Configuration for Router1

```
Router1#conf t
Router1(config)#router rip
Router1(config-router)#network 192.168.1.0
Router1(config-router)#network 10.0.0.0
Router1(config-router)#version 2
```

Listing 6.38 shows the configuration for Router2.

LISTING 6.38 Configuration for Router2

```
Router2#conf t
Router2(config)#router rip
Router2(config-router)#network 192.168.1.0
Router2(config-router)#network 192.168.2.0
Router2(config-router)#version 2
```

Listing 6.39 shows the configuration for Router3.

LISTING 6.39 Configuration for Router3

```
Router3#conf t
Router3(config)#router rip
Router3(config-router)#network 10.0.0.0
Router3(config-router)#network 192.168.2.0
Router3(config-router)#version 2
```

Listing 6.40 shows the debug messages from the routers.

LISTING 6.40 Debug messages for Router1

```
Router1#debug ip rip
00:23:43: RIP: received v2 update from 192.168.1.1 on fa0/0
00:23:44:    10.0.0.0/24 via 0.0.0.0 in 2 hops
00:23:44:    192.168.2.0/24 via 0.0.0.0 in 1 hops
00:23:50: RIP: sending v2 update to 224.0.0.9 via fa0/0  (192.168.1.2)
00:23:50: RIP: build update entries
00:23:50:    10.0.0.0/24 via 0.0.0.0, metric 1, tag 0
```

The debug messages in the Listing 6.40 for Router1 indicate that Router1 is advertising the major class network id (summarized network id) in its routing update, which is sent via the interface connected to a different major network id. Router1 also receives the routing update from Router2, which would carry the same summarized route but sent by Router3 to Router2. It will ignore the route information about network 10.0.0.0 sent by Router2, because it already has its interface connected to the 10.0.0.0 network.

Listing 6.41 shows the debug messages for Router3.

LISTING 6.41 Debug Messages for Router3

```
Router3#debug ip rip
00:23:43: RIP: received v2 update from 192.168.2.1 on fa0/0
00:23:44:    10.0.0.0/24 via 0.0.0.0 in 2 hops
00:23:44:    192.168.1.0/24 via 0.0.0.0 in 1 hops
00:23:50: RIP: sending v2 update to 224.0.0.9 via fa0/0  (192.168.2.2)
00:23:50: RIP: build update entries
00:23:50:    10.0.0.0/24 via 0.0.0.0, metric 1, tag 0
```

Similarly, Router3 sends the summarized major network id in its update and receives the same from the update advertised by Router2. It will ignore the route information about network 10.0.0.0 sent by Router2, because it already has its interface connected to the 10.0.0.0 network. To understand the debug messages, let us look at the routing tables of the routers in Listings 6.42 and 6.43.

LISTING 6.42 Routing Table for Router1

```
Router1#sh ip route
Codes: C - connected, S - static, I - IGRP, R - RIP,
       M - mobile, B - BGP
       D - EIGRP, EX - EIGRP external, O - OSPF,
       IA - OSPF inter area
       E1 - OSPF external type 1,
       E2 - OSPF external type 2, E - EGP
       i - IS-IS, L1 - IS-IS level-1, L2 - IS-IS level-2,
       - candidate default
       U - per-user static route

Gateway of last resort is not set
     10.0.0.0 is subnetted, 2 subnets
  C  10.0.1.0/24 is directly connected, Loopback1
  C  10.0.2.0/24 is directly connected, Loopback2
```

```
     C  192.168.1.0/24 is directly connected, FastEthernet0/0
     R  192.168.2.0/24 [120/1] via 192.168.1.1, 00:06:29, FastEthernet0/0
```

Listing 6.43 shows the routing table for Router3.

LISTING 6.43 Routing Table for Router3

```
Router1#sh ip route
Codes: C - connected, S - static, I - IGRP, R - RIP,
       M - mobile, B - BGP
       D - EIGRP, EX - EIGRP external, O - OSPF,
      IA - OSPF inter area
       E1 - OSPF external type 1,
       E2 - OSPF external type 2, E - EGP
       i - IS-IS, L1 - IS-IS level-1, L2 - IS-IS level-2,
       - candidate default
       U - per-user static route

Gateway of last resort is not set
     10.0.0.0 is subnetted, 3 subnets
  C  10.0.3.0/24 is directly connected, Loopback1
  C  10.0.4.0/24 is directly connected, Loopback2
  C  10.0.5.0/24 is directly connected, Loopback2
  C  192.168.2.0/24 is directly connected, FastEthernet0/0
  R  192.168.1.0/24 [120/1] via 192.168.2.1, 00:06:29, FastEthernet0/0
```

While sending the routing update to their neighbors, both Router1 and Router3 send major classful route information about network 10.0.0.0. These routers send this information because the interface used to send these updates belongs to a different major network than the advertised networks. Due to this, Router1 and Router3 assume that network boundaries for 10.0.0.0 end with them (classful routing property). Therefore, if you try and ping the IP address 10.0.1.1 or 10.0.2.1 from Router3, it is not successful.

Listing 6.44 shows the pinging IP address 10.0.1.1 for Router3.

LISTING 6.44 Pinging IP Address 10.0.1.1 for Router3

```
Router3#ping 10.0.1.1
Type escape sequence to abort.
Sending 5, 100-byte ICMP Echos to 10.0.1.1, timeout is 2 seconds:
.....
Success rate is 0 percent (0/5), round-trip min/avg/max = 1/2/4 ms
```

The debug messages indicate that the routing updates sent by Router1 and Router3 for networks like 10.0.1.0, 10.0.2.0, and so on are summarized to 10.0.0.0. As a result, routers do not maintain a proper routing table, which they should in case of RIPv2 protocol. To rectify this, configure every router with the command no auto-summary to avoid summarization of the routes. Configure this change in the router configuration mode. After the changes are made, the routers will receive the routing updates with the right subnet mask information and update the routing table. After the configuration changes, Router3 will communicate with the 10.0.1.0 and 10.0.2.0 networks. Listing 6.45 shows successful pinging of IP address 10.0.1.1 for Router3.

LISTING 6.45 Pinging IP Address 10.0.1.1 for Router3 Successfully

```
Router3#ping 10.0.1.1
Type escape sequence to abort.
Sending 5, 100-byte ICMP Echos to 10.0.1.1, timeout is 2 seconds:
!!!!!
Success rate is 0 percent (0/5), round-trip min/avg/max = 1/2/4 ms
```

Compatibility Between RIPv1 and RIPv2

As per the scenario depicted in Figure 6.6, Router2 is running RIPv2 and Router1 is running RIPv1. The problem with this topology is that Router2 and Router1 are not learning their partner routes. Both these routers are ignoring the routing updates advertised by the peer router because they are different versions of routing protocols. This is more evident on observing the debug messages from these routers.

Listing 6.46 shows the output of the debug ip rip command.

LISTING 6.46 Debugging Router1 Using the debug ip rip Command

```
Router1#debug ip rip
RIP protocol debugging is on
Router1#
00:23:43: RIP:  ignored v2 packet from 192.168.1.1 on fa0/0
00:23:50: RIP:  sending v1 update to 255.255.255.255 via fa0/1
               (192.168.1.2)
               Subnet 11.0.0.0, metric 1
```

Listing 6.47 shows the output of the debug ip rip command.

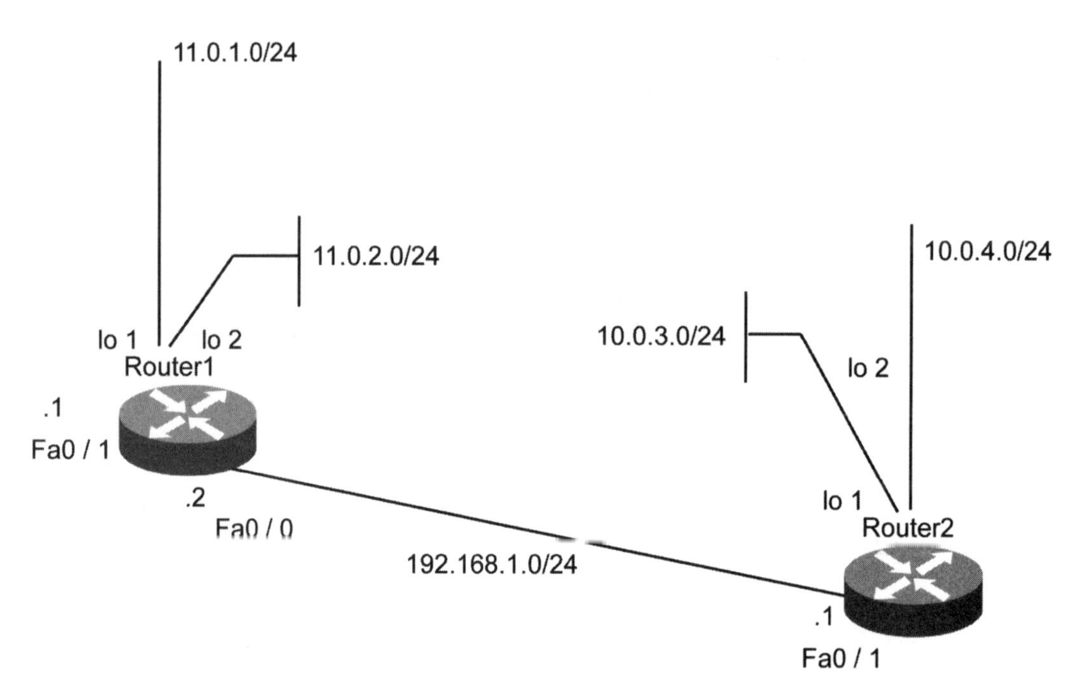

FIGURE 6.6 Scenario in which Router1 is configured with RIPv1 and Router2 is configured with RIPv2.

LISTING 6.47 Debugging Router2 using the debug ip rip Command

```
Router2#debug ip rip
RIP protocol debugging is on
Router2#
00:23:43: RIP:    ignored v1 packet from 192.168.1.2 on fa0/0
00:23:50: RIP: sending v2 update to 224.0.0.9 via fa0/0
               (192.168.1.2)
               Subnet 10.0.0.0, metric 1
```

These debug messages make it very clear that the routers are ignoring the routing updates sent by each other. To resolve this issue, configure one of the routers to advertise in the format the peer router will understand and also understand the routing update sent by the peer router. Listing 6.48 shows the changes made to the configuration of Router2 to send and receive the routing updates in the RIPv1 format while continuing to send and receive the original configured version, that is, the RIPv2 routing updates.

LISTING 6.48 Reconfiguration for Router2

```
Router2#conf t
Router2(config)#router rip
```

```
Router2(config-router)#network 10.0.0.0
Router2(config-router)#network 192.168.1.0
Router2(config-router)#exit
Router2(config)#interface fa0/0
Router2(config-if)#ip rip send version 1 2
Router2(config-if)#ip rip receive version 1 2
```

The configuration shown in Listing 6.48 ensures that both the routers will communicate and understand the other routing updates. Commands that can resolve compatibility issues between RIPv1 and RIPv2 are described in Table 6.4.

TABLE 6.4 Commands to Resolve Compatibility Issues Between RIPv1 and RIPv2

Command	Utility
Router(config-router) #version {1 | 2}	Configures the software to receive and send only RIPv1 or only RIPv2 packets.
Router(config-if)#ip rip send version 1	Configures an interface to send only RIPv1 packets.
Router(config-if)#ip rip send version 2	Configures an interface to send only RIPv2 packets.
Router(config-if)#ip rip send version 1 | 2	Configures an interface to send RIPv1 and RIPv2 packets.
Router(config-if)# ip rip receive version 1	Configures an interface to accept only RIPv1 packets.
Router(config-if)# ip rip receive version 2	Configures an interface to accept only RIPv2 packets.
Router(config-if)# ip rip receive version 1|2	Configures an interface to accept either RIPv1 or RIPv2 packets.

TIMER PROBLEM

Timers are very critical to the RIP routing protocol, because they dictate and influence the RIP routing calculations and its operating behavior.

When an RIP-enabled router starts, it sends the entire routing table as response packets to all directly attached hosts through its RIP-configured interfaces. It uses broadcast 255.255.255.255 as the transmission method. The frequency of this broadcast is governed by the update timer. The update timer value can be modified as per the requirement, but any kind of modification to the default values is generally not recommended. To modify the default values of the timer, use the command timers basic update invalid hold-down flush. To view the current configuration of these timers, use the command show ip protocols. Listing 6.49 shows the output of the command show ip protocols highlighting the timer values.

LISTING 6.49 Output of show ip protocols Command

```
show ip protocols
Routing Protocol is "rip"
  Sending updates every 30 seconds, next due in 31 seconds
  Invalid after 180 seconds, hold down 180, flushed after 240
  Outgoing update filter list for all interfaces is
  Incoming update filter list for all interfaces is
  Redistributing:   rip
  Default version control: send version 1, receive any version
    Interface        Send  Recv   Key-chain
    FastEthernet0      1    1 2
    FastEthernet0      1    1 2
  Routing for Networks:
    192.168.1.0
    192.168.2.0
  Routing Information Sources:
    192.168.1.2           120       00:00:03
    192.168.2.2           120       00:00:06
    Distance: (default is 120)
```

When a router does not receive any updates for the same route for 180 seconds, it automatically designates the route as unreachable. RIP marks a particular route as unreachable by increasing the hop count to 16. Routers wait for an update for six times (180 seconds) the update period before marking the route unreachable.

Consider a scenario showing absence of the update packets from Router1 to Router2. Under such conditions, Router2 will wait for a time equivalent to six times the update timer and designate the route as unreachable, as shown in the routing table of Router2. Figure 6.7 depicts this scenario.

Listing 6.50 shows the routing table of Router2.

LISTING 6.50 Routing Table of Router2

```
Router2#sh ip route
Codes: C - connected, S - static, I - IGRP, R - RIP,
       M - mobile, B - BGP
```

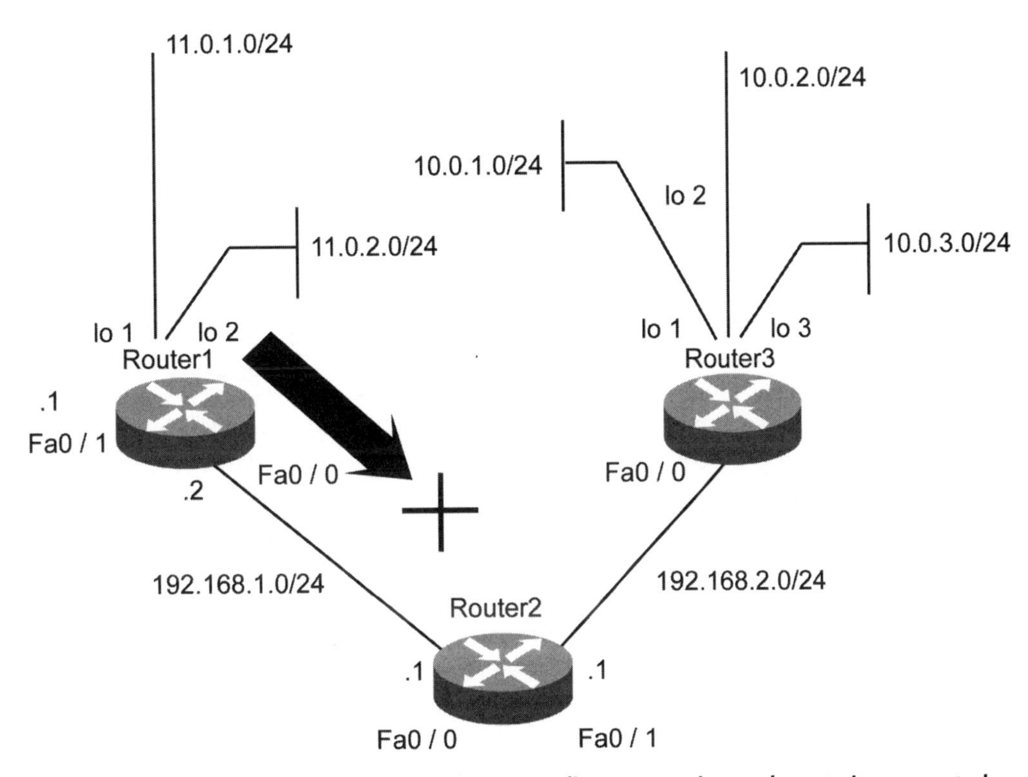

FIGURE 6.7 Scenario in which Router1 is not sending any routing updates to its connected neighbor.

```
        D - EIGRP, EX - EIGRP external, O - OSPF,
        IA - OSPF inter area
        E1 - OSPF external type 1,
        E2 - OSPF external type 2, E - EGP
        i - IS-IS, L1 - IS-IS level-1, L2 - IS-IS level-2,
        - candidate default
        U - per-user static route

Gateway of last resort is not set
  C  192.168.1.0/24 is directly connected, FastEthernet0/0
  C  192.168.2.0/24 is directly connected, FastEthernet0/1
  R  11.0.1.0/24 is possibly down
  routing via 192.168.1.2, 00:08:20, FastEthernet0/0
  R  11.0.2.0/24 is possibly down
  routing via 192.168.1.2, 00:01:42, FastEthernet0/0
  R  10.0.1.0/24 [120/1] via 192.168.2.2, 00:02:22, FastEthernet0/1
  R  10.0.2.0/24 [120/1] via 192.168.2.2, 00:01:24, FastEthernet0/1
  R  10.0.3.0/24 [120/1] via 192.168.2.2, 00:07:38, FastEthernet0/1
```

Even though the routes are marked as unreachable by the routers, they are still maintained in the routing tables for a certain period of time. This timer is known as the *flush timer*. Each route in the routing table maintains its own value of the flush timer. After the flush timer gets activated and then expires, the route is cleaned from the routing table. During the period after the invalid timer and before the completion of the flush timer, these unreachable routes also become a part of the routing updates sent by the routers.

The flush timer value should not be more than the invalid timer value or it can result in unpredictable routing calculation.

Hold-down timer is the settling duration provided to the routes that are learned and registered in the routing table of the router. Until the hold-down timer expires, any updates about that particular route are suppressed without any modification. This holds true only if the routing updates carry a higher metric value and are then installed. To change the value of this timer, command 'timers basic update invalid update invalid hold-down flush' can be used. Listing 6.51 displays the output of the show ip protocol command that is used to verify timers.

LISTING 6.51 RIP Timers Verified with show ip protocol

```
router9> show ip protocol
Routing Protocol is "rip"
Sending updates every 30 seconds, next due in 3 seconds
 Invalid after 180 seconds, hold down 180, flushed after 240
 Outgoing update filter list for all interfaces is
 Incoming update filter list for all interfaces is
 Redistributing: rip
 Default version control: send version 1, receive any version
  Interface       Send Recv Triggered RIP Key-chain
  Ethernet0       1    1 2
  Serial0         1    1 2
 Automatic network summarization is in effect
 Routing for Networks:
  172.16.0.0
 Routing Information Sources:
  Gateway      Distance   Last Update
  172.16.4.2       120    00:00:00
  172.16.1.2       120    00:00:07
 Distance: (default is 120)
```

To change the default values of the timers, you can use the command timers basic. To reset the timers to the default settings, use the command no timers basic.

If the timer value of a router is changed, the timer value of all the routers in the same RIP domain should be changed. Therefore, it is not recommended to modify the default values of these timers.

LOOPING

The information of a distance vector algorithm is said to be based on rumors. This is also known as *routing by rumor*. Because RIP is a distance vector protocol, it builds its routing table based on the information received from its directly attached neighbors after checking if the information received is genuine and error-free. To understand the looping characteristics of the RIP routing protocol and its workaround, it is important to understand the routing procedure of the RIP protocol. So you may understand the RIP routing behavior, its topology is shown in Figure 6.8.

In the topology shown in Figure 6.8, assume that the router link from Router1 to Router4 has just been activated. Therefore, none of the routers have any information about the network topology to which they are connected or the next hop

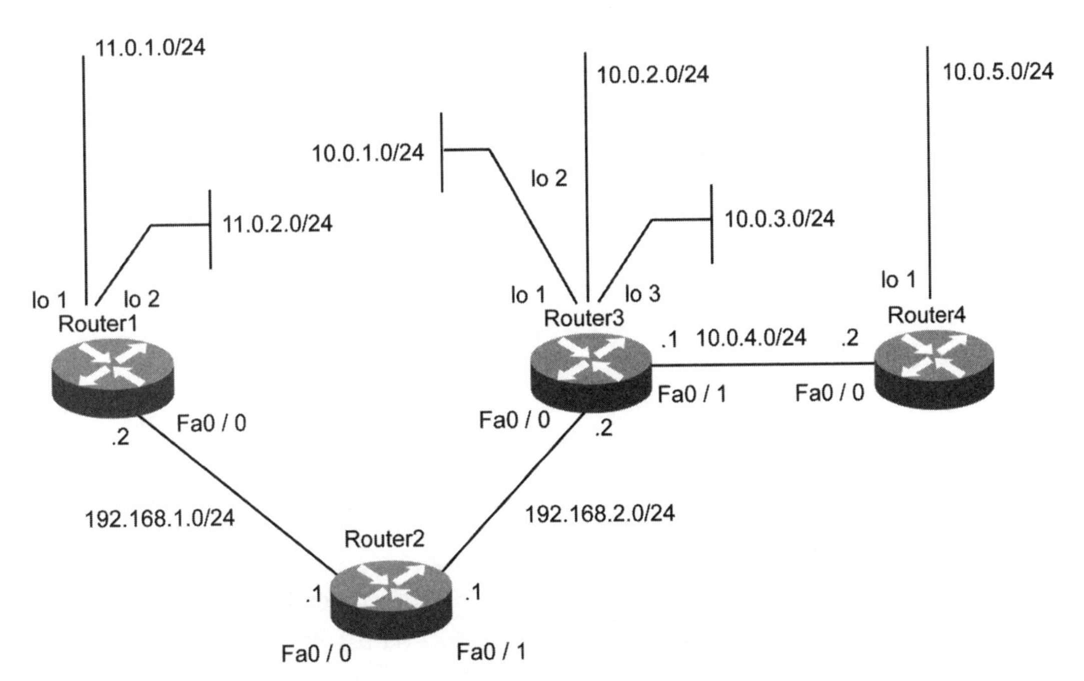

FIGURE 6.8 Scenario to describe the time required to build a consistent routing table across four routers.

router to which they should forward the traffic. The only information they possess at this point is about the directly connected networks.

Table 6.5 shows the details of routing calculation for the RIP routing protocol.

TABLE 6.5 Routing Details for RIP

Situation	Router	Routes	Metric
Before any update, just after the router startup	Router1	11.0.1.0 -> Directly connected	
		11.0.2.0 -> Directly connected	
		192.168.1.0->Directly connected	
	Router2	192.168.1.0->Directly connected	
		192.168.2.0->Directly connected	
	Router3	192.168.2.0->Directly connected	
		10.0.1.0 -> Directly connected	
		10.0.2.0 -> Directly connected	
		10.0.3.0 -> Directly connected	
		10.0.4.0 -> Directly connected	
	Router4	10.0.4.0 -> Directly connected	
		10.0.5.0 -> Directly connected	
Just after receiving the first update	Router1	11.0.2.0 -> Directly connected	
		11.0.2.0 -> Directly connected	
		192.168.1.0->Directly connected	1
		192.168.2.0 -> via 192.168.1.1	
	Router2	192.168.1.0->Directly connected	
		192.168.2.0->Directly connected	1
		11.0.0.0 -> via 192.168.1.2	1
		10.0.0.0 -> via 192.168.2.2	
		192.168.2.0->Directly connected	
		10.0.1.0 -> Directly connected	
		10.0.2.0 -> Directly connected	
	Router3	10.0.3.0 -> Directly connected	
		10.0.4.0 -> Directly connected	
		192.168.1.0 ->via 192.168.2.1	1
		10.0.5.0 -> via 10.0.4.2	1
		10.0.4.0 -> Directly connected	
		10.0.5.0 -> Directly connected	
	Router4	192.168.2.0 -> via 10.0.4.1	1
		10.0.1.0 -> via 10.0.4.1	1

TABLE 6.5 *(continued)*

Situation	Router	Routes	Metric
		10.0.2.0 -> via 10.0.4.1	1
		10.0.3.0 -> via 10.0.4.1	1
Just after receiving the second update	Router1	11.0.1.0 -> Directly connected	
		11.0.2.0 -> Directly connected	
		192.168.1.0->Directly connected	
		192.168.2.0 -> via 192.168.1.1	1
		10.0.0.0 -> via 192.168.1.1	2
	Router2	192.168.1.0->Directly connected	
		192.168.2.0->Directly connected	
		11.0.0.0 -> via 192.168.1.2	1
		10.0.0.0 -> via 192.168.2.2	1
		192.168.2.0->Directly connected	
		10.0.1.0 -> Directly connected	
		10.0.2.0 -> Directly connected	
	Router3	10.0.3.0 -> Directly connected	
		10.0.4.0 -> Directly connected	
		192.168.1.0 ->via 192.168.2.1	1
		10.0.5.0 -> via 10.0.4.2	1
		11.0.0.0 -> via 192.168.2.1	2
	Router4	10.0.4.0 -> Directly connected	
		10.0.5.0 -> Directly connected	
		192.168.2.0 -> via 10.0.4.1	1
		10.0.1.0 -> via 10.0.4.1	1
		10.0.2.0 -> via 10.0.4.1	1
		10.0.3.0 -> via 10.0.4.1	1
Just after receiving the third update	Router1	11.0.1.0 -> Directly connected	
		11.0.2.0 -> Directly connected	
		192.168.1.0->Directly connected	
		192.168.2.0 -> via 192.168.1.1	1
		10.0.0.0 -> via 192.168.1.1	2
	Router2	192.168.1.0->Directly connected	
		192.168.2.0->Directly connected	
		11.0.0.0 -> via 192.168.1.2	1
		10.0.0.0 -> via 192.168.2.2	1
	Router3	192.168.2.0->Directly connected	
		10.0.1.0 -> Directly connected	

(continued)

TABLE 6.5 *(continued)*

Situation	Router	Routes	Metric
		10.0.2.0 -> Directly connected	
		10.0.3.0 -> Directly connected	
		10.0.4.0 -> Directly connected	
		192.168.1.0 ->via 192.168.2.1	1
		10.0.5.0 -> via 10.0.4.2	1
		11.0.0.0 -> via 192.168.2.1	2
	Router4	10.0.4.0 -> Directly connected	
		10.0.5.0 -> Directly connected	
		192.168.2.0 -> via 10.0.4.1	1
		10.0.1.0 -> via 10.0.4.1	1
		10.0.2.0 -> via 10.0.4.1	1
		10.0.3.0 -> via 10.0.4.1	1
		11.0.0.0 -> via 10.0.4.1	3

Table 6.5 gives a clear understanding of the operation of RIP. After the four routing update broadcasts by each of the routers, the routers converge to the entire network topology. In this case, the routers do not understand the actual state of the network indicated by their routing paths, that is, whether the routing entries in their routing table can actually reach the directed destination network. For example, Router4 learns about network 11.0.0.0 from Router3 without knowing the actual state of networks 11.0.1.0 and 11.0.2.0. Router3 learns from Router2 without checking the network state. Therefore, it is observed that RIP populates its routing table using secondhand information.

Because of this typical attribute and the time taken for the convergence router, the routers encounter a problem called *routing loop*. Assuming that when the routers are fully converged and one of the interfaces of the router goes down or there is a topology change, the affected router becomes the source for the changed information and forces other communicating routers to reconverge. If the router itself goes down or is unavailable, the information is not synchronized with the existing routers' routing tables. As a result, the invalid timer has an important role in maintaining the validity of the existing routes in the routing table. To accommodate various routing loops occurring within certain network topologies, RIP has implemented a number of features to counter looping effects, including:

- Hop count
- Split horizon
- Route poisoning
- Triggered updates
- Hold-down timers

Hop Count

Also referred to as *counting to infinity*, hop count sets the maximum number of routers that a particular path can cross. This parameter helps control the endless routing loops in the network. According to this control, the maximum number of hop counts a path can cross is 15, and a hop count of 16 and onwards is called an invalid path or an unreachable path.

Split Horizon

According to the distance vector functionalities, every router sends its entire routing table information as an update to the directly connected neighbors. The split horizon deviates from this routine activity by restricting the routers from sending information about the network interface from which it was learned. The route entry pointing back to the router from which that particular route was learned is known as the reverse route, which becomes the major cause for maturing of routing loops. The split horizon eliminates the existence of reverse routes and helps in avoiding routing loops. There are two types of split horizon methods:

- Simple split horizon
- Split horizon with poisoned reverse

Simple Split Horizon

This is defined as sending updates out of an interface, excluding the networks that were learned from the updates received on that interface. For example, as per Figure 6.8, when Router3 sends its routing update to Router4 via its interface fa0/0, the information sent will be as shown in Table 6.6.

TABLE 6.6 Routing Update Information for Router3

Routes	Metric	Status
192.168.2.0 -> Directly connected		Will be sent
10.0.1.0 -> Directly connected		Will be sent
10.0.2.0 -> Directly connected		Will be sent
10.0.3.0 -> Directly connected		Will be sent
10.0.4.0 -> Directly connected	1	Will not be sent
192.168.1.0 ->via 192.168.2.1	1	Will be sent
10.0.5.0 -> via 10.0.4.2	2	Will not be sent
11.0.0.0 -> via 192.168.2.1		Will be sent

The update packet broadcasted from Router4 on its interface fa0/0 will not carry any information about the networks learned from Router3. This is shown in Table 6.7.

TABLE 6.7 Routing Update Information for Router4

Routes	Metric	Status
10.0.4.0 -> Directly connected		Will not be sent
10.0.5.0 -> Directly connected		Will be sent
192.168.2.0 -> via 10.0.4.1	1	Will not be sent
10.0.1.0 -> via 10.0.4.1	1	Will not be sent
10.0.2.0 -> via 10.0.4.1	1	Will not be sent
10.0.3.0 -> via 10.0.4.1	1	Will not be sent
11.0.0.0 -> via 10.0.4.1	3	Will not be sent

To check the split horizon configuration status for an interface, look at the highlighted line in the output of the command show ip interface fa0/0 from Router4, as shown in Listing 6.52.

LISTING 6.52 Output of show ip interface fa0/0 Command from Router4

```
Router4#show ip interface fa0/0
FastEthernet0/0 is up, line protocol is up
  Internet address is 10.4.0.2/16
  Broadcast address is 255.255.0.0
  MTU 1500 bytes,
  Helper address is not set
  Directed broadcast forwarding is disabled
  Outgoing access list is not set
  Inbound  access list is not set
  Proxy ARP Is Enabled
  Security Level Is Default
  Split horizon Is Enabled
  ICMP redirects are always sent
  ICMP unreachables are always sent
  ICMP mask replies are never sent
  IP fast switching is enabled
  IP fast switching on the same interface is enabled
  IP Null turbo vector
  IP multicast fast switching is enabled
  IP multicast distributed fast switching is disabled
  router Discovery Is disabled
  IP output packet accounting is disabled
```

```
IP access violation accounting is disabled
TCP/IP header compression is disabled
RTP/IP header compression is disabled
Probe proxy name replies are disabled
Policy routing Is disabled
Network address translation is disabled
WCCP Redirect outbound is disabled
WCCP Redirect exclude is disabled
BGP Policy Mapping is disabled
```

To enable and disable the split horizon, use the command shown in Listing 6.53 from the interface configuration mode.

LISTING 6.53 Command to Enable and Disable Split Horizon

```
Router#conf t
Router(config)#int fa0/0
Router(config-if)#ip split-horizon     =  Enables
Router(config-if)#no ip split-horizon  =  Disables
```

Split Horizon with Poisoned Reverse

Split horizon and split horizon with poisoned reverse have common objectives of avoiding reverse routes and ensuring a loop-free routing environment. The difference is in the way they craft their update packets. In case of split horizon with poisoned reverse, the router resends the information about the networks on the interface, including the networks that were learned from the updates received on that interface. This is done by marking the route unreachable when the information is sent back to the direction from which it was received.

Table 6.8 shows the split horizon with poisoned reverse update packet format broadcasted by Router3 on its interface fa0/1.

TABLE 6.8 Split Horizon with Poisoned Reverse Update Packet for Router3

Routes	Metric	Status
192.168.2.0 -> Directly connected		Will be sent
10.0.1.0 -> Directly connected		Will be sent
10.0.2.0 -> Directly connected		Will be sent
10.0.3.0 -> Directly connected	Unreachable	Will be sent
10.0.4.0 -> Directly connected	1	Will be sent
192.168.1.0 ->via 192.168.2.1	Unreachable	Will be sent
10.0.5.0 -> via 10.0.4.2	2	Will be sent
11.0.0.0 -> via 192.168.2.1		Will be sent

The update packet broadcast from Router4 on its interface fa0/0 will carry information about all the networks. Only the networks that were learned from the update packets sent by Router3 will be marked as unreachable. This is shown in Table 6.9.

TABLE 6.9 Router4 Update Packet Status

Routes	Metric	Status
10.0.4.0 -> Directly connected	Unreachable	Will be sent
10.0.5.0 -> Directly connected		Will be sent
192.168.2.0 -> via 10.0.4.1	Unreachable	Will be sent
10.0.1.0 -> via 10.0.4.1	Unreachable	Will be sent
10.0.2.0 -> via 10.0.4.1	Unreachable	Will be sent
10.0.3.0 -> via 10.0.4.1	Unreachable	Will be sent
11.0.0.0 -> via 10.0.4.1	Unreachable	Will be sent

Route Poisoning

Route poisoning reduces the convergence time when a particular route fails. As per the standard response, if a particular route to a destination fails, it is immediately removed from the routing table. Similarly, other neighboring routers in the network will remove the route when they note the failure of the route via the routing update packet received by them. Route poisoning sets the distance of the failed route to infinity on the next routing update, and all the neighbor routers immediately track that the route has become inaccessible.

Depending on the network size, it can take more time to converge all the connected routers.

Figure 6.9 shows unavailability of one of the networks connected to Router3 for the understanding of route poisoning.

As shown in Figure 6.9, if network 10.0.3.0 connected to Router3 fails, instead of removing this entry from the routing table, Router3would mark that particular route as inaccessible and update its directly connected routers. This ensures faster convergence in the network.

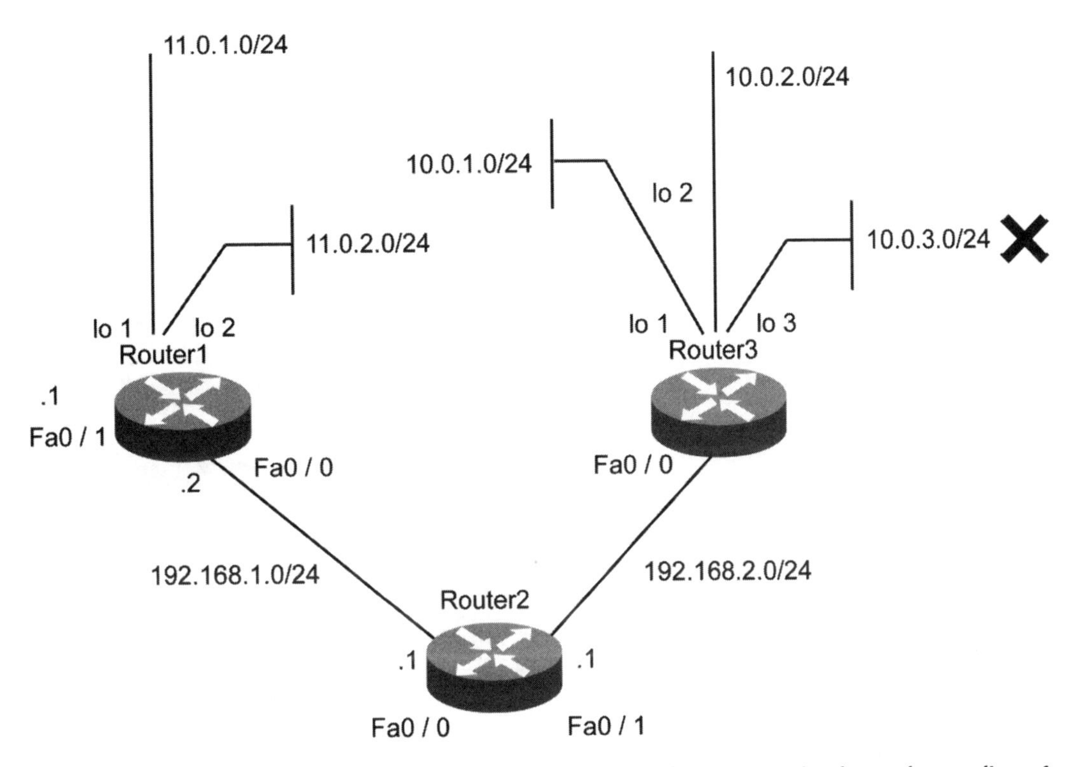

FIGURE 6.9 Unavailability of one of the networks connected to Router3 for the understanding of route poisoning.

Triggered Updates

Using this technique, whenever there is an addition or a modification (increase or decrease in the metric value) in the existing routes of the routing table, routers will immediately broadcast their updates without waiting for their update timer to expire. The triggered updates differ from the normal updates by not waiting for the timer to expire. They just send the changed information rather than the entire routing table. This significantly reduces the time required for reconvergence and does not affect the normal timed updates. Figure 6.10 depicts the mechanism of the triggered update.

As shown in Figure 6.10, if network 10.5.0.0 connected to Router4 fails, Router4 will immediately send this update on interface fa0/0. After Router3 receives this information, it will update its routing table and send the update via its interface fa0/0 without waiting for the update timer to expire and will go on until Router1 synchronizes its routing table. This technique reduces the processing time and optimizes the bandwidth use.

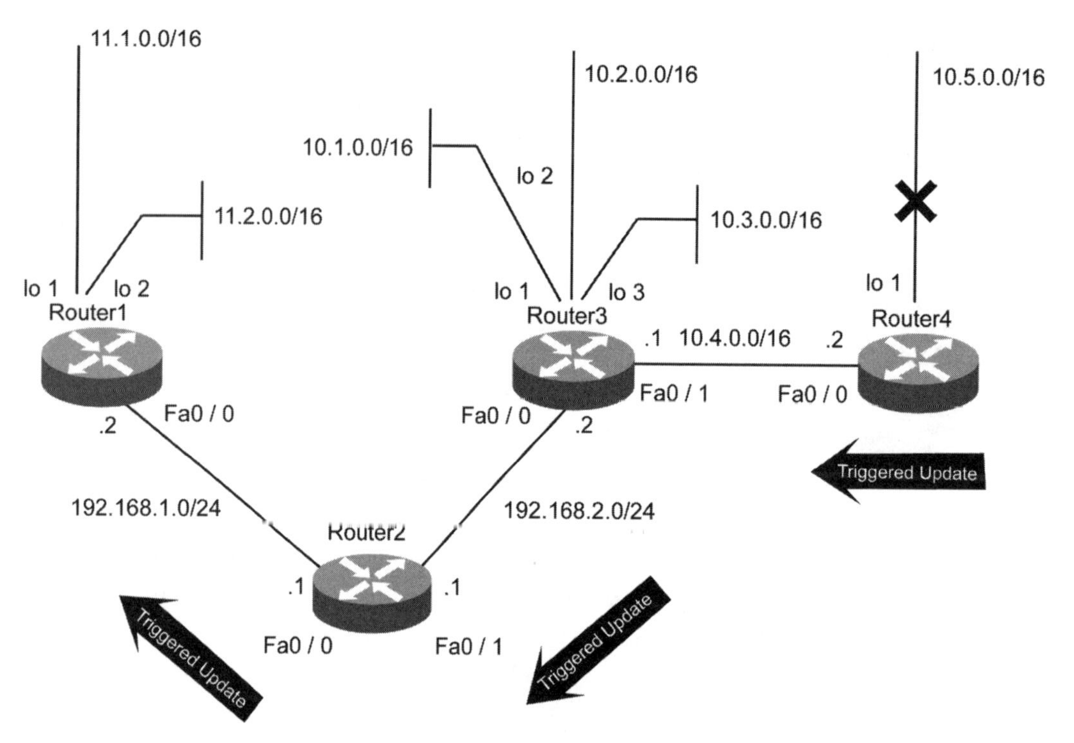

FIGURE 6.10 Mechanism of the triggered update.

Hold-down Timers

Unlike triggered updates, the hold-down timer shows resistance to the immediate topology change. This method ensures some consistency of the routing information by reducing the acceptance of any inconsistent routing information. The hold-down timer is active for a particular route in which the router receives the update for that route with an increased metric value. After the hold-down timer expires, the router will accept the route with the increased metric value and update its routing table. The advantage of this timer is that it prevents routes from being advertised as having failed when the interface is flapping and, therefore, prevents other routers from recalculating their routing table. In addition, this introduces some amount of latency in the total convergence time. This prevents any false information from being propagated in the network.

The command to configure the hold-down timer is shown in Listing 6.54.

LISTING 6.54 Command to Configure the Hold-down Timer

```
Router#conf t
Router(config-router)# timers basic hold-down <time in seconds>
```

SUMMARY

In this chapter, we learned about the common problems encountered in RIP environments. We also reminded you about isolating these problems and their resolutions. In the next chapter, we move on to troubleshooting IGRP environments.

POINTS TO REMEMBER

- RIP uses hop count as a metric to calculate the best routing path to a destination network.
- RIP uses the Bellman-Ford algorithm to compute the metric used for routing path decision.
- RIPv1 is a classful routing protocol, does not carry subnet information, and is incapable of supporting Variable Length Subnet Masking (VLSM).
- RIPv2 is a classless routing protocol and supports VLSM and authentication.
- The common problems encountered in RIP environments are misconfiguration, classful routing, classless routing, timers, looping, and version incompatibility issues.
- RIP is activated using the command router rip.
- The syntax for assigning the network is network network number.
- Routing convergence means ensuring that the routing information about the entire network is registered with all the participating routers.
- To verify the routing protocol configuration on a router, use the command show ip protocols.
- The command to view the routing table of a router is show ip route.
- The command to check router configuration with a passive interface is show running-config.
- To create equal cost path routes for a router, use the command maximum-path <path #>.
- The process of consolidating all subnet routes of a major network while crossing the boundaries of another major network is called autosummarization.
- To verify a RIPv2 protocol configuration for different routers, use the command show ip protocols.
- To debug a router RIP message, use the command debug ip rip.
- RIPv2 cannot reach discontiguous networks by default.
- Timers dictate and influence the RIP routing calculations and their operating behavior.
- Four types of timers associated with RIP are update, invalid, flush, and hold-down.
- Invalid timer is also referred to as an expiration timer.

7 Troubleshooting IGRP Routing Environments

IN THIS CHAPTER

- Features of IGRP
- Problem Isolation in IGRP
- Misconfiguration in IGRP
- Timer problem in IGRP

Interior Gateway Routing Protocol (IGRP) is a distance vector protocol suitable for medium and large-sized networks. IGRP uses a composite metric that is calculated by a combination of delay, bandwidth, reliability, load, and Maximum Transmission Unit (MTU).

This chapter discusses various problems, their effective isolation, and different methods to resolve these problems in IGRP routing environments.

FEATURES OF IGRP

A thorough understanding of the IGRP protocol is required to successfully isolate and troubleshoot problems pertaining to IGRP networks. The different features of IGRP are shown in Table 7.1.

TABLE 7.1 Features of the IGRP Routing Protocol

Features	IGRP Values
Protocol type	Distance vector
Subnet information	Classful
Metric	Composite
Count to infinity	100 by default
Routing updates mode	Broadcast
Flash or triggered updates	Yes
Load balancing	Up to four paths, by default
Algorithm	Bellman-Ford

IGRP requires timers for different routing operations. These IGRP timers are set to default values that can be changed, if required. Table 7.2 lists the default timer values used by IGRP.

TABLE 7.2 IGRP Timers

Timer	Function	Default Value
Update	Specifies how frequently routing update messages should be sent.	90 seconds.
Invalid	Specifies how long a router should wait in the absence of routing updates about a route before declaring it invalid.	270 seconds. Three times update timer.
Hold-down	Specifies duration given to routes learned and registered in the routing table of the router.	280 seconds. 10 seconds more than invalid timer.
Flush	Specifies how much time should pass before a route should be flushed from the routing table.	630 seconds. Seven times update timer.

The IGRP metrics used to calculate the best path to reach a destination are:

Bandwidth: Difference between highest and lowest frequencies available for network signals. Bandwidth value ranges from 1.2 kbps to 10 gbps.

Delay: Time taken for packets to move from source to destination. Delay is expressed in units of 10 microseconds.

Load: Amount of router resource used. Load is measured in a scale of 1 to 255.

Reliability: Ratio of expected keepalives to received keepalives from a link. Reliability is expressed in a scale of 1 to 225.

MTU: Maximum packet size that an interface can handle. MTU is expressed in bytes.

Interior route: Directly connected to a router interface.

System route: Learned by other IGRP enabled routers configured with the same Autonomous System Number (ASN).

Exterior route: Learned or redistributed by other IGRP enabled routers configured with a different ASN.

PROBLEM ISOLATION IN IGRP

There are a number of commands available in the Cisco IOS with which to isolate and troubleshoot problems in IGRP networks. Table 7.3 lists the different show command options for IGRP.

TABLE 7.3 Show Commands and Description

Show Command	Description
show ip route	Displays entries in the IP routing table for the router in which this command is issued.
show ip protocols	Displays filters, parameters, and network information details corresponding to the active routing protocol that is configured in the router.
show ip route igrp *Number*	Displays only IGRP routes in the routing table. It takes one parameter, *Number*, which is the number of the IGRP AS whose routes are being verified using this command.
show interface	Displays information about network interfaces within the IGRP network.
show running-config	Displays the current updated operating configuration for IGRP in the router where this command is issued.

The show commands are used to verify the IGRP routing protocol configuration on a router. These commands verify the configuration or any customizations performed in IGRP networks. The output of show commands displays details of routing protocols that are configured in routers.

Table 7.4 lists different debug commands that are used for thorough packet level troubleshooting for problems in IGRP networks.

TABLE 7.4 Debug Commands and Descriptions

Debug Command	Description
debug ip igrp transaction	Lists detailed contents of both sent and received IGRP updates.

(continued)

TABLE 7.4 *(continued)*

Debug Command	Description
debug ip igrp events	Lists summary contents of both sent and received IGRP updates. Only shows the sending or receiving of IGRP packets and the number of routes in each update.
ping ip-addr/hostname	Tests the network connectivity with the specified network address. It accepts one parameter as ip-address or host name of the router that we need to verify for network connectivity. This command sends test packets to the specified IP address or hostname and verifies the status of the network connectivity.
undebug all	Sets the IGRP debugging system off for the current router.

Consider the example depicted in Figure 7.1. Part of an IGRP network with five routers R1, R2, R3, R4, and R5 is connected as shown in the figure.

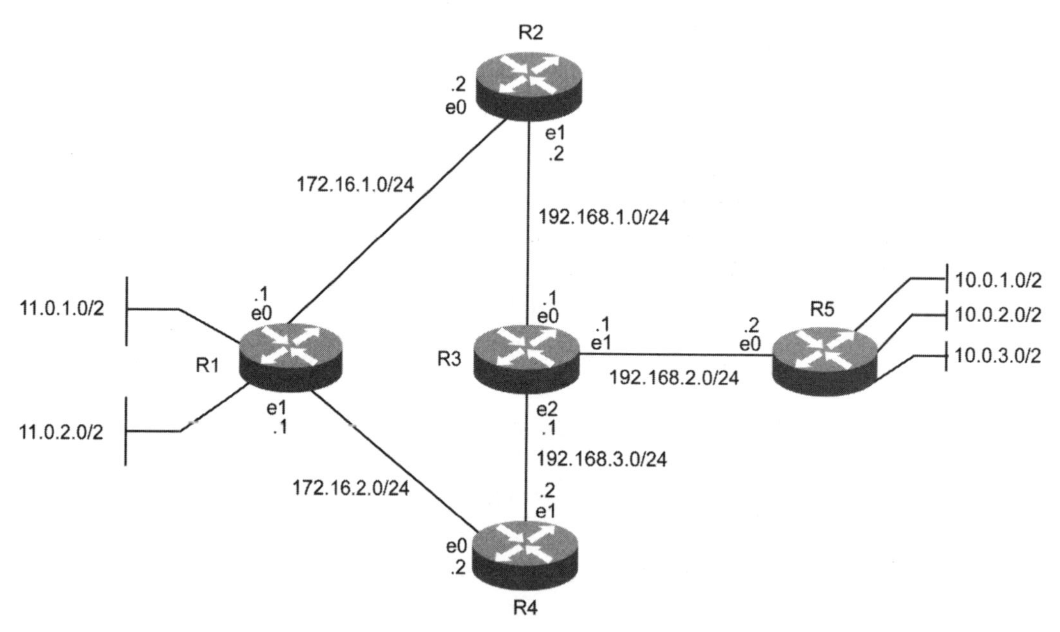

FIGURE 7.1 A network configured with the IGRP routing protocol.

In this IGRP scenario, R1 is connected to two other networks with IPs 11.0.1.0 and 11.0.2.0, respectively. R5 connects three networks, 10.0.1.0, 10.0.2.0, and 10.0.3.0.

In Figure 7.1, the show and debug commands are issued at R2. The configuration of R1 is shown in Listing 7.1.

LISTING 7.1 Configuration of R1

```
R1#conf t
R1(config)#ip routing
R1(config)#router igrp 100
R1(config-router)#network 11.0.1.0
R1(config-router)#network 11.0.2.0
R1(config-router)#exit
R1(config)#interface ethernet 0
R1(config-if)#ip address 172.16.1.1 255.255.255.0
R1(config-if)#exit
R1(config)#interface ethernet 1
R1(config-if)#ip address 172.16.2.1 255.255.255.0
R1(config-if)#exit
R1(config)#exit
R1
```

The configuration of R2 is shown in Listing 7.2.

LISTING 7.2 Configuration of R2

```
R2#conf t
R2(config)#ip routing
R2(config)#router igrp 100
R2(config-router)#exit
R2(config)#interface ethernet 0
R2(config-if)#ip address 172.16.1.2 255.255.255.0
R2(config-if)#exit
R2(config)#interface ethernet 1
R2(config-if)#ip address 192.168.1.2 255.255.255.0
R2(config-if)#exit
R2(config)#exit
R2#
```

The configuration of R3 is shown in Listing 7.3.

LISTING 7.3 Configuration of R3

```
R3#conf t
R3(config)#ip routing
```

```
R3(config)#router igrp 100
R3(config-router)#exit
R3(config)#interface ethernet 0
R3(config-if)#ip address 192.168.1.1 255.255.255.0
R3(config-if)#exit
R3(config)#interface ethernet 1
R3(config-if)#ip address 192.168.2.1 255.255.255.0
R3(config-if)#exit
R3(config)#interface ethernet 2
R3(config-if)#ip address 192.168.3.1 255.255.255.0
R3(config-if)#exit
R3(config)#exit
R3#
```

The configuration of R4 is shown in Listing 7.4.

LISTING 7.4 Configuration of R4

```
R4#conf t
R4(config)#ip routing
R4(config)#router igrp 100
R4(config-router)#exit
R4(config)#interface ethernet 0
R4(config-if)#ip address 172.16.2.2 255.255.255.0
R4(config-if)#exit
R4(config)#interface ethernet 1
R4(config-if)#ip address 192.168.3.2 255.255.255.0
R4(config-if)#exit
R4(config)#exit
R4#
```

The configuration of R5 is shown in Listing 7.5.

LISTING 7.5 Configuration of R5

```
R5#conf t
R5(config)#ip routing
R5(config)#router igrp 100
R5(config-router)#network 10.0.1.0
R5(config-router)#network 10.0.2.0
R5(config-router)#network 10.0.3.0
R5(config-router)#exit
R5(config)#interface ethernet 0
R5(config-if)#ip address 192.168.2.2 255.255.255.0
```

```
R5(config-if)#exit
R5(config)#exit
R5#
```

Output of the show ip protocols command at R2 is as shown in Listing 7.6.

LISTING 7.6 Output of the show ip protocols Command at R2

```
R2#show ip protocols
Routing Protocol is “igrp 100”
  Sending updates every 90 seconds, next due in 32 seconds
  Invalid after 270 seconds, hold down 280, flushed after 630
  Outgoing update filter list for all interfaces is not set
  Incoming update filter list for all interfaces is not set
  Default networks flagged in outgoing updates
  Default networks accepted from incoming updates
  IGRP metric weight K1=1, K2=0, K3=1, K4=0, K5=0
  IGRP maximum hopcount 100
  IGRP maximum metric variance 1
  Redistributing: igrp 100
  Routing for Networks:
    172.16.0.0
    192.168.1.0
  Routing Information Sources:
    172.16.1.1            100       00:00:06
    192.168.1.1           100       00:00:09
  Distance: (default is 100)
```

The output of the show ip route command for R2 is as shown in Listing 7.7.

LISTING 7.7 Output of the show ip route Command at R2

```
R2#sh ip route
Codes: C  - connected, S - static, I - IGRP, R - RIP,
       M  - mobile, B - BGP
       D  - EIGRP, EX - EIGRP external, O - OSPF,
       IA - OSPF inter area
       E1 - OSPF external type 1,
       E2 - OSPF external type 2, E - EGP
       i  - IS-IS, L1 - IS-IS level-1, L2 - IS-IS level-2,
       *  - candidate default
       U  - per-user static route

Gateway of last resort is not set
  172.16.0.0 is subnetted, 2 subnets
```

```
C  172.16.1.0/24 is directly connected, Ethernet0
I  172.16.2.0/24 [100/273] via 172.16.1.1, 00:08:23, Ethernet0
C  192.168.1.0/24 is directly connected, Ethernet1
I  11.0.0.0/8 [100/273] via 172.16.1.1, 00:04:20, Ethernet0
I  192.168.2.0/24 [100/273] via 192.168.1.1, 00:02:30, Ethernet1
I  192.168.3.0/24 [100/273] via 192.168.1.1, 00:06:32, Ethernet1
I  10.0.0.0/8 [100/437] via 192.168.1.1, 00:03:15, Ethernet1
```

Output of the show ip route igrp command for R2 is as shown in Listing 7.8.

LISTING 7.8 Output of the show ip route igrp Command at R2

```
R2#show ip route igrp 100
Codes: C - connected, S - static, I - IGRP, R - RIP,
       M - mobile, B - BGP
       D - EIGRP, EX - EIGRP external, O - OSPF,
       IA - OSPF inter area
       N1 - OSPF NSSA external type 1,
       N2 - OSPF NSSA external type 2
       E1 - OSPF external type 1,
       E2 - OSPF external type 2, E - EGP
       i - IS-IS, L1 - IS-IS level-1, L2 - IS-IS level-2,
       *  - candidate
       U - per-user static route, o - ODR

Gateway of last resort is not set
  C  172.16.1.0/24 is directly connected, Ethernet0
  C  192.168.1.0/24 is directly connected, Ethernet0
```

The output of the debug ip igrp transactions command for R1 is as shown in Listing 7.9.

LISTING 7.9 Output of the debug ip igrp transactions Command at R1

```
R1#debug ip igrp transactions
IGRP protocol debugging is on
R1#
IGRP: sending update to 255.255.255.0 via Ethernet0 (172.16.1.1)
      network 11.0.1.0, metric=8476
      network 11.0.2.0, metric=8576
IGRP: sending update to 255.255.255.0 via Ethernet1 (172.16.2.1)
      network 11.0.1.0, metric=8476
      network 11.0.2.0, metric=8576
```

Output of the debug ip igrp events command for R2 is as shown in Listing 7.10.

LISTING 7.10 Output of the debug ip igrp events Command at R2

```
R2#debug ip igrp events
IGRP event debugging is on
R2#
IGRP: sending update to 255.255.255.0 via Ethernet0 (172.16.1.2)
IGRP: Update contains 0 interior, 2 system, and 0 exterior routes.
IGRP: Total routes in update: 2
IGRP: sending update to 255.255.255.0 via Ethernet1
 (192.168.1.2)
IGRP: Update contains 0 interior, 2 system, and 0 exterior routes.
IGRP: Total routes in update: 2
```

Output of the ping ip-address command for R2 is as shown in Listing 7.11.

LISTING 7.11 Output of the ping ip-address Command

```
R2#ping 172.16.0.2
Sending 5, 64 byte ICMP echoes to 172.16.0.2, timeout is 2 seconds:
36 bytes from 172.16.0.2: icmp_seq=0. time=0. ms
36 bytes from 172.16.0.2: icmp_seq=1. time=0. ms
36 bytes from 172.16.0.2: icmp_seq=2. time=0. ms
36 bytes from 172.16.0.2: icmp_seq=3. time=0. ms
36 bytes from 172.16.0.2: icmp_seq=4. time=0. Ms
100% success rate (5/5), round-trip min/avg/max = 0/0/0 ms
```

The output of the undebug all command for R2 is as shown in Listing 7.12.

LISTING 7.12 Output of the undebug all Command for R2

```
R2#undebug all
IGRP event debugging is off
```

MISCONFIGURATION IN IGRP

Misconfiguration can be caused either by manual or system errors. You need to take care while configuring IGRP in a router, because even a small configuration error can affect the traffic of the whole network. Misconfiguration problems in IGRP can occur when:

- IGRP routes are missing from the routing table.
- IGRP is not installing all possible equal cost paths.
- ASN is misconfigured.

IGRP Routes Missing from Routing Table

IGRP routes may be missing from the routing table when:

- Network statement is missing or misconfigured.
- Layer 1 or 2 is down.
- Distribute list is blocking the route.
- Access list is blocking IGRP source address.
- Access list is blocking IGRP broadcast.
- Discontiguous network occurs.
- Invalid source of route.
- Switch, Frame Relay, and other Layer 2 media problems occur.
- AS mismatch of sender occurs.

Check the router configuration specific to routing. When a network command is added to the routing configuration of the router:

- The IGRP routing engine is activated to send and receive routing updates.
- The network is advertised in update packets.

Consider the example shown in Figure 7.1. The problems that arise when IGRP routes are missing from the routing table are:

- R1 is unable to ping any IP address.
- Delay in routing convergence.
- R1 is unable to reach any hosts in networks.

Unable to Ping IP Address

In Figure 7.1, R1 is unable to ping any IP address that belongs to network 10.0.0.0. Any hosts, including router interfaces, belonging to network 10.0.0.0 are unreachable from R1. Listing 7.13 shows the current state of the routing table of R1. This output is obtained by executing the sh ip route command.

LISTING 7.13 Output of the sh ip route Command at R1

```
R1#sh ip route
Codes: C - connected, S - static, I - IGRP, R - RIP,
```

```
       M  - mobile, B - BGP
       D  - EIGRP, EX - EIGRP external, O - OSPF,
       IA - OSPF inter area
       E1 - OSPF external type 1,
       E2 - OSPF external type 2, E - EGP
       i  - IS-IS, L1 - IS-IS level-1, L2 - IS-IS level-2,
       *  - candidate default
       U  - per-user static route

Gateway of last resort is not set
     11.0.0.0 is subnetted, 2 subnets
  C  11.0.1.0/24 is directly connected, Loopback1
  C  11.0.2.0/24 is directly connected, Loopback2
     172.16.0.0 is subnetted, 2 subnets
  C  172.16.1.0/24 is directly connected, Ethernet0
  C  172.16.2.0/24 is directly connected, Ethernet1
  I  192.168.1.0/24 [100/273] via 172.16.1.2, 00:05:21, Ethernet0
  I  192.168.3.0/24 [100/273] via 172.16.2.2, 00:05:28, Ethernet1
  I  192.168.2.0/24 [100/437] via 172.16.1.2, 00:01:36, Ethernet0
```

Listing 7.13 shows that R1 does not have knowledge about networks 10.0.1.0, 10.0.2.0, and 10.0.3.0. This problem occurs because R1 is not receiving routing updates about the mentioned networks.

You must check if R5 is advertising the 10.0.0.0 network in the routing update. This is done by executing the show ip protocols command. Output of the show ip protocols command at R5 is as shown in Listing 7.14.

LISTING 7.14 Output of the show ip protocols Command at R5

```
R5#sh ip protocols
Routing Protocol is "igrp 100"
  Sending updates every 90 seconds, next due in 44 seconds
  Invalid after 270 seconds, hold down 280, flushed after 630
  Outgoing update filter list for all interfaces is not set
  Incoming update filter list for all interfaces is not set
  Default networks flagged in outgoing updates
  Default networks accepted from incoming updates
  IGRP metric weight K1=1, K2=0, K3=1, K4=0, K5=0
  IGRP maximum hopcount 100
  IGRP maximum metric variance 1
  Redistributing: igrp 100
  Routing for Networks:
    192.168.2.0
  Routing Information Sources:
```

```
    192.168.2.1  100  00:00:03
  Distance: (default is 100)
```

In Listing 7.14, it's clear that the IGRP in R1 is only active for network 192.168.2.0 and not for network 10.0.0.0. To troubleshoot this problem, the commands issued at R5 are:

```
R5#conf t
R5(global)#router igrp 100
R5(global-router)#network 10.0.0.0
```

The new configuration can be checked using the show ip protocols command at R5, and the show ip route command at R1. Output of the show ip protocols command in R5 is as shown in Listing 7.15.

LISTING 7.15 Output of the show ip protocols Command at R5

```
R5#sh ip protocols
Routing Protocol is "igrp 100"
  Sending updates every 90 seconds, next due in 44 seconds
  Invalid after 270 seconds, hold down 280, flushed after 630
  Outgoing update filter list for all interfaces is not set
  Incoming update filter list for all interfaces is not set
  Default networks flagged in outgoing updates
  Default networks accepted from incoming updates
  IGRP metric weight K1=1, K2=0, K3=1, K4=0, K5=0
  IGRP maximum hopcount 100
  IGRP maximum metric variance 1
  Redistributing: igrp 100
  Routing for Networks:
    10.0.0.0
    192.168.2.0
  Routing Information Sources:
    192.168.2.1           100       00:00:03
  Distance: (default is 100)
```

Output of the show ip route command at R1 is as shown in Listing 7.16.

LISTING 7.16 Output of the show ip route Command at R1

```
R1#sh ip route
Codes: C - connected, S - static, I - IGRP, R - RIP,
       M - mobile, B - BGP
       D - EIGRP, EX - EIGRP external, O - OSPF,
```

```
        IA - OSPF inter area
        E1 - OSPF external type 1,
        E2 - OSPF external type 2, E - EGP
        i  - IS-IS, L1 - IS-IS level-1, L2 - IS-IS level-2,
        *  - candidate default
        U  - per-user static route

Gateway of last resort is not set
     11.0.0.0 is subnetted, 2 subnets
  C  11.0.1.0/24 is directly connected, Loopback1
  C  11.0.2.0/24 is directly connected, Loopback2
     172.16.0.0 is subnetted, 2 subnets
  C  172.16.1.0/24 is directly connected, Ethernet0
  C  172.16.2.0/24 is directly connected, Ethernet1
  I  192.168.1.0/24 [100/273] via 172.16.1.2, 00:05:21, Ethernet0
  I  192.168.3.0/24 [100/273] via 172.16.2.2, 00:05:28, Ethernet1
  I  192.168.2.0/24 [100/437] via 172.16.1.2, 00:01:36, Ethernet0
  I  10.0.0.0/8 [100/437] via 172.16.1.2, 00:01:34, Ethernet0
     [100/437] via 172.16.2.2, 00:02:42, Ethernet1
```

Delay in Convergence

Sometimes, it may happen that configuration changes do not affect the routing tables. This problem is because of a delay in routing convergence. The router will send updates only after 90 seconds. To enforce an immediate effect on the routing table of a router, use the command clear ip route *. This command refreshes the current routing table.

Passive Interface

The configuration of passive interface in IGRP networks also causes the unavailability of routes. In Figure 7.1, R1 is unable to reach any hosts in networks 10.0.1.0, 10.0.2.0, and 10.0.3.0. This problem can be caused by the configuration of passive interface. The passive interface command prevents the interface from sending routing updates. However, this interface can access the routing update broadcasts on the link. It can also receive routing packets from other routers and accordingly modify the routing table.

Consider the example shown in Figure 7.2. The figure shows a network with passive interface configuration at R5.

In the IGRP scenario depicted in Figure 7.2, there are five routers: R1, R2, R3, R4, and R5. The interface that connects R1 and R2 has a bandwidth of 10 mbps. R1 is connected to two other networks with IPs 11.0.1.0 and 11.0.2.0. The interface between R1 and R4 has a bandwidth of 100 mbps. R2 is connected to R3, which in turn is connected to R4 using interfaces that have bandwidths 100 mbps each. R3

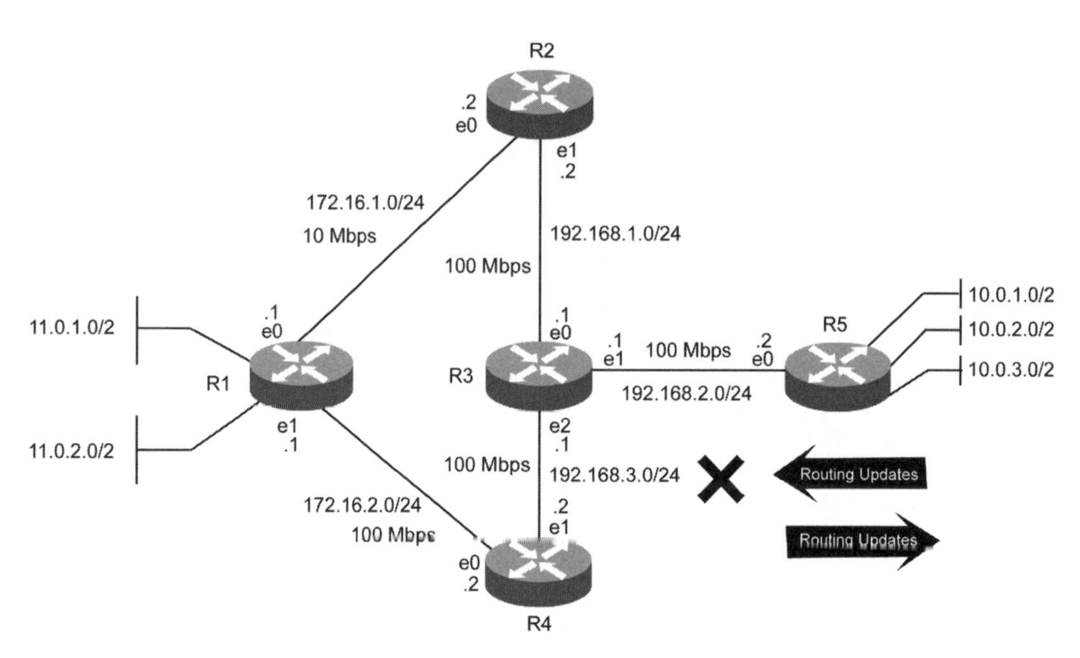

FIGURE 7.2 An IGRP network showing passive interface configuration at R5.

is connected to R5 using an interface with bandwidth 100 mbps. R5 is configured with a passive interface.

In Figure 7.2, R5 is configured with passive interface. The configuration of R5 is shown in Listing 7.17.

LISTING 7.17 Configuration of R5

```
R5#show run
router igrp 100
  network 192.168.2.0
  network 10.0.0.0
  passive-interface e0
```

In Listing 7.17, the interface has been configured as passive. As a result, interface e0 of R5 will only be able to listen to the IGRP routing updates and hold all its outgoing IGRP routing updates via the configured interface.

If the interface is removed from passive mode, it will start sending its routing table to the neighboring routers and will reach R1. The routing table of R1 will be modified after receiving updates. R1 will now be able to reach any host in networks 10.0.1.0, 10.0.2.0, and 10.0.3.0. The no passive-interface command removes an interface from passive mode. The output of the no passive-interface command at R5 is shown in Listing 7.18.

LISTING 7.18 Output of the no passive-interface Command at R5

```
R5#conf t
R5(config)#router rip
R5(config-router)#no passive-interface e0
```

IGRP Does Not Install All Possible Equal Cost Paths

This section will focus only on problems pertaining to equal cost paths. Equal cost path refers to multiple routes to a common destination with same metric values. By default, IGRP can use up to four equal cost paths. Misconfiguration can lead to missing of equal cost routes in the routing table. Consider the example depicted in Figure 7.3.

In the IGRP scenario depicted in Figure 7.3, there are five routers: R1, R2, R3, R4, and R5. The interface that connects R1 and R2 has a bandwidth of 100 Mbps. R1 is connected to two other networks with IPs 11.0.1.0 and 11.0.2.0. The interface between R1 and R4 has a bandwidth of 100 mbps. R2 is connected to R3, which in turn is connected to R4 using interfaces that have bandwidths 100 mbps each. R3 is connected to R5 using an interface with bandwidth 100 mbps. R5 is configured with an active interface.

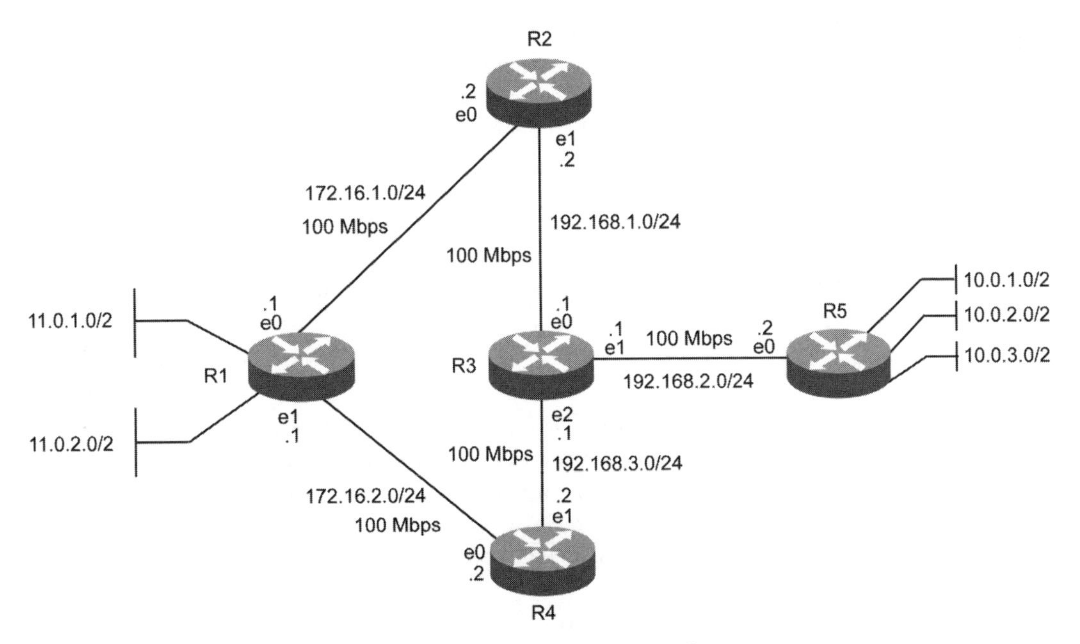

FIGURE 7.3 An IGRP routing scenario depicting equal cost paths.

In Figure 7.3, R1 can reach networks 10.0.1.0, 10.0.2.0, and 10.0.3.0 via two paths:

- R1 to R2 to R3 to R5
- R1 to R4 to R3 to R5

Both paths have the same cost to the destination. Suppose the routing table of R1 shows only one path for the destination networks 10.0.1.0, 10.0.2.0, and 10.0.3.0. This scenario is shown in the output for the show ip route command at R1. Output for the show ip route command is shown in Listing 7.19.

LISTING 7.19 Output of the show ip route Command at R1

```
R1#sh ip route
Codes: C - connected, S - static, I - IGRP, R - RIP,
       M - mobile, B - BGP
       D - EIGRP, EX - EIGRP external, O - OSPF,
       IA - OSPF inter area
       E1 - OSPF external type 1,
       E2 - OSPF external type 2, E - EGP
       i  - IS-IS, L1 - IS-IS level-1, L2 - IS-IS level-2,
       *  - candidate default
       U  - per-user static route

Gateway of last resort is not set
     11.0.0.0 is subnetted, 2 subnets
  C  11.0.1.0/24 is directly connected, Loopback1
  C  11.0.2.0/24 is directly connected, Loopback2
     172.16.0.0 is subnetted, 2 subnets
  C  172.16.1.0/24 is directly connected, e0
  C  172.16.2.0/24 is directly connected, e1
  I  192.168.1.0/24 [100/273] via 172.16.1.2, 00:05:21, e0
  I  192.168.3.0/24 [100/273] via 172.16.2.2, 00:05:28, e1
  I  192.168.2.0/24 [100/437] via 172.16.1.2, 00:01:36, e0
  I  10.0.0.0/8 [100/437] via 172.16.1.2, 00:01:34, e0
```

To troubleshoot this problem, you need to understand the igrp updates that are being received by R1. Use the command debug ip igrp at R1 to see the routing updates. Output of the debug ip igrp command is as shown in Listing 7.20.

LISTING 7.20 Output of the debug ip igrp Command at R1

```
R1#debug ip igrp transactions
IGRP protocol debugging is on
```

```
R1#
IGRP : sending update to 255.255.255.255. via e0  (172.16.1.1)
       network 11.0.1.0, metric=xxx
       network 11.0.2.0, metric=xxx
IGRP : sending update to 255.255.255.255. via e1  (172.16.2.1)
       network 11.0.1.0, metric=xxx
       network 11.0.2.0, metric=xxx

IGRP : received update from 172.16.1.2 on e0
       network 192.168.1.0, metric xxx (neighbor xxx)
       network 192.168.2.0, metric xxx (neighbor xxx)
       network 192.168.3.0, metric xxx (neighbor xxx)
       network 10.0.0.0, metric xxx (neighbor xxx)

IGRP : received update from 172.16.2.2 on e1
       network 192.168.1.0, metric xxx (neighbor xxx)
       network 192.168.2.0, metric xxx (neighbor xxx)
       network 192.168.3.0, metric xxx (neighbor xxx)
       network 10.0.0.0, metric xxx (neighbor xxx)
```

Listing 7.20 shows that R1 is receiving updates about network 10.0.0.0 from two different sources. Even after receiving the updates from two sources, R1 is registering it once. To troubleshoot this problem, look into the configuration of R1 using the show run command. The output of the show run command is shown in Listing 7.21.

LISTING 7.21 Output of the show run Command at R1

```
R1#show run
router IGRP 100
  network 172.16.0.0
  network 11.0.0.0
  maximum-path 1
```

IGRP routers can install up to four equal cost paths to a common destination. In the scenario discussed, the router is configured to install only one route to a destination. Routers can be configured with the command maximum-path 1, when only one path is required for routing.

In the example depicted in Figure 7.3, the solution is to enter the command maximum-path 6 in R1 to maintain a maximum of six equal cost paths for a destination. R1 now shows two equal path cost routes for 10.0.0.0 networks. This can be verified by looking at the routing table of R1, as given in Listing 7.22.

LISTING 7.22 Routing Table of R1

```
R1#sh ip route
Codes: C - connected, S - static, I - IGRP, R - RIP,
       M - mobile, B - BGP
       D - EIGRP, EX - EIGRP external, O - OSPF,
       IA - OSPF inter area
       E1 - OSPF external type 1,
       E2 - OSPF external type 2, E - EGP
       i  - IS-IS, L1 - IS-IS level-1, L2 - IS-IS level-2,
       *  - candidate default
       U - per-user static route

Gateway of last resort is not set
     11.0.0.0 is subnetted, 2 subnets
  C  11.0.1.0/24 is directly connected, Loopback1
  C  11.0.2.0/24 is directly connected, Loopback2
     172.16.0.0 is subnetted, 2 subnets
  C  172.16.1.0/24 is directly connected, e0
  C  172.16.2.0/24 is directly connected, e1
  I  192.168.1.0/24 [100/273] via 172.16.1.2, 00:05:21, e0
  I  192.168.3.0/24 [100/273] via 172.16.2.2, 00:05:28, e1
  I  192.168.2.0/24 [100/437] via 172.16.1.2, 00:01:36, e0
  I  10.0.0.0/8 [100/437] via 172.16.1.2, 00:01:34, e0
     [100/437] via 172.16.2.2, 00:02:42, e1
```

Misconfigured ASN

Consider the example depicted in Figure 7.4. In the figure, there are two different ASs that run IGRP with ids 100 and 1000, respectively. There are five routers: R1, R2, R3, R4, and R5. R1, R2, R3, and R4 are in the same AS as id 100 whereas R5 is in a different AS with id 1000.The interface that connects R1 and R2 has a bandwidth of 10 mbps. R1 is connected to two other networks with IPs 11.0.1.0 and 11.0.2.0. The interface between R1 and R4 has a bandwidth of 100 mbps. R2 is connected to R3, which in turn is connected to R4 using interfaces that have bandwidths of 100 mbps each. R3 is connected to R5 using an interface with a bandwidth of 100 mbps. Three networks with network IDs 10.0.1.0, 10.0.2.0, and 10.0.3.0 are connected to R5.

In Figure 7.4, the routing table for R3 is shown in Listing 7.23.

LISTING 7.23 Routing Table of R3

```
R3#sh ip route
Codes: C - connected, S - static, I - IGRP, R - RIP,
```

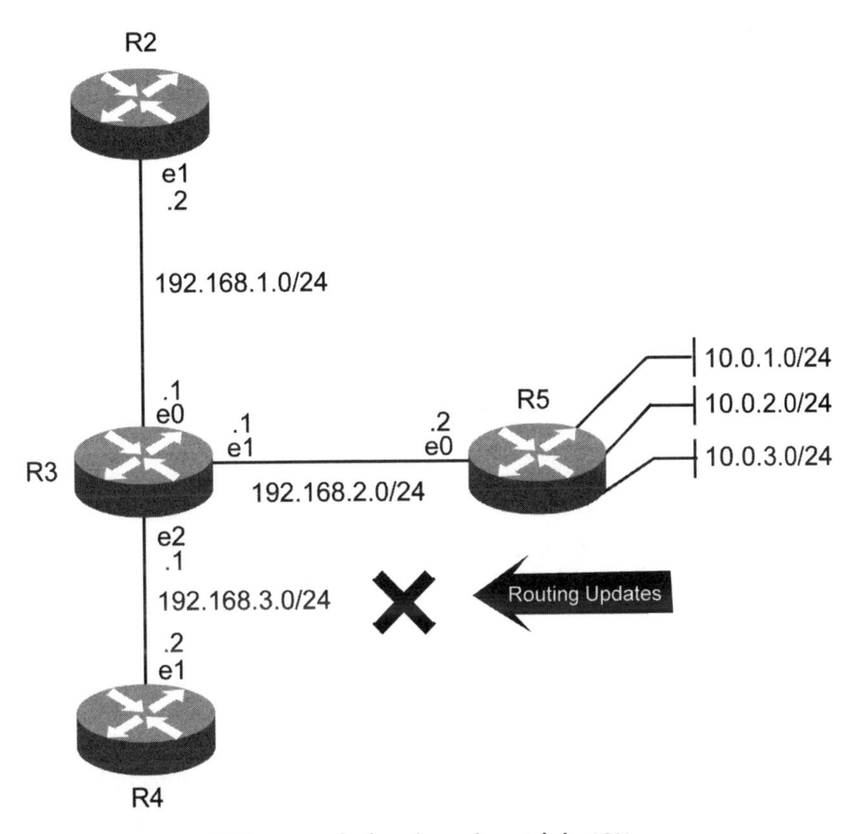

FIGURE 7.4 An IGRP network showing mismatch in ASN.

```
        M - mobile, B - BGP
        D - EIGRP, EX - EIGRP external, O - OSPF,
        IA - OSPF inter area
        E1 - OSPF external type 1,
        E2 - OSPF external type 2, E - EGP
        i  - IS-IS, L1 - IS-IS level-1, L2 - IS-IS level-2,
        *  - candidate default
        U  - per-user static route

Gateway of last resort is not set
  C  192.168.1.0/24 is directly connected, Ethernet0
  C  192.168.2.0/24 is directly connected, Ethernet1
  C  192.168.3.0/24 is directly connected, Ethernet2
  I  172.16.1.0/24 [100/273] via 192.168.1.2, 00:08:41, Ethernet0
  I  11.0.1.0/24 [100/437] via 192.168.1.2, 00:07:37, Ethernet0
  I  11.0.2.0/24  [100/437] via 192.168.3.2, 00:03:17, Ethernet2
  I  172.16.2.0/24 [100/437] via 192.168.1.2, 00:03:18, Ethernet0
```

In Listing 7.23, it is clear that the route information about network 10.0.0.0 is missing. The routing table of R5 is shown in Listing 7.24.

LISTING 7.24 Routing Table of R5

```
R5#sh ip route
Codes: C - connected, S - static, I - IGRP, R - RIP,
       M - mobile, B - BGP
       D - EIGRP, EX - EIGRP external, O - OSPF,
       IA - OSPF inter area
       E1 - OSPF external type 1,
       E2 - OSPF external type 2, E - EGP
       i  - IS-IS, L1 - IS-IS level-1, L2 - IS-IS level-2,
       * - candidate default
       U - per-user static route

Gateway of last resort is not set
  C   10.0.1.0/24 is directly connected, Loopback1
  C   10.0.2.0/24 is directly connected, Loopback2
  C   10.0.3.0/24 is directly connected, Loopback3
  C   192.168.2.0/24 is directly connected, Ethernet0
```

Listing 7.24 indicates that R5 is cut off from the network, because the routing table displays only information about the directly connected routes. This could happen if the routing process is disabled in R5 or if it is misconfigured, assuming that the connectivity is error free. Compare the configuration of R5 with that of other routers in the network. The configuration of R5 is as shown in Listing 7.25.

LISTING 7.25 Configuration of R5

```
R5#sh ip protocols
Routing Protocol is "igrp 1000"
  Sending updates every 90 seconds, next due in 44 seconds
  Invalid after 270 seconds, hold down 280, flushed after 630
  Outgoing update filter list for all interfaces is not set
  Incoming update filter list for all interfaces is not set
  Default networks flagged in outgoing updates
  Default networks accepted from incoming updates
  IGRP metric weight K1=1, K2=0, K3=1, K4=0, K5=0
  IGRP maximum hopcount 100
  IGRP maximum metric variance 1
  Redistributing: igrp 1000
  Routing for Networks:
  10.0.0.0
```

```
    192.168.2.0
  Routing Information Sources:
    192.168.2.1          100      00:00:03
  Distance: (default is 100)
```

In Listing 7.25, the code highlighted in bold should be checked for the proper functioning of the routing protocol. The configuration of R5 is compared with that of R3. The configuration of R3 is as shown in Listing 7.26.

LISTING 7.26 Configuration of R3

```
R3#show ip protocols
Routing Protocol is "igrp 100"
  Sending updates every 90 seconds, next due in 54 seconds
  Invalid after 270 seconds, hold down 280, flushed after 630
  Outgoing update filter list for all interfaces is not set
  Incoming update filter list for all interfaces is not set
  Default networks flagged in outgoing updates
  Default networks accepted from incoming updates
  IGRP metric weight K1=1, K2=0, K3=1, K4=0, K5=0
  IGRP maximum hopcount 100
  IGRP maximum metric variance 1
  Redistributing: igrp 100
  Routing for Networks:
    192.168.1.0
    192.168.2.0
    192.168.3.0
  Routing Information Sources:
    192.168.1.2          100      00:00:00
    192.168.3.2          100      00:00:00
    192.168.2.2          100      00:00:00
  Distance: (default is 100)
```

A comparison of the configuration of R5 with R3 shows that the IGRP routing process of R5 is using ASN 1000, and the routing process of R3 is using ASN 100. This is a mismatch caused by configuration error. Only those routers carrying the same ASN are allowed to share and exchange routing table information among them.

To troubleshoot this problem, reconfigure R5 by reenabling the IGRP routing process with the correct process id. Listing 7.27 displays the output of the routing protocol configuration of R5 with the correct process id.

LISTING 7.27 Reconfiguring R5

```
R5#conf t
R5(config)#no router igrp 1000
R5(config)#router igrp 100
```

```
R5(config-router)#network 10.0.0.0
R5(config-router)#network 192.168.3.0
R5(config-router)#exit
```

To confirm the correct functioning of the routers, check the routing tables of R3 and R5 using the show ip route command. The routing table of R3 is shown in Listing 7.28.

LISTING 7.28 Routing Table for R3

```
R3#sh ip route
Codes: C - connected, S - static, I - IGRP, R - RIP,
       M - mobile, B - BGP
       D - EIGRP, EX - EIGRP external, O - OSPF,
       IA - OSPF inter area
       E1 - OSPF external type 1,
       E2 - OSPF external type 2, E - EGP
       i  - IS-IS, L1 - IS-IS level-1, L2 - IS-IS level-2,
       *  - candidate default
       U  - per-user static route

Gateway of last resort is not set
  C  192.168.1.0/24 is directly connected, Ethernet0
  C  192.168.2.0/24 is directly connected, Ethernet1
  C  192.168.3.0/24 is directly connected, Ethernet2
  I  172.16.1.0/24 [100/273] via 192.168.1.2, 00:08:41, Ethernet0
  I  11.0.1.0/24 [100/437] via 192.168.1.2, 00:07:37, Ethernet0
     [100/437] via 192.168.3.2, 00:03:17, Ethernet2
  I  172.16.2.0/24 [100/437] via 192.168.1.2, 00:03:18, Ethernet0
  I  10.0.0.0/8 [100/273] via 192.168.2.2, 00:01:27, Ethernet1
```

The routing table of R5 is shown in Listing 7.29.

LISTING 7.29 Routing Table for R5

```
R5#sh ip route
Codes: C - connected, S - static, I - IGRP, R - RIP,
       M - mobile, B - BGP
       D - EIGRP, EX - EIGRP external, O - OSPF,
       IA - OSPF inter area
       E1 - OSPF external type 1,
       E2 - OSPF external type 2, E - EGP
       i  - IS-IS, L1 - IS-IS level-1, L2 - IS-IS level-2,
       *  - candidate default
       U  - per-user static route
```

```
Gateway of last resort is not set
  C  10.0.1.0/24 is directly connected, Loopback1
  C  10.0.2.0/24 is directly connected, Loopback2
  C  10.0.3.0/24 is directly connected, Loopback3
  C  192.168.2.0/24 is directly connected, Ethernet0
  I  192.168.1.0/24 [100/273] via 192.168.2.1, 00:05:31, Ethernet0
  I  192.168.3.0/24 [100/273] via 192.168.2.1, 00:01:28, Ethernet0
  I  172.16.0.0/16 [100/437] via 192.168.2.1, 00:09:34, Ethernet0
  I  11.0.0.0/8 [100/437] via 192.168.2.1, 00:05:14, Ethernet0
```

The routing tables of R3 and R5 show that all the routers—R1, R2, R3, R4, and R5—exchange routing table information. R3 can now learn the routes of network 10.0.0.0, and R5 can learn the routes that are connected to the other routers.

TIMER PROBLEM IN IGRP

Timers govern the frequency at which the IGRP routers learn and update routing tables. These timers also affect the time required by routers to attain convergence. Timers are critical to the IGRP protocol, because they influence routing calculations and operating behavior. Different timers used in IGRP are update, invalid, flush, and hold-down. The default values of all these timers can be modified as per the requirement. However, modifications to default values should be performed with extreme caution.

You can use the timers basic update invalid holddown flush command to modify the default values of timers. The show ip protocols command is used to view the current configuration of these timers. The output of the show ip protocols command is shown in Listing 7.30.

LISTING 7.30 Output of the show ip protocols Command

```
show ip protocols
Routing Protocol is "igrp 100"
  Sending updates every 90 seconds, next due in 76 seconds
  Invalid after 270 seconds, hold down 280, flushed after 630
  Outgoing update filter list for all interfaces is not set
  Incoming update filter list for all interfaces is not set
  Default networks flagged in outgoing updates
  Default networks accepted from incoming updates
  IGRP metric weight K1=1, K2=0, K3=1, K4=0, K5=0
  IGRP maximum hopcount 100
  IGRP maximum metric variance 1
  Redistributing: igrp 100
  Routing for Networks:
```

```
    11.0.0.0
    172.16.0.0
  Routing Information Sources:
    172.16.1.2              100        00:00:03
    172.16.2.2              100        00:00:03
  Distance: (default is 100)
```

When an IGRP router learns about a new routing path to a destination, it is registered in the routing table with the invalid timer set to the default value of 270 seconds. Consequently, when the same route is seen in an update packet, it resets the invalid timer to 270 seconds. Assuming that the router does not receive any updates for the same route for 270 seconds, the route is designated as an unreachable route. Routers wait for 270 seconds before marking the route unreachable. To understand this concept, consider the scenario depicted in Figure 7.5, which shows an IGRP network that consists of routers R1, R2, R3, R4, and R5.

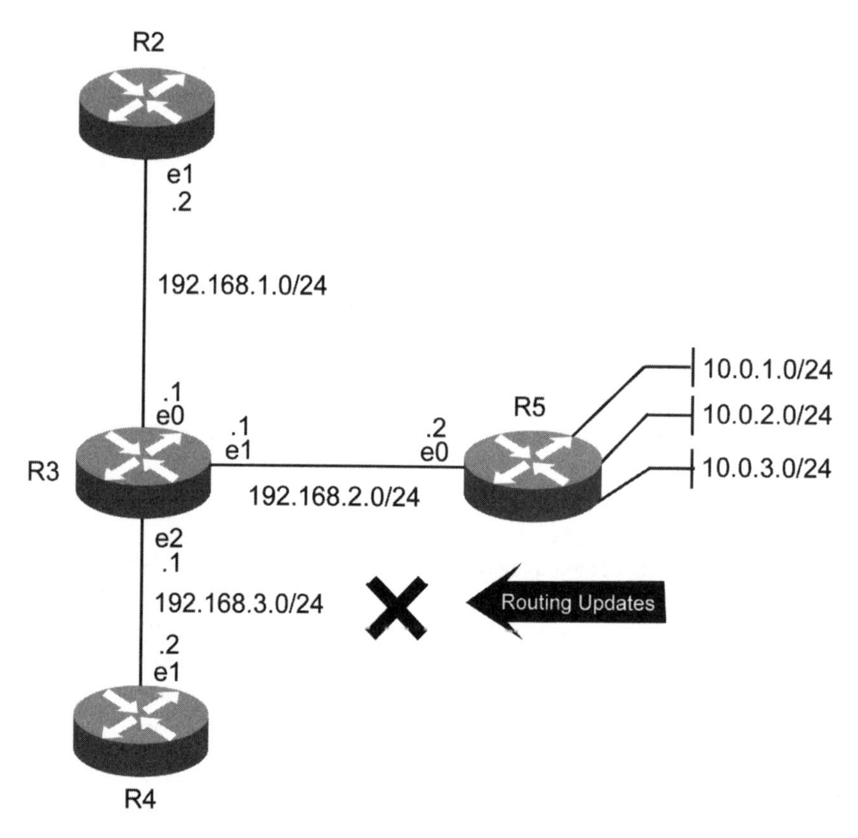

FIGURE 7.5 An IGRP network depicting a scenario of missing update packets.

Figure 7.5 shows the absence of the updates packets from R5 to R3 and to R2 and R4. R3 will wait for a duration that equals three times the update timer before designating the route as unreachable, as shown in the routing table of R3. The routing table of R3 is as shown in Listing 7.31.

LISTING 7.31 Routing Table of R3

```
R3#sh ip route
Codes: C - connected, S - static, I - IGRP, R - RIP,
       M - mobile, B - BGP
       D - EIGRP, EX - EIGRP external, O - OSPF,
       IA - OSPF inter area
       E1 - OSPF external type 1,
       E2 - OSPF external type 2, E - EGP
       i - IS-IS, L1 - IS-IS level-1, L2 - IS-IS level-2,
       - candidate default
       U - per-user static route

Gateway of last resort is not set
  C  192.168.1.0/24 is directly connected, Ethernet0
  C  192.168.2.0/24 is directly connected, Ethernet1
  C  192.168.3.0/24 is directly connected, Ethernet2
  I  172.16.1.0/24 [100/273] via 192.168.1.2, 00:08:41, Ethernet0
  I  11.0.1.0/24 [100/437] via 192.168.1.2, 00:07:37, Ethernet0
     [100/437] via 192.168.3.2, 00:03:17, Ethernet2
  I  172.16.2.0/24 [100/437] via 192.168.1.2, 00:03:18, Ethernet0
  I  10.0.0.0/8 is possibly down
  routing [100/273] via 192.168.2.2, 00:01:27, Ethernet1
```

Listing 7.31 indicates that network 10.0.0.0 is unreachable because R3 is not receiving routing updates from R5. Even though routes are marked as unreachable, it is still maintained in routing tables for a certain period of time that is determined by the flush timer. To make invalid timer value less than flush timer value, use the command:

```
Invalid after 270 seconds, hold down 280, flushed after 630
```

Set the flush timer value of R3 to be less than invalid timer value as shown:

```
R3#timers basic update invalid holddown flush 250
```

This will solve the routing update problem of R3.

Flush timer value should not be more than the invalid timer value; otherwise, it will result in unpredictable routing calculation.

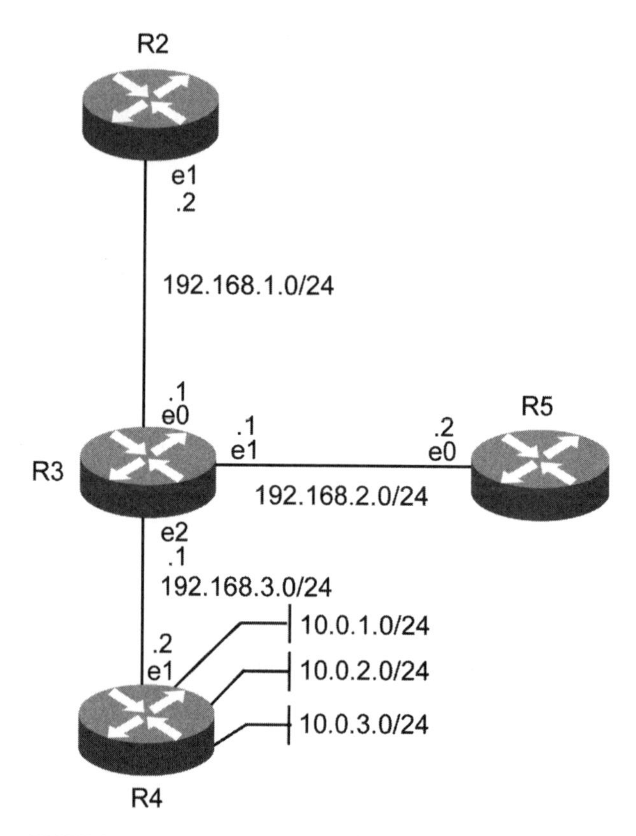

FIGURE 7.6 An IGRP case study.

CASE STUDY

Consider the scenario depicted in Figure 7.6. There are four routers—R2, R3, R4, and R5—in the same IGRP AS. R4 has networks connected to it with network ids 10.0.1.0, 10.0.2.0, and 10.0.3.0.

In Figure 7.6, the configuration of R4 is as shown in Listing 7.32.

LISTING 7.32 Configuration of R4

```
R4#conf t
R4(config)#ip routing
R4(config)#router igrp 100
R4(config-router)#network 10.0.1.0
R4(config-router)#network 10.0.3.0
R4(config-router)#exit
R4(config)#interface ethernet 1
```

```
R4(config-if)#ip address 192.168.3.2 255.255.255.0
R4(config-if)#exit
R4(config)#exit
R4#
```

The problem is that routing packets are not reaching R4 from network 10.0.2.0, which is connected to R4. To troubleshoot this problem, verify the IP route in R4 using the show ip route command. This command output shows only two network ids connected. The third network with network id 10.1.2.0 is not connected. The output of the show ip route command is shown in Listing 7.33.

LISTING 7.33 Output of the show ip route Command

```
R4#show ip route
Codes: C - connected, S - static, I - IGRP, R - RIP,
       M - mobile, B - BGP
       D - EIGRP, EX - EIGRP external, O - OSPF,
       IA - OSPF inter area
       E1 - OSPF external type 1,
       E2 - OSPF external type 2, E - EGP
       i  - IS-IS, L1 - IS-IS level-1, L2 - IS-IS level-2,
       *  - candidate default
       U  - per-user static route
Gateway of last resort is not set
  C  10.0.1.0/24 is directly connected, Loopback1
  C  10.0.3.0/24 is directly connected, Loopback3
  C  192.168.3.1/24 is directly connected, Ethernet0
```

Reconfigure R4 with the configuration for network 10.1.2.0 as shown in Listing 7.34.

LISTING 7.34 Reconfiguration of R4

```
R4#conf t
R4(config)#ip routing
R4(config)#router igrp 100
R4(config-router)#network 10.0.1.0
R4(config-router)#network 10.1.2.0
R4(config-router)#network 10.0.3.0
R4(config-router)#exit
R4(config)#interface ethernet 1
R4(config-if)#ip address 192.168.3.2 255.255.255.0
R4(config-if)#exit
R4(config)#exit
R4#
```

Verify routing information again using the show ip route command. The output for the command is shown in Listing 7.35.

LISTING 7.35 Output of the show ip route Command

```
R4#show ip route
Codes: C - connected, S - static, I - IGRP, R - RIP,
       M - mobile, B - BGP
       D - EIGRP, EX - EIGRP external, O - OSPF,
       IA - OSPF inter area
       E1 - OSPF external type 1,
       E2 - OSPF external type 2, E - EGP
       i  - IS-IS, L1 - IS-IS level-1, L2 - IS-IS level-2,
       *  - candidate default
       U  - per-user static route

Gateway of last resort is not set
  C  10.0.1.0/24 is directly connected, Loopback1
  C  10.0.2.0/24 is directly connected, Loopback1
  C  10.0.3.0/24 is directly connected, Loopback3
  C  192.168.3.1/24 is directly connected, Ethernet0
```

In Listing 7.35, the text in bold shows that network 10.0.2.0 is correctly configured.

Now all networks connected to R4 are correctly configured, and routing updates are sent properly to R4.

SUMMARY

In this chapter, we learned about troubleshooting IGRP environments, which included problem isolation in IGRP environments. We also reminded you about resolution of problems in IGRP environments. In the next chapter, we move on to troubleshooting EIGRP routing environments.

POINTS TO REMEMBER

- IGRP uses a composite metric that is calculated by a combination of delay, bandwidth, reliability, load, and MTU.
- The show ip route command displays entries in the IP routing table for the router in which this command is issued.
- The show ip protocols command displays filters, parameters, and network information details corresponding to the active routing protocol configured in the router.

- The show ip route igrp *Number* command displays only the IGRP routes in the routing table.
- The show interface command displays information about network interfaces within the IGRP network.
- The show running-config command displays the current updated operating configuration for IGRP in the router where this command is executed.
- The debug ip igrp transaction command lists detailed contents of both sent and received IGRP updates.
- The debug ip igrp events command lists summary contents of both sent and received IGRP updates.
- The ping ip-addr/hostname command tests the network connectivity with a specified network address.
- The undebug all command sets the IGRP debugging system off for the current router.
- Misconfiguration problems in IGRP can occur when IGRP routes are missing from the routing table, IGRP is not installing all possible equal cost paths, or if ASN is misconfigured.
- The clear ip route * command refreshes the current routing table.
- The passive interface command prevents the interface from sending routing updates.
- Equal cost path refers to multiple routes to a common destination with same metric values.

8 Troubleshooting EIGRP Routing Environments

IN THIS CHAPTER

- Features of EIGRP
- Problem Isolation in EIGRP
- Misconfiguration in EIGRP
- EIGRP Neighbor Formation Problem
- EIGRP Route Problem
- EIGRP Metric Problem
- Stuck in Active State
- Redistribution Problem

Enhanced Interior Routing Gateway Protocol (EIGRP) is a Cisco-proprietary Interior Gateway Protocol (IGP). It is a highly advanced distance vector routing protocol that scales well to larger internetworks. EIGRP uses the Diffused Update Algorithm (DUAL) to calculate the best path to a destination network, thereby reducing overheads associated with other IGPs. The network resource utilization is reduced as well, because only Hello packets are exchanged between neighbors. In case of a change in the routing table, only the change is propagated instead of the whole routing table, ensuring minimum bandwidth utilization. The EIGRP algorithm also ensures fast convergence in case of any changes in network topology.

EIGRP is widely deployed around the world in large Cisco-only networks. This chapter discusses the various problems, their effective isolation, and the different methods to resolve problems.

FEATURES OF EIGRP

The features of EIGRP routing protocol are shown in Table 8.1.

TABLE 8.1 Features of EIGRP

Features	EIGRP Values
Protocol type	Hybrid
Subnet information	Classless
Routing updates	Sends routing updates when there is a change in topology
Metric	Composite (bandwidth, delay, reliability, load)
Count to infinity	224
Routing updates mode	Multicast; uses address 224.0.0.10
Algorithm	DUAL

EIGRP uses different data packets during the routing process. They are:

Hello: Sent by a router to its neighbor during the discovery process. These packets are sent using the multicast unreliable delivery method.

Acknowledgment: Sent by an EIGRP router on receipt of Hello packets. The Unicast delivery protocol is used to send Acknowledgment packets back to the sender of Hello packets.

Routing Updates: Sent in the form of Reliable Transport Protocol (RTP) when there is a change in the network topology. They are sent using multicast or unicast, depending upon the number of routers to which data is to be sent.

Queries: Sent by EIGRP using multicast or unicast reliable delivery protocol.

Replies: Sent by EIGRP using unicast.

These terms and concepts will help you understand EIGRP:

Adjacency: Virtual link between two routers over which the routing information is sent and received. Adjacency is formed by exchanging Hello packets between neighboring routers. Routers exchange routing updates after adjacency is established. A router adds the distance metric to the network as advertised by its neighbor to the cost of the link. This enables the router to access its neighbor. Routers perform this calculation for every route to a destination network.

Feasible Distance: The lowest calculated metric among the different alternate paths to reach a destination network.

Advertised Distance: Distance advertised by a neighboring router.

Feasible Successor: Neighboring router that has an advertised distance that is less than the feasible distance. The feasible successor is closer to the destination than the router. A router can have multiple feasible successors for reaching a destination network.

Successor: The feasible successor that has the lowest metric to reach a destination.

PROBLEM ISOLATION IN EIGRP

There are a number of commands available in the Cisco IOS, to isolate problems in EIGRP networks. The basic TCP/IP troubleshooting commands like ping, traceroute, and show ip route are also used in problem isolation in EIGRP environments. Table 8.2 lists the various show command options for EIGRP.

TABLE 8.2 Show Commands and Explanations

Command	Explanation
show ip eigrp interfaces	Shows list of interfaces in which EIGRP is enabled
show ip eigrp neighbors	Shows the EIGRP enabled routers that are on directly connected networks of a router
show ip eigrp topology	Shows topology of the entire EIGRP routing domain, including all the networks that are declared by the EIGRP routers
show ip eigrp traffic	Shows statistics pertaining to the traffic generated by EIGRP

Consider the example shown in Figure 8.1. Part of an EIGRP network with routers B1, B2, and B3 is shown. The show commands are issued at B2.

We will discuss the output of the show commands at B2. Listing 8.1 displays the output of the show ip eigrp interfaces command.

LISTING 8.1 Output for show ip eigrp interfaces Command in B2

```
B2#show ip eigrp interfaces
IP-EIGRP interfaces for process 10
Xmit        Queue  Mean          Pacing Time          Multicast   Pending
Interface   Peers  Un/Reliable   SRTT   Un/Reliable   Flow Timer  Routes
```

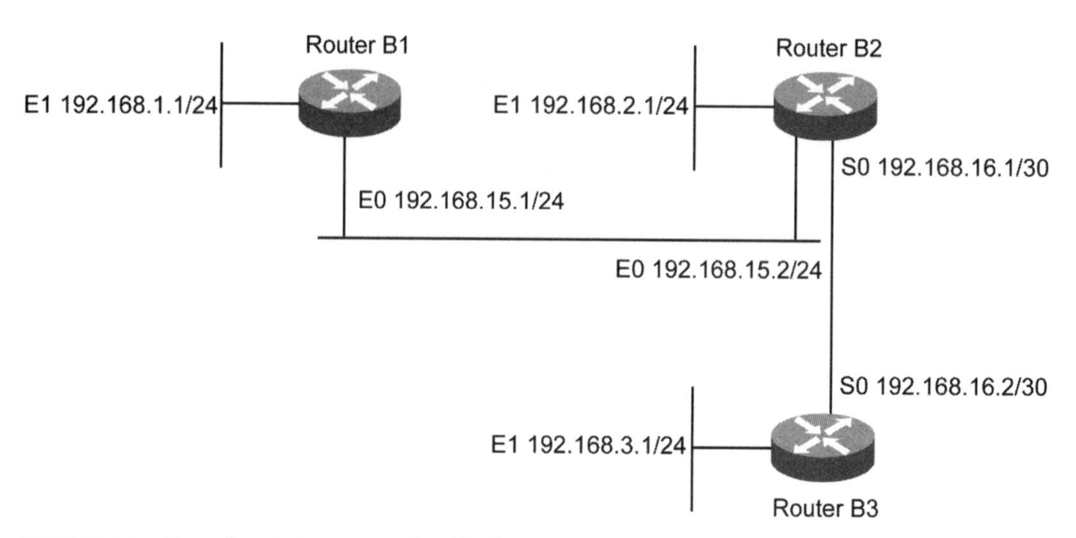

FIGURE 8.1 Part of an EIGRP network with show commands at B2.

```
Se0   1   0/0   214  10/380  1232   0
Et0   1   0/0   3    0/10    50     0
Et1   0   0/0   3    0/10    50     0
```

Listing 8.2 displays the output of the show ip eigrp neighbors command in B2.

LISTING 8.2 Output of show ip eigrp neighbors Command in B2

```
B2#show ip eigrp neighbors
IP-EIGRP neighbors for process 10
H   Address          Interface        Hold Uptime   SRTT   RTO  Q     Seq
                                      (sec)                (ms) Cnt   Num
1   192.168.16.2     Se0              14  1w6d      48     288  0     3471
0   192.168.15.1     Et0              14  1w6d      3      200  0     8558
```

Listing 8.3 displays the output of the show ip eigrp topology command in B2.

LISTING 8.3 Output of show ip eigrp topology Command in B2

```
B2#show ip eigrp topology
IP-EIGRP Topology Table for AS(10)/ID(192.168.16.1)
Codes: P - Passive, A - Active, U - Update, Q - Query, R - Reply,
       r - Reply status
       P 192.168.2.0/24, 1 successors, FD is 281600
       via Connected, Ethernet1
       P 192.168.15.0/24, 1 successors, FD is 281600
```

```
via Connected, Ethernet0
P 192.168.3.0/24, 1 successors, FD is 2195456
via 192.168.16.2 (2195456/281600), Serial0
P 192.168.3.0/24, 1 successors, FD is 2195456
via 192.168.15.1 (2195456/2169856), Ethernet0
```

Listing 8.4 displays the output of the show ip eigrp traffic command in B2.

LISTING 8.4 Output of show ip eigrp traffic Command in B2

```
B2#show ip eigrp traffic
IP-EIGRP Traffic Statistics for process 300
  Hellos sent/received: 1210064/1207881
  Updates sent/received: 4914/3832
  Queries sent/received: 1849/774
  Replies sent/received: 775/1819
  Acks sent/received: 5830/6697
  Input queue high water mark 8, 0 drops
```

Table 8.3 lists different debug command options for EIGRP along with the explanation for each command.

TABLE 8.3 debug Commands and Explanations

Command	Explanation
debug eigrp neighbors	Monitors the status of EIGRP neighbors and reflects when a neighbor router comes up or goes down
debug eigrp packets	Monitors all EIGRP packets and reports each time an EIGRP routing packet is sent or received by the router
debug eigrp transmit	Tracks all EIGRP transmission events, such as ack, packetize, startup, peerdown, link, build, strange, sia, and detail
debug eigrp fsm	Displays EIGRP DUAL algorithm functioning and the various stages involved in building the topology table

Different types of EIGRP packets can be monitored using the debug eigrp packets command. Different EIGRP packets are Update, Request, Query, Reply, Hello,

Ipxsap, Probe, Ack, Stub, SIAquery, and SIAreply. Each type of EIGRP packet can be monitored separately using the commands:

debug eigrp packets SIAquery: Monitors EIGRP SIA-Query packets
debug eigrp packets SIAreply: Monitors EIGRP SIA-Reply packets
debug eigrp packets ack: Monitors EIGRP ack packets
debug eigrp packets hello: Monitors EIGRP hello packets
debug eigrp packets ipxsap: Monitors EIGR ipxsap packets
debug eigrp packets probe: Monitors EIGRP probe packets
debug eigrp packets query: Monitors EIGRP query packets
debug eigrp packets reply: Monitors EIGRP reply packets
debug eigrp packets request: Monitors EIGRP request packets
debug eigrp packets retry: Monitors EIGRP retransmissions
debug eigrp packets stub: Monitors EIGRP stub packets
debug eigrp packets terse: Monitors all EIGRP packets except Hello packets
debug eigrp packets update: Monitors EIGRP update packets
debug eigrp packets verbose: Displays all EIGRP packets

Consider part of an EIGRP network showing routers B1, B2, B3, B4, B5, B6, and B7. The debug commands are issued at router B2. This scenario is depicted in Figure 8.2.

First, the debugging of EIGRP neighbor is carried out. We are considering a stable network. The B2#cle ip eigrp neighbors 172.18.1.52 command is issued to simulate the going down of one of the EIGRP networks. EIGRP neighbor debugging is

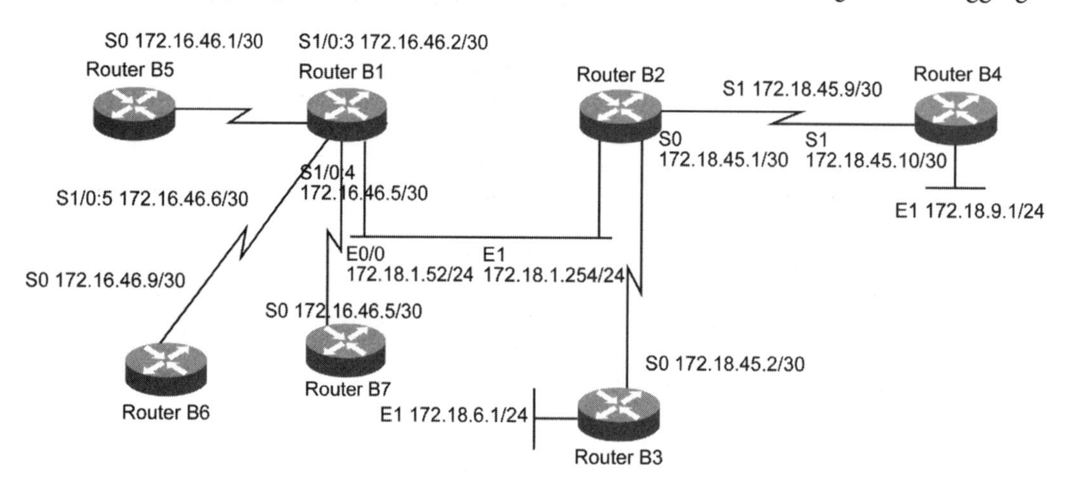

FIGURE 8.2 Part of an EIGRP network with debug commands at B2.

enabled with the B2#debug eigrp neighbors command. Listing 8.5 displays the output of the debug command as observed in B2.

LISTING 8.5 Output of debug eigrp neighbors Command at B2

```
B2#cle ip eigrp neighbors 172.18.1.52
B2#debug eigrp neighbors
5w5d: EIGRP: Holdtime expired
5w5d: EIGRP: Neighbor 172.18.1.52 went down on Ethernet1
5w5d: EIGRP: New peer 172.18.1.52
```

Second, the EIGRP packets are debugged. Listing 8.6 displays the movement of EIGRP hello packets in particular across the EIGRP network, as observed at B2.

LISTING 8.6 Output of debug eigrp packets hello Command at B2

```
B2#debug eigrp packets hello
EIGRP Packets debugging is on
  (HELLO)
5w6d: EIGRP: Sending HELLO on Ethernet1
5w6d:    AS 300, Flags 0x0, Seq 0/0 idbQ 0/0 iidbQ un/rely 0/0
5w6d: EIGRP: Sending HELLO on Serial1
5w6d:    AS 300, Flags 0x0, Seq 0/0 idbQ 0/0 iidbQ un/rely 0/0
5w6d: EIGRP: Sending HELLO on Serial0
5w6d:    AS 300, Flags 0x0, Seq 0/0 idbQ 0/0 iidbQ un/rely 0/0
5w6d: EIGRP: Received HELLO on Serial1 nbr 172.18.45.10
5w6d:    AS 300, Flags 0x0, Seq 0/0 idbQ 0/0 iidbQ un/rely 0/0 peerQ
  un/rely 0/0
5w6d: EIGRP: Received HELLO on Serial0 nbr 172.18.45.2
5w6d:    AS 300, Flags 0x0, Seq 0/0 idbQ 0/0 iidbQ un/rely 0/0 peerQ
  un/rely 0/0
5w6d: EIGRP: Received HELLO on Ethernet1 nbr 172.18.1.52
5w6d:    AS 300, Flags 0x0, Seq 0/0 idbQ 1/0 iidbQ un/rely 0/0 peerQ
  un/rely 0/0
5w6d: EIGRP: Sending HELLO on Ethernet1
5w6d:    AS 300, Flags 0x0, Seq 0/0 idbQ 0/0 iidbQ un/rely 0/0
5w6d: EIGRP: Sending HELLO on Serial1
5w6d:    AS 300, Flags 0x0, Seq 0/0 idbQ 0/0 iidbQ un/rely 0/0
5w6d: EIGRP: Sending HELLO on Serial0
5w6d:    AS 300, Flags 0x0, Seq 0/0 idbQ 0/0 iidbQ un/rely 0/0
5w6d: EIGRP: Received HELLO on Serial1 nbr 172.18.45.10
5w6d:    AS 300, Flags 0x0, Seq 0/0 idbQ 0/0 iidbQ un/rely 0/0 peerQ
  un/rely 0/0
```

```
5w6d: EIGRP: Received HELLO on Serial0 nbr 172.18.45.2
5w6d:   AS 300, Flags 0x0, Seq 0/0 idbQ 0/0 iidbQ un/rely 0/0 peerQ
un/rely 0/0
5w6d: EIGRP: Received HELLO on Ethernet1 nbr 172.18.1.52
5w6d:   AS 300, Flags 0x0, Seq 0/0 idbQ 0/0 iidbQ un/rely 0/0 peerQ
un/rely 0/0
5w6d: EIGRP: Sending HELLO on Ethernet1
5w6d:   AS 300, Flags 0x0, Seq 0/0 idbQ 0/0 iidbQ un/rely 0/0
5w6d: EIGRP: Sending HELLO on Serial1
5w6d:   AS 300, Flags 0x0, Seq 0/0 idbQ 0/0 iidbQ un/rely 0/0
5w6d: EIGRP: Sending HELLO on Serial0
5w6d:   AS 300, Flags 0x0, Seq 0/0 idbQ 0/0 iidbQ un/rely 0/0
```

All EIGRP transmission events are tracked across the EIGRP network, as displayed in Listing 8.7. The command clear ip eigrp 300 neighbors is used to trigger flow of transmission events.

LISTING 8.7 Output of the debug eigrp transmit Command at B2

```
B2#debug eigrp transmit
B2#clear ip eigrp 300 neighbors
EIGRP Transmission Events debugging is on
  (ACK, PACKETIZE, STARTUP, PEERDOWN, LINK, BUILD, STRANGE, SIA, DETAIL)
5w5d: Peer 172.18.1.52 going down
5w5d: DNDB QUERY 192.9.201.0/24, serno 17692 to 17751, refcount 3
5w5d:   Anchoring Serial1, starting Serial1 timer
5w5d:   Anchoring Serial0, starting Serial0 timer
5w5d:   Anchoring Ethernet1, starting Ethernet1 timer
5w5d: DNDB QUERY 192.9.212.0/24, serno 17714 to 17752, refcount 3
5w5d: DNDB QUERY 192.9.214.0/24, serno 17674 to 17753, refcount 3
5w5d: Last peer deleted from Ethernet1
5w5d:   Dropping refcount on 192.9.201.0/24, refcount now 2
5w5d:   Dropping refcount on 192.9.212.0/24, refcount now 2
5w5d: Peer 172.18.45.2 going down
5w5d: DNDB QUERY 172.18.6.0/24, serno 15520 to 17829, refcount 2
5w5d: Last peer deleted from Serial0
5w5d:   Dropping refcount on 172.18.6.0/24, refcount now 1
5w5d: Peer 172.18.45.10 going down
5w5d: Last peer deleted from Serial1
5w5d:   Dropping refcount on 172.18.6.0/24, refcount now 0
5w5d: Packetizing timer expired on Serial1
5w5d: Packetizing timer expired on Serial0
5w5d: Packetizing timer expired on Ethernet1
5w5d: New peer 172.18.45.2 on Serial0
```

```
5w5d:    No IIDB anchor
5w5d:    Packetized serno 1-15519
5w5d: New peer 172.18.45.10 on Serial1
5w5d:    No IIDB anchor
5w5d:    Packetized serno 1-15519
5w5d: Building STARTUP packet for 172.18.45.2, serno 1-15519
5w5d:    Items:  1 15518 S15519
5w5d: Building STARTUP packet for 172.18.45.10, serno 1-15519
5w5d:    Items:  1 S15518 15519
5w5d: DNDB UPDATE 172.18.9.0/24, serno 0 to 17830, refcount 2
5w5d:    Anchoring Serial1, starting Serial1 timer
5w5d:    Anchoring Serial0, starting Serial0 timer
5w5d: Packetizing timer expired on Serial1
5w5d: Packets pending on Serial1
5w5d: Intf Serial1 packetized UPDATE 17830-17830
5w5d:    Interface is now quiescent
5w5d: Packetizing timer expired on Serial0
5w5d: Packets pending on Serial0
5w5d: Intf Serial0 packetized UPDATE 17830-17830
5w5d:    Interface is now quiescent
5w5d: Building STARTUP packet for 172.18.45.2, serno 1-15519
5w5d:    Items:  1 15518 S15519
5w5d: Building STARTUP packet for 172.18.45.10, serno 1-15519
5w5d:    Items:  1 S15518 15519
5w5d: Packet acked from 172.18.45.10 (Serial1), serno 1-15519
5w5d: Startup update acked from 172.18.45.10, serno 1-15519, thread
ends 15519
5w5d: Building MULTICAST UPDATE packet for Serial1, serno 17830-17830
5w5d:    Items:  U17830
5w5d: Packet acked from 172.18.45.10 (Serial1), serno 17830-17830
5w5d: Flow blocking cleared on Serial1
5w5d: Multicast acked from Serial1, serno 17830-17830
5w5d:    Found serno 17830, refcount now 1
5w5d: New peer 172.18.1.52 on Ethernet1
5w5d:    No IIDB anchor
5w5d:    Packetized serno 1-17830
5w5d: Building STARTUP packet for 172.18.1.52, serno 1-17830
5w5d:    Items:  S1 15518 15519 17830
5w5d: DNDB UPDATE 172.18.6.0/24, serno 0 to 17831, refcount 3
5w5d:    Anchoring Ethernet1, starting Ethernet1 timer
5w5d:    Anchoring Serial0, starting Serial0 timer
5w5d:    Anchoring Serial1, starting Serial1 timer
5w5d: Packetizing timer expired on Ethernet1
5w5d: Packets pending on Ethernet1
```

```
5w5d: Intf Ethernet1 packetized UPDATE 17831-17831
5w5d:   Interface is now quiescent
5w5d: Packetizing timer expired on Serial0
5w5d: Packets pending on Serial0
5w5d:   Interface not flow-ready
5w5d: Packetizing timer expired on Serial1
5w5d: Packets pending on Serial1
5w5d: Intf Serial1 packetized UPDATE 17831-17831
5w5d:   Interface is now quiescent
5w5d: Building MULTICAST UPDATE packet for Ethernet1, serno 17831-17831
5w5d:   Items:  17831
5w5d: Building MULTICAST UPDATE packet for Serial1, serno 17831-17831
5w5d:   Items:  17831
5w5d: DNDB UPDATE 192.9.19.0/24, serno 0 to 17832, refcount 3
5w5d:   Anchoring Serial1, starting Serial1 timer
5w5d:   Anchoring Ethernet1, starting Ethernet1 timer
5w5d: DNDB UPDATE 192.9.214.0/24, serno 0 to 17833, refcount 3
5w5d: Packet acked from 172.18.45.10 (Serial1), serno 17831-17831
5w5d: Packets pending on Serial1
5w5d:   Interface not flow-ready
5w5d: Flow blocking cleared on Serial1
5w5d: Multicast acked from Serial1, serno 17831-17831
5w5d:   Found serno 17831, refcount now 2
5w5d: Packets pending on Serial1
5w5d: Intf Serial1 packetized UPDATE 17832-17857
5w5d:   Interface is now quiescent
5w5d: Packetizing timer expired on Serial1
5w5d: Packetizing timer expired on Ethernet1
5w5d: Packets pending on Ethernet1
5w5d:   Interface not flow-ready
5w5d: Building MULTICAST UPDATE packet for Serial1, serno 17832-17857
5w5d:   Items:  17832 17833 17834 17835 17836 17837 17838 17839 17840
  17841 1784
2 17843 17844 17845 17846 17847 17848 17849 17850 17851 17852 17853
  17854 17855
17856 17857
5w5d: Packet acked from 172.18.45.10 (Serial1), serno 17832-17857
5w5d: Flow blocking cleared on Serial1
5w5d: Multicast acked from Serial1, serno 17832-17857
5w5d:   Found serno 17832, refcount now 2
5w5d:   Found serno 17833, refcount now 2
5w5d: Building STARTUP packet for 172.18.1.52, serno 1-17830
5w5d:   Items:  S1 15518 15519 17830
5w5d: Packet acked from 172.18.1.52 (Ethernet1), serno 1-17830
```

```
5w5d: Startup update acked from 172.18.1.52, serno 1-17830, thread ends
  17830
5w5d: Packets pending on Ethernet1
5w5d:   Interface not flow-ready
5w5d: DNDB UPDATE 172.19.2.0/24, serno 0 to 17858, refcount 3
5w5d:   Anchoring Serial1, starting Serial1 timer
5w5d: Building MULTICAST UPDATE packet for Ethernet1, serno 17831-17831
5w5d:   Items:  17831
5w5d: Packetizing timer expired on Serial1
5w5d: Packets pending on Serial1
5w5d: Intf Serial1 packetized UPDATE 17858-17887
5w5d:   Interface is now quiescent
```

Finally, Listing 8.8 displays the working of DUAL algorithm as captured by the debug eigrp fsm command. The command clear ip eigrp neighbors serial 1 is used to initiate some activity on the DUAL front.

LISTING 8.8 Output of debug eigrp fsm Command at B2

```
B2#debug eigrp fsm
B2#clear ip eigrp neighbors serial 1
EIGRP FSM Events/Actions debugging is on
5w5d: DUAL: linkdown: start - 172.18.45.10 via Serial1
5w5d: DUAL: Destination 192.9.201.0/24
5w5d: DUAL: Destination 192.9.212.0/24
5w5d: DUAL: Destination 192.9.214.0/24
5w5d: DUAL: Destination 172.16.46.0/30
5w5d: DUAL: Destination 172.16.46.4/30
5w5d: DUAL: Destination 172.16.46.8/30
5w5d: DUAL: Destination 172.18.9.0/24
5w5d: DUAL: Find FS for dest 172.18.9.0/24. FD is 2195456, RD is
  2195456
5w5d: DUAL: 172.18.45.10 metric 4294967295/4294967295 not found
  Dmin is 4294
967295
5w5d: DUAL: Dest 172.18.9.0/24 entering active state.
5w5d: DUAL: Set reply-status table. Count is 2.
5w5d: DUAL: Not doing split horizon
5w5d: DUAL: Destination 172.18.6.0/24
5w5d: DUAL: linkdown: finish
5w5d: DUAL: dest(172.18.9.0/24) active
5w5d: DUAL: rcvreply: 172.18.9.0/24 via 172.18.45.2 metric
4294967295/4294967295
5w5d: DUAL: reply count is 2
```

```
5w5d: DUAL: Clearing handle 0, count now 1
5w5d: DUAL: Removing dest 172.18.9.0/24, nexthop 172.18.45.2
5w5d: DUAL: dest(172.18.9.0/24) active
5w5d: DUAL: rcvquery: 172.18.9.0/24 via 172.18.1.52 metric
  4294967295/4294967295, RD is 4294967295
5w5d: DUAL: send REPLY(r1/n1) about 172.18.9.0/24 to 172.18.1.52
5w5d: DUAL: rcvreply: 172.18.9.0/24 via 172.18.1.52 metric
4294967295/4294967295
5w5d: DUAL: reply count is 1
5w5d: DUAL: Clearing handle 2, count now 0
5w5d: DUAL: Freeing reply status table
5w5d: DUAL: Find FS for dest 172.18.9.0/24. FD is 4294967295, RD is
4294967295 found
5w5d: DUAL: Removing dest 172.18.9.0/24, nexthop 172.18.45.10
5w5d: DUAL: Removing dest 172.18.9.0/24, nexthop 172.18.1.52
5w5d: DUAL: No routes.  Flushing dest 172.18.9.0/24
5w5d: DUAL: dest(172.18.9.0/24) not active
5w5d: DUAL: rcvupdate: 172.18.9.0/24 via 172.18.45.10 metric
  2195456/281600
5w5d: DUAL: Find FS for dest 172.18.9.0/24. FD is 4294967295, RD is
  4294967295 found
5w5d: DUAL: RT installed 172.18.9.0/24 via 172.18.45.10
5w5d: DUAL: Send update about 172.18.9.0/24.  Reason: metric chg
5w5d: DUAL: Send update about 172.18.9.0/24.  Reason: new if
```

MISCONFIGURATION IN EIGRP

Misconfiguration is one of the root causes of problems in any network. Misconfiguration can be caused either by manual or system errors. You need to take extreme care while configuring an EIGRP routing protocol in a router, because even a small configuration error can affect the network traffic. Consider the scenario depicted in Figure 8.3. There are three routers, B1, B2, and B3, in an EIGRP network.

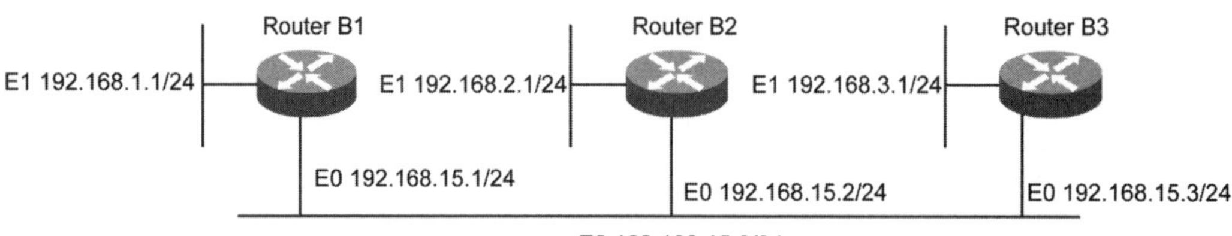

FIGURE 8.3 An EIGRP network with possibility for misconfigurations.

In Figure 8.3, simulate a few possible misconfigurations and define methods to solve them. The misconfigurations could be due to:

- Network not declared in EIGRP
- Same Autonomous System (AS) not defined in all routers in EIGRP domain

Network Not Declared in EIGRP

Consider a network segment 192.168.1.0/24 containing the secondary DNS server of the company with IP 192.168.1.10. The misconfiguration is caused by not declaring the network 192.168.1.0/24 in the EIGRP configurations. Table 8.4 shows the configurations of the three routers, B1, B2, and B3.

TABLE 8.4 Configurations of EIGRP Routers

Router	Configuration
B1	interface Ethernet 0 ip address 192.168.15.1 255.255.255.0 ! interface Ethernet 1 ip address 192.168.1.1 255.255.255.0 ! router eigrp 10 network 192.168.15.0 !
B2	interface Ethernet 0 ip address 192.168.15.2 255.255.255.0 ! interface Ethernet 1 ip address 192.168.2.1 255.255.255.0 ! router eigrp 10 network 192.168.2.0 network 192.168.15.0 !

Observation

The end users connected to the LAN segments 192.168.2.0/24 and 192.168.3.0/24 are not able to browse the Internet because of the misconfiguration.

Problem Isolation

The steps to isolating the problem caused due to not declaring a network in the EIGRP domain are:

1. Note if the problem occurs when the primary DNS server, 192.168.2.4, is down.
2. Require users to contact the secondary DNS server, 192.168.1.1.
3. Issue a ping command to 192.168.1.1 from B2 and B3 to check connectivity to this server. A 100% failure is noted. The ping command is executed as shown in Listings 8.9 and 8.10.

LISTING 8.9 Ping Command at B2

```
B2#ping 192.168.1.1
Type escape sequence to abort.
Sending 5, 100-byte ICMP Echos to 192.168.1.1, timeout is 2 seconds:
.....
Success rate is 0 percent (0/5)
```

LISTING 8.10 Ping Command at B3

```
B3#ping 192.168.1.1
Type escape sequence to abort.
Sending 5, 100-byte ICMP Echos to 192.168.1.1, timeout is 2 seconds:
.....
Success rate is 0 percent (0/5)
```

4. Issue the commands:

```
B2#show ip route 192.168.1.0
% Network not in table
B3#show ip route 192.168.1.0
% Network not in table
```

5. Check if B1 is an EIGRP neighbor of B3 and B2 using the show ip eigrp neighbors command. The output for the show ip eigrp neighbors command is shown in Listing 8.11 and Listing 8.12. This output confirms that B1, B2, and B3 are EIGRP neighbors.

LISTING 8.11 Output of show ip eigrp neighbors Command at B2

```
B2#show ip eigrp neighbors
IP-EIGRP neighbors for process 10
```

```
H    Address           Interface    Hold Uptime   SRTT    RTO    Q     Seq
                                    (sec)                 (ms)   Cnt   Num
1    192.168.15.3      Se0          14  1w6d      48      288    0     3471
0    192.168.15.1      Et0          14  1w6d      3       200    0     8558
```

LISTING 8.12 Output of show ip eigrp neighbors Command at B3

```
B3#show ip eigrp neighbors
IP-EIGRP neighbors for process 10
H    Address           Interface    Hold Uptime   SRTT    RTO    Q     Seq
                                    (sec)                 (ms)   Cnt   Num
1    192.168.15.2      Se0          14  1w6d      48      288    0     3471
0    192.168.15.1      Et0          14  1w6d      3       200    0     8558
```

6. Check the availability of route to network 192.168.1.0 at B1 using the show ip route 192.168.1.0 command. The output for the show ip route 192.168.1.0 command showing that 192.168.1.0 is a connected network is shown in Listing 8.13.

LISTING 8.13 Output for show ip route 192.168.1.0 Command at B2

```
B2#show ip route 192.168.1.0
Routing entry for 192.168.1.0/24
Known via "connected", distance 0, metric 0 (connected, via interface)
  Routing Descriptor Blocks:
  * directly connected, via Ethernet1
  Route metric is 0, traffic share count is 1
```

7. Check if EIGRP is enabled for this network in B1 using the show ip protocols command. The output for the show ip protocols command is shown in Listing 8.14.

LISTING 8.14 Output for show ip protocols Command at B1

```
B1#show ip protocols
Routing Protocol is "eigrp 10"
  Outgoing update filter list for all interfaces is not set
  Incoming update filter list for all interfaces is not set
  Default networks flagged in outgoing updates
  Default networks accepted from incoming updates
  EIGRP metric weight K1=1, K2=0, K3=1, K4=0, K5=0
  EIGRP maximum hopcount 100
  EIGRP maximum metric variance 1
```

```
Redistributing: static, eigrp 300
Automatic network summarization is in effect
Maximum path: 4
Routing for Networks:
 192.168.15.0
Passive Interface(s):
  Routing Information Sources:
  Gateway         Distance   Last Update
  192.168.15.2    90         00:05:35
  192.168.15.3    90         00:05:20
 Distance: internal 90 external 170
```

Listing 8.14 shows the networks for which EIGRP routing takes place. EIGRP routing does not occur for network 192.168.1.0. The corrected configuration of B1 should be as shown in Listing 8.15.

LISTING 8.15 Correct Configuration of B1

```
!
interface Ethernet 0
ip address 192.168.15.1 255.255.255.0
!
interface Ethernet 1
ip address 192.168.1.1 255.255.255.0
!
router eigrp 10
network 192.168.15.0
network 192.168.1.0
!
```

Same AS Not Defined in EIGRP Routers

Consider the same example depicted in Figure 8.3. Table 8.5 lists the configuration of the routers in the EIGRP network.

TABLE 8.5 Configuration of Routers

Router	Configuration
B1	interface Ethernet 0 ip address 192.168.15.1 255.255.255.0 ! interface Ethernet 1 ip address 192.168.1.1 255.255.255.0

TABLE 8.5 *(continued)*

Router	Configuration
	! router eigrp 10 network 192.168.15.0 !
B2	interface Ethernet 0 ip address 192.168.15.2 255.255.255.0 ! interface Ethernet 1 ip address 192.168.2.1 255.255.255.0 ! router eigrp 20 network 192.168.2.0 network 192.168.15.0 !
B3	interface Ethernet 0 ip address 192.168.15.3 255.255.255.0 ! interface Ethernet 1 ip address 192.168.3.1 255.255.255.0 ! router eigrp 10 network 192.168.3.0 network 192.168.15.0 !

Observation

The users connected to the LAN segments 192.168.1.0/24 and 192.168.3.0/24 are not able to connect to the internal Web server, 192.168.2.5. This problem occurs because the same AS is not defined in all the routers in the EIGRP routing domain.

Problem Isolation

The steps to isolate the problem are:

1. Issue a ping command to network 192.168.2.5 from B1 and B3 to check connectivity using the ping 192.168.2.5 command. The output of the ping 192.168.2.5 command is shown in Listings 8.16 and 8.17.

LISTING 8.16 Output of ping 192.168.2.5 Command at B1

```
B1#ping 192.168.2.5
Type escape sequence to abort.
Sending 5, 100-byte ICMP Echos to 192.168.2.5, timeout is 2 seconds:
.....
Success rate is 0 percent (0/5)
```

LISTING 8.17 Output of ping 192.168.2.5 Command at B3

```
B3#ping 192.168.2.5
Type escape sequence to abort.
Sending 5, 100-byte ICMP Echos to 192.168.2.5, timeout is 2 seconds:
.....
Success rate is 0 percent (0/5)
```

The listings show a 100% failure of the ping command.

2. Issue the commands:

```
B1#show ip route 192.168.2.0
% Network not in table
B3#show ip route 192.168.2.0
% Network not in table
```

The output of these commands shows there is no route available to the network.

3. Check if B2 is an EIGRP neighbor of B1 and B3 using the show ip eigrp neighbors command. Listings 8.18 and 8.19 show the output for the show ip eigrp neighbors command.

LISTING 8.18 Output of show ip eigrp neighbors Command at B1

```
B1#show ip eigrp neighbors
IP-EIGRP neighbors for process 10
H    Address           Interface    Hold Uptime   SRTT    RTO   Q    Seq
                                    (sec)                 (ms)  Cnt  Num
1    192.168.15.3      Se0          14 1w6d       48      288   0    3471
```

LISTING 8.19 Output of show ip eigrp neighbors Command at B3

```
B3#show ip eigrp neighbors
IP-EIGRP neighbors for process 10
```

```
H   Address          Interface   Hold Uptime  SRTT   RTO   Q    Seq
                                 (sec)               (ms)  Cnt  Num
1   192.168.15.1     Et0         14 1w6d      3      200   0    8558
```

The listings show that B2 is not a neighbor of B1 and B3.

4. Check the availability of route to network 192.168.2.0 at B2 using the show ip route 192.168.1.0 command at B2. The output shows that 192.168.2.0 is a connected network
5. Check if EIGRP is enabled for this network in B2 using the show ip protocols command. The output for the command is shown in Listing 8.20.

LISTING 8.20 Output for show ip protocols Command at B2

```
B2#show ip protocols
Routing Protocol is “eigrp 20”
  Outgoing update filter list for all interfaces is not set
  Incoming update filter list for all interfaces is not set
  Default networks flagged in outgoing updates
  Default networks accepted from incoming updates
  EIGRP metric weight K1=1, K2=0, K3=1, K4=0, K5=0
  EIGRP maximum hopcount 100
  EIGRP maximum metric variance 1
  Redistributing: static, eigrp 300
  Automatic network summarization is in effect
  Maximum path: 4
  Routing for Networks:
   192.168.2.0
   192.168.15.0
  Passive Interface(s):
    Routing Information Sources:
    Gateway          Distance      Last Update
   Distance: internal 90 external 170
```

Listing 8.20 shows that the Autonomous System Number (ASN) in B2 is 20 and is 10 in B1 and B3. The corrected configuration of B2 is shown in Listing 8.21.

LISTING 8.21 Correct Configuration of B2

```
interface Ethernet 0
ip address 192.168.15.2 255.255.255.0
!
interface Ethernet 1
ip address 192.168.2.1 255.255.255.0
```

```
!
router eigrp 20
network 192.168.2.0
network 192.168.15.0
!
```

EIGRP NEIGHBOR FORMATION PROBLEM

The primary requirement for an EIGRP routing environment is to establish adjacencies between neighbor routers by exchanging Hello packets. Troubleshooting is relevant when an EIGRP network is being setup, or when a new node is being added to the existing network. There may be situations in which initially existing neighbor relations are not present any more and need to be resolved. Look at the example of troubleshooting an EIGRP network. Figure 8.4 depicts an EIGRP network in which routers A2 and A3 are directly connected to router A1.

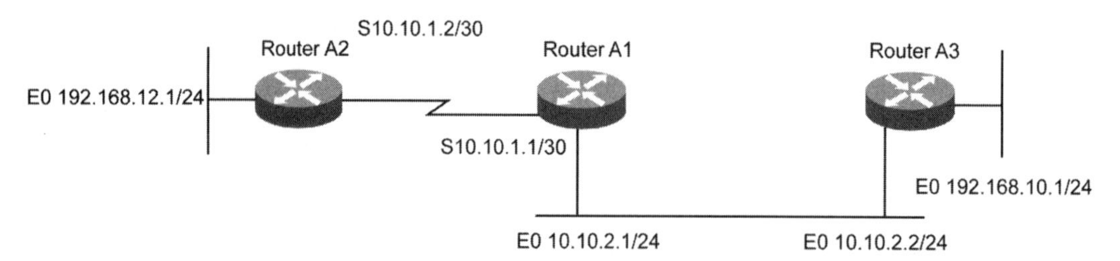

FIGURE 8.4 An EIGRP network depicting neighbor formation problems.

Observation

In Figure 8.4, A1 has no route to network 192.168.12.0/24. This network is also missing from the topology table. An attempt to check neighbors does not reflect A2 as a neighbor of A1. Listing 8.22 shows the output of the show ip eigrp neighbors command at A1.

LISTING 8.22 Output of the show ip eigrp neighbors Command at A1

```
A1#show ip eigrp neighbors
IP-EIGRP neighbors for process 100
H    Address        Interface     Hold Uptime   SRTT     RTO   Q     Seq
                                  (sec)                  (ms)  Cnt   Num
0    10.10.2.2      Et0           14 1w6d       3        200   0     8558
```

Problem Isolation

The output of the debug command debug eigrp packets hello in Listing 8.6 shows that no hello packets are sent or received at interface Serial 0, resulting in the inability to form neighbor relationships. The steps to isolating the neighbor formation problems are:

1. Check if network 10.10.1.0/30 is the primary network defined in the interface Serial 0 of A1 and A2. The commands used are:

```
A1#show running-config
A1#show ip interface brief
A2#show running-config
A2#show ip interface brief
```

No neighbor relationship is formed by EIGRP if network 10.10.1.0/30 is not the primary network. The network topology needs to be reworked.

2. Check if the routers can ping each other with packet sizes starting from minimum possible packet size, up to a packet size equal to the interface MTU size. In case of failure in ping response, there could be two problems:

- Mismatch in MTU size at the interface at both sides. This can be checked using show interface Serial0 commands. The output for the show interface Serial0 command at A1 is shown in Listing 8.23.

LISTING 8.23 Output of the show interface Serial0 at A1

```
A1#show interface Serial0
Serial0 is up, line protocol is up
  Hardware is HD64570
  Description: WAN link to router A2
  Internet address is 10.10.1.1/30
  MTU 1500 bytes, BW 1544 Kbit, DLY 20000 usec,
     reliability 255/255, txload 1/255, rxload 1/255
  Encapsulation HDLC, loopback not set
  Keepalive set (10 sec)
  Last input 00:00:00, output 00:00:00, output hang never
  Last clearing of “show interface” counters never
  Input queue: 0/75/3134350/0 (size/max/drops/flushes); Total output
   drops: 11
  Queueing strategy: weighted fair
  Output queue: 0/1000/64/0 (size/max total/threshold/drops)
     Conversations  0/11/256 (active/max active/max total)
```

```
Reserved Conversations 0/0 (allocated/max allocated)
5 minute input rate 13000 bits/sec, 29 packets/sec
5 minute output rate 2000 bits/sec, 3 packets/sec
115001192 packets input, 4155732628 bytes, 272 no buffer
Received 425130 broadcasts, 0 runts, 0 giants, 0 throttles
228 input errors, 227 CRC, 68 frame, 0 overrun, 0 ignored, 34 abort
21650119 packets output, 3170761599 bytes, 0 underruns
0 output errors, 0 collisions, 1 interface resets
0 output buffer failures, 0 output buffers swapped out
10 carrier transitions
DCD=up  DSR=up  DTR=up  RTS=up  CTS=up
```

The MTU value is 1500 as shown in Listing 8.23. In case of a mismatch, the value can be adjusted to make it equal at both ends, using the command:

```
A1(config)#interface Serial0
A1(config-if)#mtu "size in bytes"
```

- A layer-2 level connectivity issue that needs to be checked and rectified. In the case of a WAN link, there may be a transmission media, hardware, or cable problem. In the case of LAN, it may be a problem of switch port or cabling.

3. Check the configuration of routers to see if the ping response is OK. Check if EIGRP is enabled for the network in both routers by using the show running-config command. If EIGRP is enabled, the output for the command for both A1 and A2 will be as shown in Listing 8.24.

LISTING 8.24 Output for show running-config Command at A1

```
A1#show running-config
router eigrp 100
network  10.10.1.0
```

The show ip protocols command can also be used to check if EIGRP routing protocol has been enabled in the routers. If it is enabled, the output at the routers would look like Listing 8.25.

LISTING 8.25 Output for show ip protocols Command at A1

```
A1#show ip protocols
Routing Protocol is "eigrp 100"
  Outgoing update filter list for all interfaces is not set
```

```
Incoming update filter list for all interfaces is not set
Default networks flagged in outgoing updates
Default networks accepted from incoming updates
EIGRP metric weight K1=1, K2=0, K3=1, K4=0, K5=0
EIGRP maximum hopcount 100
EIGRP maximum metric variance 1
Redistributing: static, eigrp 300
Automatic network summarization is in effect
Maximum path: 4
Routing for Networks:
  10.10.2.0
  10.10.1.0
Passive Interface(s):
Routing Information Sources:
Gateway     Distance    Last Update
10.10.2.1   90          00:05:35
Distance: internal 90 external 170
```

The output Routing Protocol is "eigrp 100" indicates that EIGRP is enabled for the network 10.10.1.0.

4. Check the access lists at interfaces if the commands are present in both routers. The access lists are checked by using the show ip interface Serial 0 command at both routers. Listing 8.26 shows the output of show ip interface Serial 0 command at A1.

LISTING 8.26 Output of the show ip interface Serial 0 Command at A1

```
A1#show ip interface Serial 0
Serial0 is up, line protocol is up
  Internet address is 10.10.1.1/30
  Broadcast address is 255.255.255.255
  Address determined by non-volatile memory
  MTU is 1500 bytes
  Helper address is not set
  Directed broadcast forwarding is disabled
  Multicast reserved groups joined: 224.0.0.10
  Outgoing access list is not set
  Inbound  access list is not set
  Proxy ARP is enabled
  Security level is default
  Split horizon is enabled
  ICMP redirects are never sent
  ICMP unreachables are always sent
  ICMP mask replies are never sent
```

```
IP fast switching is enabled
IP fast switching on the same interface is enabled
IP Flow switching is disabled
IP Feature Fast switching turbo vector
IP multicast fast switching is enabled
IP multicast distributed fast switching is disabled
IP route-cache flags are Fast
Router Discovery is disabled
IP output packet accounting is disabled
IP access violation accounting is disabled
TCP/IP header compression is disabled
RTP/IP header compression is disabled
Probe proxy name replies are disabled
Policy routing is disabled
Network address translation is enabled, interface in domain inside
WCCP Redirect outbound is disabled
WCCP Redirect exclude is disabled
BGP Policy Mapping is disabled
```

The part of the command output Outgoing access list is not set Inbound access list is not set shows the position where the access list setting appears. Access lists can be modified suitably to allow EIGRP packets to pass through.

EIGRP ROUTE PROBLEM

In certain cases, not all EIGRP routes may be available in the routing table. This may be caused due to nonformation of neighbor relationships.

Look at the example depicted in Figure 8.5. Routers A1 and A2 are part of the EIGRP routing environment represented by a cloud.

Observation

In Figure 8.5, the route for network 10.10.1.0/24 is not available in the routing table of A1. This is checked using the show ip route 10.10.1.0/24 command.

FIGURE 8.5 An EIGRP network depicting route problems.

Problem Isolation

The steps to isolate the route problem are:

1. Check if any EIGRP routes are available in A1 by using the A1#show ip route eigrp command.
2. Check the EIGRP topology table by using the A1#show ip eigrp topology command, if there are no routes available at A1.
3. Check and verify the neighbor relationships of A1 if an empty topology table is viewed. The neighbor relationships can be checked as discussed in the previous section.
4. Check and fix the distribute lists in both inbound and outbound directions if neighbor relationships are OK. The distribute lists are checked by using the show ip protocols command. The output of the show ip protocols command is shown in Listing 8.27.

LISTING 8.27 Output of show ip protocols Command at A1

```
A1#show ip protocols
Routing Protocol is "eigrp 100"
Outgoing update filter list for all interfaces is not set
Incoming update filter list for all interfaces is not set
Default networks flagged in outgoing updates
Default networks accepted from incoming updates
EIGRP metric weight K1=1, K2=0, K3=1, K4=0, K5=0
EIGRP maximum hopcount 100
EIGRP maximum metric variance 1
Redistributing: static, eigrp 300
Automatic network summarization is in effect
Maximum path: 4
Routing for Networks:
10.10.1.0
Passive Interface(s):
Ethernet0
Routing Information Sources:
Gateway          Distance       Last Update
10.10.2.1        90             00:05:35
Distance: internal 90 external 170
```

The output indicates whether any routing update filter is set or not.

5. Check if the local router has the same Router ID (RID) as the originating router. In that case, RID needs to be changed accordingly, and the EIGRP process is restarted.

EIGRP METRIC PROBLEM

EIGRP uses a composite metric composed of bandwidth, delay, reliability, load, and MTU. The metric is calculated using the formula:

$$\text{Metric} = [(\text{K1} \times \text{Bw}) + (\text{K2} \times \text{Bw}) + (\text{K3} \times \text{Delay})] + \text{K5}\,(256 - \text{Load})(\text{Reliability} + \text{K4})$$

in which Bw refers to bandwidth.

In EIGRP, K2 = K4 = K5 = 0 and K1 = K3 = 1, by default. As a result, the formula reduces to:

$$\text{Metric} = \text{Bw} + \text{Delay}$$

$$\text{Bw} = 10{,}000{,}000 \times \frac{1}{\text{Bandwidth in Kbps}}.$$

$$\text{Delay} = 100 \times (\text{delay in ms})$$

The final metric used in EIGRP is:

$$\text{Metric} = 10{,}000{,}000 \times \frac{1}{\text{Bandwidth in Kbps}} + 100\ (\text{delay in ms})$$

Bandwidth and delay are configurable parameters at the router interfaces with default values. For example, the default bandwidth value of a serial interface is that of a T1 link. Consider the example depicted in Figure 8.6 to detect and troubleshoot

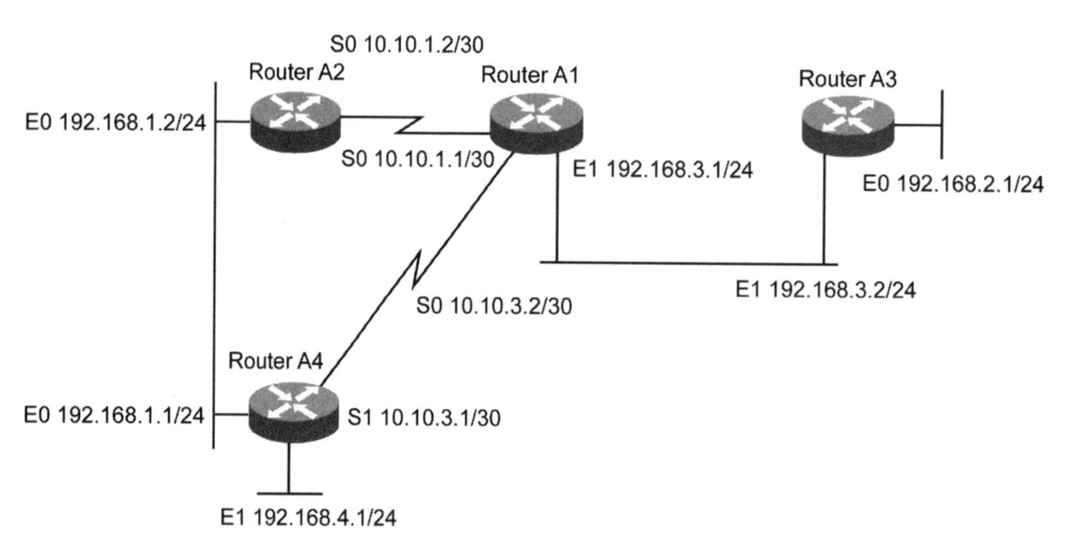

FIGURE 8.6 An EIGRP network depicting metric problems.

issues related to the EIGRP metric. In Figure 8.6, routers A2, A3, and A4 are directly connected to router A1.

Observation

Data traffic moving between A1 and A4 is erratic, with some packets getting dropped from time to time. This erratic behavior was reported by users accessing the Web server 192.168.4.2. The Web server is directly connected to A4.

Problem Isolation

The steps to isolate the metric problem are:

1. Check if there is any problem in the route availability in the routing table and the topology table. This is verified by using the show ip route 192.168.4.0 and show ip eigrp topology | inc 192.168.4.0 commands. The output for the show ip route 192.168.4.0 command is shown in Listing 8.28.

LISTING 8.28 Output for the show ip route 192.168.4.0 Command at A1

```
A1#show ip route 192.168.4.0
Routing entry for 192.168.4.0/24
  Known via "eigrp 300", distance 90, metric 1787392, type internal
  Redistributing via eigrp 100
  Last update from 10.10.3.1 on Serial0, 1d06h ago
  Routing Descriptor Blocks:
  * 10.10.3.1, from 10.10.3.1, 1d06h ago, via Serial0
      Route metric is 1787392, traffic share count is 1
      Total delay is 21000 microseconds, minimum bandwidth is 2048 Kbit
      Reliability 254/255, minimum MTU 1500 bytes
      Loading 1/255, Hops 1
```

The output for the show ip eigrp topology | inc 192.168.4.0 command at A1 is:

```
A1#show ip eigrp topology | inc 192.168.4.0
P 192.168.4.0/24, 1 successors, FD is 1787392
```

2. Make sure there is no neighbor relationship problem by using the A1#show ip eigrp neighbors command.
3. Make sure the transmission link connecting the rest of the EIGRP network with A4, which is terminated at interface Serial 0, has no errors. This is verified by using the show interface Serial 0 command. Listing 8.29 shows the output for the show interface Serial 0 command for A1.

LISTING 8.29 Output of the show interface Serial 0 Command for A1

```
A1#show interface Serial 0
Serial0 is up, line protocol is up
  Hardware is HD64570
  Description: WAN link for Router A4
  Internet address is 10.10.3.2/30
  MTU 1500 bytes, BW 1544 Kbit, DLY 20000 usec,
     reliability 255/255, txload 1/255, rxload 1/255
  Encapsulation HDLC, loopback not set
  Keepalive set (10 sec)
  Last input 00:00:00, output 00:00:00, output hang never
  Last clearing of "show interface" counters never
  Input queue: 0/75/3134350/0 (size/max/drops/flushes); Total output
    drops: 11
  Queueing strategy: weighted fair
  Output queue: 0/1000/64/0 (size/max total/threshold/drops)
  Conversations  0/11/256 (active/max active/max total)
  Reserved Conversations 0/0 (allocated/max allocated)
  5 minute input rate 13000 bits/sec, 29 packets/sec
  5 minute output rate 2000 bits/sec, 3 packets/sec
     115001192 packets input, 4155732628 bytes, 272 no buffer
     Received 425130 broadcasts, 0 runts, 0 giants, 0 throttles
     0 input errors, 0 CRC, 0 frame, 0 overrun, 0 ignored, 0 abort
     21650119 packets output, 3170761599 bytes, 0 underruns
     0 output errors, 0 collisions, 0 interface resets
     0 output buffer failures, 0 output buffers swapped out
     0 carrier transitions
     DCD=up  DSR=up  DTR=up  RTS=up  CTS=up
```

The command output shows errors in the link. EIGRP routing updates are exchanged without any problem, but the data traffic gets hampered.

4. Check the configured bandwidth parameter at the interface by using the show interface Serial 0 command. The output for the show interface Serial 0 command for A1 is shown in Listing 8.30.

LISTING 8.30 Output for the show interface Serial 0 Command for A1

```
A1#show interface Serial 0
Serial0 is up, line protocol is up
  Hardware is HD64570
  Description: WAN link for Router A4
```

```
Internet address is 10.10.3.2/30
MTU 1500 bytes, BW 1544 Kbit, DLY 20000 usec,
   reliability 255/255, txload 1/255, rxload 1/255
Encapsulation HDLC, loopback not set
Keepalive set (10 sec)
Last input 00:00:00, output 00:00:00, output hang never
Last clearing of "show interface" counters never
Input queue: 0/75/3134350/0 (size/max/drops/flushes); Total output
  drops: 11
Queueing strategy: weighted fair
Output queue: 0/1000/64/0 (size/max total/threshold/drops)
Conversations  0/11/256 (active/max active/max total)
Reserved Conversations 0/0 (allocated/max allocated)
5 minute input rate 13000 bits/sec, 29 packets/sec
5 minute output rate 2000 bits/sec, 3 packets/sec
115001192 packets input, 4155732628 bytes, 272 no buffer
Received 425130 broadcasts, 0 runts, 0 giants, 0 throttles
0 input errors, 0 CRC, 0 frame, 0 overrun, 0 ignored, 0 abort
   21650119 packets output, 3170761599 bytes, 0 underruns
   0 output errors, 0 collisions, 0 interface resets
   0 output buffer failures, 0 output buffers swapped out
   0 carrier transitions
   DCD=up  DSR=up  DTR=up  RTS=up  CTS=up
```

In Listing 8.30, the bandwidth configured for the interface is 1544 KB. There can be two conditions of a faulty bandwidth configuration in EIGRP. They are:

- Bandwidth is not configured. EIGRP uses a maximum of 50% of the link bandwidth for the routing packets, by default. The default of a serial link is a T1 with a bandwidth of 1.5 mbps. In this case, the link is a fractional T1 with a bandwidth of 768 Kbps. The configured bandwidth being 1.5 mbps, EIGRP reserves 750 Kbps for itself, which leaves less than 50 Kbps for data traffic causing dropping of traffic. Bandwidth is configured correctly using the commands

```
A1(config)#interface S0
A1(config-if)#bandwidth 768
```

- Bandwidth is incorrectly configured. Sometimes, a value of bandwidth other than the actual value is configured in serial links to manipulate the metric value to influence route selection in routing protocols. In Figure 8.6, both serial links are 56 Kbps bandwidth. To choose the path via A2 over the path via A4 to network 192.168.1.0/24, the bandwidth parameter of the link to A4 is set at a higher value of 224 Kbps. Out of this, 50% (that is, 112 Kbps) may be used for

EIGRP traffic, which is more than the actual bandwidth of the link. This condition causes congestion of the link. To overcome this condition, without affecting the desired effect on path selection process, you can use the commands

```
A1(config)#interface Serial0
A1(config-interface)#ip bandwidth-percent eigrp 100 12
```

These commands restrict the maximum percentage of the configured value of bandwidth to be used in EIGRP to 12% (27 Kbps), which is around 50% of the actual bandwidth.

STUCK IN ACTIVE STATE

In case of any topology change in EIGRP, the successors are searched from within the topology table. If there are no feasible successors available in the topology table, the particular route is marked as active, and queries are sent out to all the neighbors regarding that route. In some cases, replies of queries may not be received from all routers to which they were sent. Such a route is said to be in Stuck-In-Active (SIA) state. The neighbor that fails to reply is removed from the neighbor table.

Consider the example depicted in Figure 8.7 to understand issues pertaining to SIA routes. Figure 8.7 shows an EIGRP network in which routers A2 and A3 are neighbors of router A1. Router A4 is a neighbor of A2.

Observation

In Figure 8.7, one of the routes is always seen in SIA mode. This is evident in the show ip eigrp topology active command in A1, as shown in Listing 8.31.

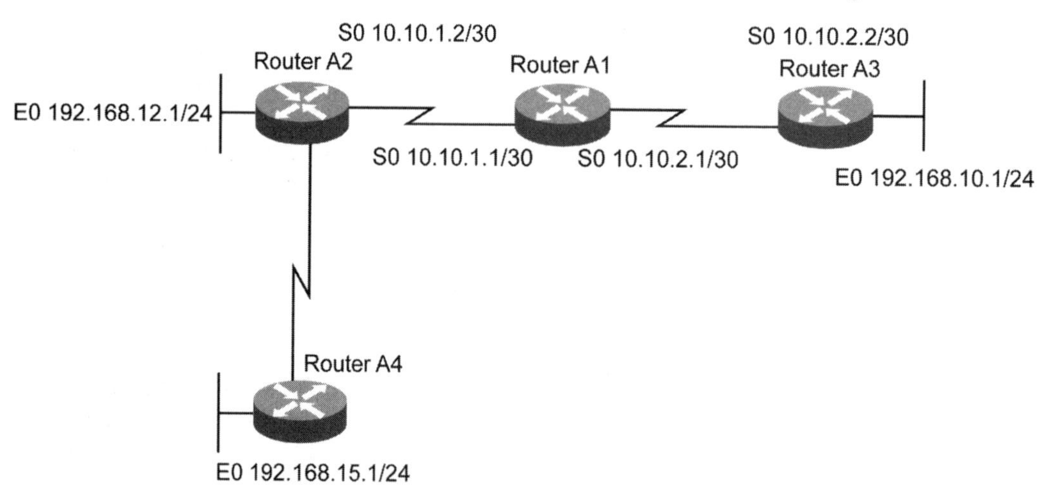

FIGURE 8.7 An EIGRP network showing the SIA state.

LISTING 8.31 Output of the show ip eigrp topology active Command in A1

```
Codes: P - Passive, A - Active, U - Update, Q - Query, R - Reply,
r - Reply status
A 192.168.15.0/24, 0 successors, FD is 512640000, Q
1 replies, active 00:00:01, query-origin: Local origin
    via 10.10.2.2 (Infinity/Infinity), Serial1
Remaining replies:
via 10.1.1.2, r, Serial0
```

Listing 8.31 shows that the neighbor 10.10.1.2 is not replying consistently, resulting in the route 192.168.15.0/24 being in the SIA state.

Problem Isolation

The steps to isolating the SIA mode problem are:

1. Identify the neighbor 10.10.1.2 to be responsible for the route being in SIA state. This is the most important step in troubleshooting SIA situations. In case of larger EIGRP networks, this process is repeated several times to reach the responsible router.
2. Examine the router that is not responding so you may find and resolve the issues. The reasons and commands used to identify the problems are listed in Table 8.6.

TABLE 8.6 Causes, Commands, and Solutions for the SIA Mode

Cause	Command	Possible Resolution
High CPU utilization	A2#show processes cpu	1. Identify the process using more of CPU. 2. Check if the process is avoidable and disable it. 3. Consider upgrading if the process is unavoidable. The syntax of this command varies in different versions.
High memory utilization	A2#show processes memory	1. Identify the process using more memory. 2. Check if the process is avoidable and disable it. 3. Consider upgrading if the process is unavoidable.

(continued)

TABLE 8.6 *(continued)*

Cause	Command	Possible Resolution
Inaccurately defined bandwidth parameter	A2#show interface Serial0 A2#show running-config	EIGRP is unable to send data packets at a proper pace if the bandwidth parameter configuration is missing or incorrect. This parameter needs to be configured as per the clock rate of the interface.
Poor link quality	A1#ping "other side WAN IP"	Packet loss and high latency is observed in the ping response. The transmission media needs to be checked thoroughly and errors should be taken care of.

In addition to the solutions listed in Table 8.6, the SIA problem can be reduced by using one of two approaches. They are:

- Modify one of the EIGRP timers to solve the problem. The time for which a router waits after sending out queries to its neighbors before declaring a route as SIA is the active time timer. This setting can be modified using the command:

```
A2(config)#router eigrp 100
A2(config-router)#timers active-time "EIGRP active-state time limit in
minutes"
```

- Reduce the query range that relates to the number of routers in the EIGRP network receiving and replying a query when there is a topology change. The more the number of routes, the longer it takes for the network to converge. Route summarization and distribution lists can be used effectively to ensure that not too many queries pass over high latency, low quality transmission links, and do not traverse routers that have higher processing tasks.

REDISTRIBUTION PROBLEM

Redistribution refers to sharing of routing information among multiple routing protocols running in the same router, which is not shared by default. You should visualize the changes that will take place in the routing table, due to the implementation of redistribution.

Consider the example depicted in Figure 8.8. Company A has both RIP and EIGRP routing protocols running in the network. Figure 8.8 shows part of the network with three routers. Redistribution occurs in router A1. A3 is running EIGRP and A2 is running RIP.

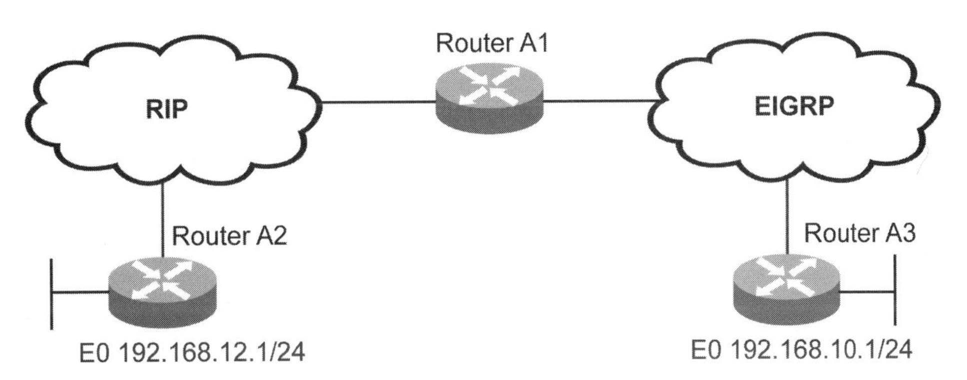

FIGURE 8.8 An EIGRP network showing redistribution problems.

Observation

Users in the network segment 192.168.10.0/24 in the EIGRP domain are not able to download pages from the Web server located in the network segment 192.168.12.0/24, which is a part of the RIP domain. The IP address of the Web server is 192.168.12.2.

Problem Isolation

The steps to isolating redistribution problems are:

1. Receive negative ping and traceroute responses to IP 192.168.12.2 at A3.
2. Run the show ip route 192.168.12.2 command in A3, which shows that no such route is available.

These two steps ascertain that the route 192.168.12.0/24 is not propagated in EIGRP.

3. Issue show ip route 192.168.12.2 at A1 where it shows an available RIP route. This ascertains that there is a problem in redistribution in EIGRP from RIP, at A1.
4. Check EIGRP routing configuration to ensure that redistribution is configured for RIP routes.

5. Configure default metric for routes redistributed into EIGRP.
6. Ensure that there is no route filter implemented in redistribution that is causing the blockage of propagation of the route 192.168.12.0/24.

The commands for configuring redistribution into EIGRP from RIP are shown in Listing 8.32.

LISTING 8.32 Commands to Configure Redistribution from RIP into EIGRP

```
router eigrp process-id
network ..................
network ..................
network ..................
redistribute rip
default-metric bandwidth delay reliability load mtu
```

CASE STUDY

Consider the network with an EIGRP routing environment as shown in Figure 8.9. The EIGRP enabled routers are A1, A2, A3, and A4. RIP is running as a part of the

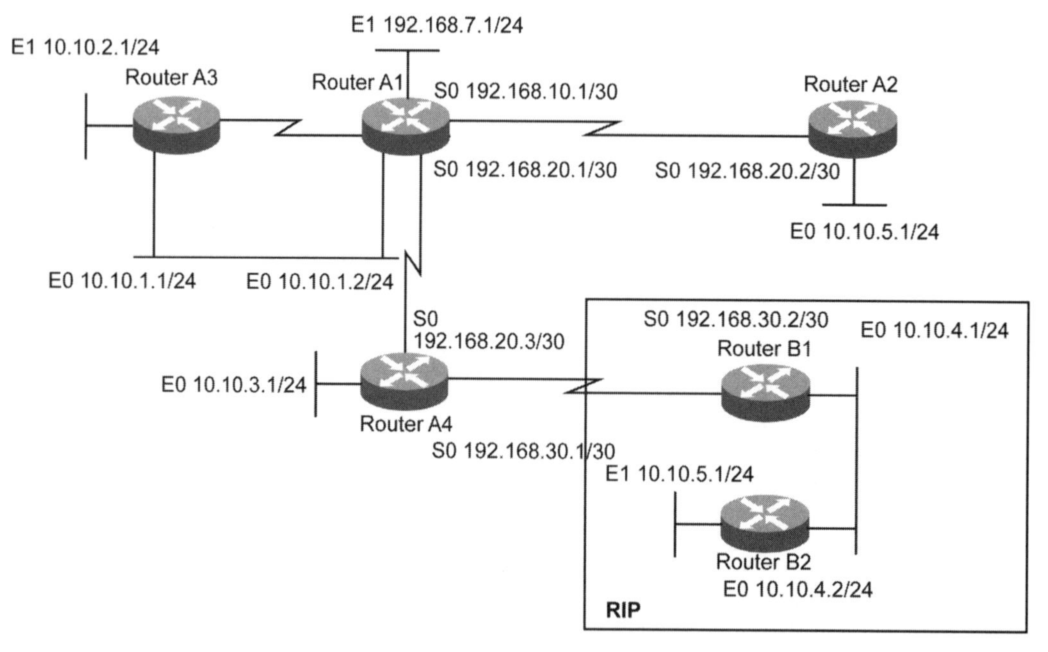

FIGURE 8.9 An EIGRP routing environment.

network in routers B1 and B2. Router A4 is running both EIGRP and RIP and is redistributing between the two routing protocols.

The problems with the network in Figure 8.9 are:

- The network segment 10.10.2.0/24 is not reachable from the network segment 10.10.6.0/24. This is an EIGRP route problem.
- The network segment 10.10.5.0/24 is not reachable from the network segment 10.10.7.0/24. This is an EIGRP redistribution problem.

To resolve these problems, consider the problem resolution approach.

EIGRP Route Problem

When the network segment 10.10.2.0/24 is not reachable from the network segment 10.10.6.0/24, it refers to an EIGRP route problem. The steps to detect and solve this problem are:

1. Check for the network segment 10.10.2.0 in the routing table and topology table of A2. The network segment 10.10.2.0/24 is connected to A3, and the network segment 10.10.6.0/24 is connected to A2. The route for the segment 10.10.2.0 is checked at A2 using the A2#show ip route 10.10.2.0 command. The topology table at A2 is checked using the A2#show ip eigrp topology command. The subject route is not found in the routing table or the topology table of A2.
2. Check for the network segment 10.10.2.0 at the neighbor router (A1) using the A1#show ip route 10.10.2.0 command. The topology table at A2 is checked using the A1#show ip eigrp topology command. The subject route is not present in the routing table or the topology table of A1. The network 10.10.2.0/24 is connected to A3, which is on a directly connected Ethernet link with A1. A neighbor check is necessary to ensure that neighbor relationship is stable between A1 and A3, because the route from A3 is not available at A1.
3. Check the neighbor relationship at A1 using the A1#show ip eigrp neighbors command. The output shows that A4 is the only existing neighbor of A1. The debug command A1#debug eigrp packet hello is used to confirm that there is no neighbor relationship.
4. Monitor the console or terminal output by issuing the A1#debug eigrp neighbor command. It is found that there is no exchange of hello packets via the Ethernet interface. The command output shows that the neighbor relationship with A3 does not come up.
5. Check the basic IP level connectivity from A1 to A4 by executing the A1#ping 10.10.1.1 command. The output shows 100% success with admissible latency limits.

6. Check if any access list may be used in the interface of A1 and A3. The check is carried out using these commands at both routers.

The command output at A3 shows that an outbound access list is set that denies certain packets to pass through, which hampers the establishment of an EIGRP neighbor relationship with A1. The access list is identified as:

```
access-list 100 deny tcp any any
access-list 100 deny udp any any
access-list 100 permit icmp any any
```

The access list is removed from the interface to solve the problem.

```
A4(config)#interface Ethernet0
A4(config-if)#no ip access-group 100 out
```

EIGRP Redistribution Problem

When the network segment 10.10.5.0/24 is not reachable from the network segment 10.10.7.0/24, it refers to an EIGRP redistribution problem. The steps to detect and solve this problem are:

1. Check the availability of route A4 where redistribution occurs, using the A4#show ip route 10.10.5.0 command. The output shows availability of an RIP route of two hops via the next-hop IP address 192.168.30.2.
2. Check the same route at the EIGRP neighbor router A1. This is performed using the A1#show ip route 10.10.5.0 command. The output shows that there is no route available to the destination 10.10.5.0. This implies that there was no redistribution from RIP for this route.
3. Check the routing process configuration at A4 using the A4#show running-config command. Here, redistribution is configured with an appropriate metric value specified for redistributed RIP routes. There is also no route-map set to restrict networks to redistribute.
4. Check if filtering occurs for EIGRP routing updates moving out of various interfaces. This is performed using the A4#show ip protocols command. The output shows that there is a distribute list set in the outgoing updates to filter the routing updates. The access list is:

```
access-list 50 deny 10.10.5.0 0.0.0.255
access-list 50 permit any
```

5. Remove the access list from the routing updates to solve the problem.

SUMMARY

In this chapter, we discussed the different problems encountered in the EIGRP routing protocol. We also looked at the methods to isolate and troubleshoot the problems. In the next chapter, we will discuss the methods of troubleshooting in the OSPF routing protocol.

POINTS TO REMEMBER

- TCP/IP troubleshooting commands like ping, traceroute, and show ip route are used to isolate problems in EIGRP environments.
- EIGRP packets are monitored using the debug eigrp packets command.
- The show ip eigrp interfaces command shows a list of interfaces in which EIGRP is enabled.
- The show ip eigrp neighbors command shows EIGRP enabled routers that are on directly connected networks of a router.
- The show ip eigrp topology command shows topology of an entire EIGRP domain.
- The show ip eigrp traffic command shows statistics of the EIGRP generated traffic.
- The debug eigrp neighbors command monitors status of EIGRP neighbors.
- The debug eigrp packets command monitors EIGRP packets.
- The debug eigrp transmit command tracks all EIGRP transmission events, such as ack, packetize, startup, peerdown, link, build, strange, sia, and detail.
- The debug eigrp fsm command displays EIGRP DUAL algorithm functioning and stages of building topology table.
- Misconfiguration occurs because networks are not declared or because the same AS is not defined in all routers in the EIGRP domain.
- EIGRP neighbor problems occur when a new node is added or when initially existing neighbor relations are no longer present in the network.
- Routers that fail to reply to queries are said to be in Stuck-In-Active (SIA) state.

9 Troubleshooting OSPF Routing Environments

IN THIS CHAPTER

- OSPF Terminology
- Resolution of Problems in OSPF
- Problems with Assigning Priority
- OSPF Neighbor States
- OSPF Routing Table
- NBMA Networks
- OSPF Stub Areas
- Redistribution in OSPF

Open Shortest Path First (OSPF) is an industry standard protocol that ensures interoperability between routing devices manufactured by diverse vendors. OSPF is a complex link state protocol with concepts of adjacency, areas, incremental updates, and authentication.

This chapter discusses different methods of problem isolation and troubleshooting in OSPF networks.

OSPF TERMINOLOGY

This section discusses the terminology used when discussing the OSPF routing protocol. Some of the terms are:

Link State Advertisement (LSA): OSPF packet containing source, destination, and routing information. This information is advertised to all OSPF routers in a hierarchical area.

Flooding: Periodic updating of topology and routing table information with sending of LSAs.

Adjacencies: Logical connection between OSPF router and its Designated Router (DR).

Designated Router: Router used to reduce the number of adjacencies formed in a broadcast network.

Backup Designated Router (BDR): Router that acts as a standby for DR on broadcast networks. BDR collects routing information updated from the adjacent OSPF routers and takes the role of DR when the DR goes down.

Autonomous Systems (ASs): Set of routers in the same administrative control using the same protocol for routing processes.

Multi-access/Broadcast Networks: Physical networks that support interconnection of more than two routers that can communicate directly.

Nonbroadcast Multi-access Networks (NBMAs): Interconnect routers in an OSPF network without having the broadcast capability. NBMA is discussed in detail in a later section.

Single Area (SA): Logical subdivision of the greater OSPF domain, grouping routers that run OSPF with identical topological databases.

Stub Area: Type of nonstandard OSPF area.

Totally Stubby Area (TSA): Type of nonstandard OSPF area used when few networks with limited connectivity are connected to the remaining network.

Not So Stubby Area (NSSA): Type of nonstandard OSPF area used in redistribution of routing information.

OSPF Routers: Four types of OSPF routers are Internal Router (IR), Area Border Router (ABR), Backbone Router (BR), and Autonomous System Boundary Router (ASBR).

RESOLUTION OF PROBLEMS IN OSPF

There are a host of commands with which to troubleshoot issues in OSPF networks. They are the show and debug commands.

Show Commands

The different show commands and description are listed in Table 9.1.

TABLE 9.1 Show Commands and Descriptions

Command	Description
show ip ospf 1(Process ID number)	Shows details of the OSPF process for a specified process ID.
show ip ospf border-routers	Shows information about the border and boundary routers of OSPF for which entries exist in the routing table.
show ip ospf database	Shows the total OSPF topological database with all the LSAs present in a router.
show ip ospf flood-list	Shows the link states to be flooded out of an interface.
show ip ospf interface	Lists all interfaces in the router and shows OSPF specific information for each of them. This command also shows the interface IP address, OSPF process ID, route ID, type of OSPF network, and OSPF timers. The command also shows whether the interface is passive.
show ip ospf neighbor	Lists all the neighbor relationships in OSPF. Parameters such as the IP address of the neighbor, the interface over which the relationship is established, and the status of the router in the OSPF network are specified in this command.
show ip ospf request-list	Shows list of LSAs that have been requested by the router.
show ip ospf retransmission-list	Shows list of LSAs whose retransmission has been requested by the router.
show ip ospf summary-address	Shows the summary address redistribution information.
show ip ospf virtual-links	Shows OSPF virtual links.

Consider the example shown in Figure 9.1. The figure shows an OSPF enabled network. Routers B1, B2, B3, and B4 are OSPF neighbors over a shared LAN 192.168.119.0/24.

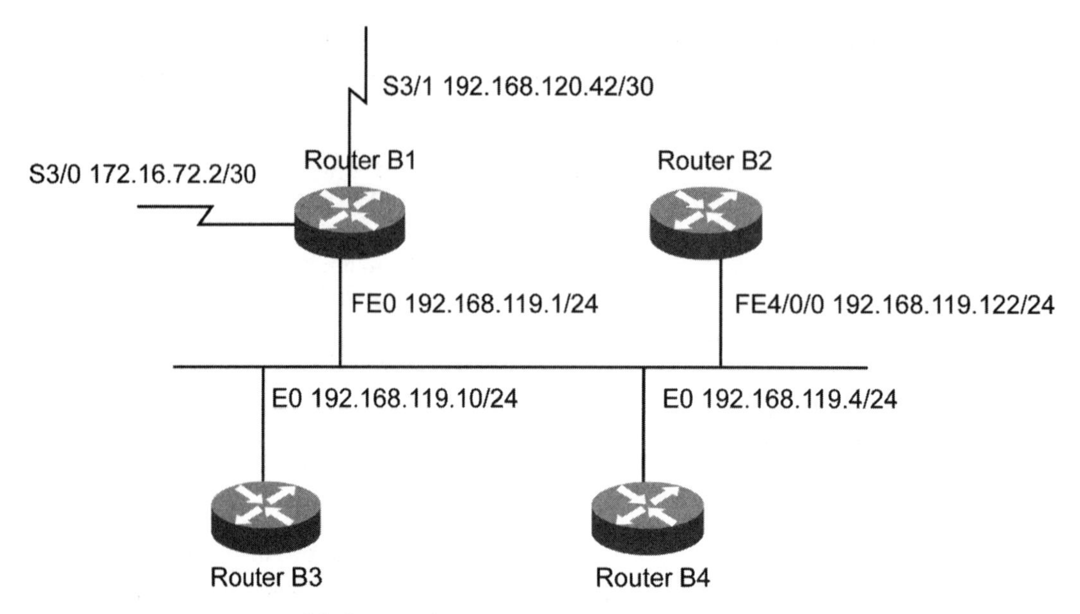

FIGURE 9.1 An OSPF enabled network.

The output of the commands listed in Table 9.1 is discussed in this section. Listing 9.1 shows the output of the show ip ospf neighbor command at B1.

LISTING 9.1 Output of the show ip ospf neighbor Command at B1

```
B1#show ip ospf neighbor
Neighbor ID      Pri     State      Dead Time      Address      Interface
0
192.168.119.122    1     2WAY/DROTHER   00:00:31      192.168.119.122
FastEthernet0/
0
192.168.120.90     1     FULL/BDR       00:00:36      192.168.119.4
FastEthernet0/
0
192.168.120.222    1     FULL/DR        00:00:35      192.168.119.10
FastEthernet0/
0
```

Listing 9.2 shows the output for the show ip ospf border-routers command at B1.

LISTING 9.2 Output of the show ip ospf border-routers Command at B1

```
B1#show ip ospf border-routers
OSPF Process 1 internal Routing Table
```

```
Codes: i - Intra-area route, I - Inter-area route
i 192.168.120.222 [1] via 192.168.119.10, FastEthernet0/0, ASBR, Area
5, SPF 939
8
i 192.168.119.122 [1] via 192.168.119.122, FastEthernet0/0, ASBR, Area 5,
  SPF 93
98
i 192.168.120.90 [1] via 192.168.119.4, FastEthernet0/0, ASBR, Area 5,
  SPF 9398
```

The output for the show ip ospf database command is shown in Listing 9.3.

LISTING 9.3 Output of the show ip ospf database Command at B1

```
B1#show ip ospf database
OSPF Router with ID (192.168.120.9) (Process ID 1)
Router Link States (Area 5)
Link ID         ADV Router      Age    Seq#       Checksum  Link count
192.168.119.122 192.168.119.122 978    0x80001C3C 0x00B6D6  10
192.168.120.90  192.168.120.90  1133   0x80001717 0x00B034  10
192.168.120.222 192.168.120.222 2      0x80005112 0x000E89  50
Net Link States (Area 5)
Link ID         ADV Router      Age    Seq#       Checksum
192.168.119.10  192.168.120.222 90     0x800000B2 0x00C83F
Type-5 AS External Link States
Link ID         ADV Router      Age    Seq#       Checksum  Tag
0.0.0.0         192.168.119.122 472    0x80000006 0x002DF3  1
24.147.216.55   192.168.119.122 716    0x800026B0 0x0047C9  0
61.3.128.49     192.168.119.122 716    0x80001C07 0x00A4E9  0
63.104.239.70   192.168.119.122 716    0x80001C07 0x002D75  0
66.218.66.240   192.168.119.122 211    0x80000006 0x00CA82  0
80.204.229.99   192.168.119.122 716    0x80001C07 0x00E534  0
150.108.77.100  192.168.119.122 716    0x800020FC 0x00656C  0
198.64.129.82   192.168.119.122 716    0x800020FC 0x007B30  0
198.172.121.204 192.168.119.122 716    0x80001C07 0x00EDD8  0
198.172.121.209 192.168.119.122 716    0x800026B3 0x0044C6  0
199.44.167.0    192.168.119.122 716    0x80001101 0x00B8BC  0
202.4.187.242   192.168.119.122 716    0x800020FC 0x0053B5  0
202.41.230.37   192.168.119.122 716    0x800026B3 0x004584  0
172.16.9.0      192.168.119.122 979    0x80001BAE 0x00CFFF  0
172.16.9.127    192.168.119.122 716    0x800000D5 0x00DAE8  0
172.16.9.140    192.168.120.222 849    0x800010CF 0x003B9C  0
172.16.9.184    192.168.120.90  1127   0x8000003A 0x003013  0
172.16.9.236    192.168.120.90  1127   0x8000000D 0x0080BB  0
```

```
172.16.52.0      192.168.119.122 716    0x80001BCE 0x00A5E1 0
172.16.52.40     192.168.120.222 1588   0x8000003C 0x00DB46 0
172.16.52.88     192.168.120.90  1128   0x8000006C 0x00B491 0
172.16.52.164    192.168.120.222 91     0x8000000C 0x005F76 0
172.16.53.0      192.168.119.122 717    0x80001BCD 0x00ABD8 0
172.16.54.16     192.168.120.90  1128   0x80000759 0x008531 0
172.16.54.64     192.168.120.222 849    0x8000053D 0x00435D 0
172.16.55.0      192.168.119.122 717    0x80001BCE 0x0093ED 0
172.16.72.36     192.168.119.122 473    0x8000004A 0x00BFCB 0
172.16.74.64     192.168.120.222 1101   0x80000002 0x008B51 0
172.16.74.128    192.168.120.222 92     0x8000004A 0x00D86B 0
172.16.74.192    192.168.120.222 850    0x80000015 0x0060E8 0
172.16.74.224    192.168.120.222 1589   0x80000169 0x00D3EF 0
172.16.75.0      192.168.120.222 1833   0x8000003D 0x00ECE3 0
172.16.75.1      192.168.120.222 1833   0x8000003D 0x0075B8 0
172.16.75.16     192.168.120.222 850    0x80000261 0x00FD9C 0
172.16.75.64     192.168.120.222 850    0x80000041 0x00417B 0
172.16.76.0      192.168.119.122 717    0x80001C07 0x00293A 0
172.16.76.24     192.168.119.122 717    0x8000158F 0x00B7D2 0
172.16.76.25     192.168.119.122 717    0x8000158F 0x00ADDB 0
172.16.77.1      192.168.120.222 92     0x8000000A 0x008D64 0
172.16.77.2      192.168.120.222 92     0x8000000A 0x00836D 0
172.16.77.3      192.168.120.222 92     0x8000000A 0x007976 0
172.16.77.128    192.168.120.222 92     0x80000068 0x005AF8 0
172.16.78.0      192.168.120.222 92     0x800000D5 0x003A0B 0
172.16.78.32     192.168.120.222 92     0x800000D5 0x0059BB 0
172.16.78.96     192.168.120.222 850    0x800010CF 0x00B218 0
172.16.78.128    192.168.120.222 850    0x800010E5 0x00454F 0
172.16.78.176    192.168.120.222 1101   0x8000008C 0x004687 0
172.16.78.192    192.168.120.222 850    0x80000261 0x00F5F0 0
172.16.78.240    192.168.120.222 92     0x800000D5 0x003113 0
172.16.115.23    192.168.119.122 718    0x80001A0C 0x009E53 0
B1#
```

The output for the show ip ospf interface command is shown in Listing 9.4.

LISTING 9.4 Output of the show ip ospf interface Command at B1

```
B1#show ip ospf interface
FastEthernet0/0 is up, line protocol is up
Internet Address  192.168.119.1/25, Area 5
Process ID 1, Router ID  192.168.120.9, Network Type BROADCAST,
  Cost: 1
Transmit Delay is 1 sec, State DROTHER, Priority 1
```

```
Designated Router (ID)    192.168.120.222, Interface address
192.168.119.10
Backup Designated router (ID)    192.168.119.90, Interface address
192.168.119.90
Timer intervals configured, Hello 10, Dead 40, Wait 40, Retransmit 5
Hello due in 00:00:02
Index 20/20, flood queue length 0
Next 0x0(0)/0x0(0)
Last flood scan length is 0, maximum is 37
Last flood scan time is 0 msec, maximum is 4 msec
Neighbor Count is 14, Adjacent neighbor count is 2
Adjacent with neighbor  192.168.120.222  (Designated Router)
Adjacent with neighbor 203.200.163.1  (Backup Designated Router)
Suppress hello for 0 neighbor(s)
Simple password authentication enabled
Serial3/0 is up, line protocol is up
Internet Address 172.16.72.2/30, Area 5
Process ID 1, Router ID  192.168.120.9, Network Type POINT_TO_POINT,
Cost: 64
Transmit Delay is 1 sec, State POINT_TO_POINT,
Timer intervals configured, Hello 10, Dead 40, Wait 40, Retransmit 5
No Hellos (Passive interface)
Index 7/7, flood queue length 0
Next 0x0(0)/0x0(0)
Last flood scan length is 0, maximum is 0
Last flood scan time is 0 msec, maximum is 0 msec
Neighbor Count is 0, Adjacent neighbor count is 0
Suppress hello for 0 neighbor(s)
Simple password authentication enabled
Serial3/1 is up, line protocol is up
Internet Address  192.168.120.42/30, Area 5
Process ID 1, Router ID   192.168.120.9, Network Type POINT_TO_POINT,
Cost: 64
Transmit Delay is 1 sec, State POINT_TO_POINT,
Timer intervals configured, Hello 10, Dead 40, Wait 40, Retransmit 5
No Hellos (Passive interface)
Index 21/21, flood queue length 0
Next 0x0(0)/0x0(0)
Last flood scan length is 0, maximum is 0
Last flood scan time is 0 msec, maximum is 0 msec
Neighbor Count is 0, Adjacent neighbor count is 0
Suppress hello for 0 neighbor(s)
Simple password authentication enabled
```

Output of the show ip ospf 1 command is shown in Listing 9.5.

LISTING 9.5 Output of the show ip ospf 1 Command at B1

```
B1#show ip ospf 1
 Routing Process "ospf 1" with ID 192.168.120.9
 Supports only single TOS(TOS0) routes
 Supports opaque LSA
 It is an autonomous system boundary router
 Redistributing External Routes from,
connected, includes subnets in redistribution
static with metric mapped to 2, includes subnets in redistribution
 SPF schedule delay 5 secs, Hold time between two SPFs 10 secs
 Minimum LSA interval 5 secs. Minimum LSA arrival 1 secs
 Number of external LSA 591. Checksum Sum 0x123BBA0
 Number of opaque AS LSA 0. Checksum Sum 0x000000
 Number of DCbitless external and opaque AS LSA 0
 Number of DoNotAge external and opaque AS LSA 0
 Number of areas in this router is 1. 1 normal 0 stub 0 nssa
 External flood list length 0
Area 5
Number of interfaces in this area is 25
Area has simple password authentication
SPF algorithm executed 9412 times
Area ranges are
192.168.119.0/24 Passive Advertise
Number of LSA 16. Checksum Sum 0x090354
Number of opaque link LSA 0. Checksum Sum 0x000000
Number of DCbitless LSA 0
Number of indication LSA 0
Number of DoNotAge LSA 0
    Flood list length 0
```

The output of the show ip protocols command is shown in Listing 9.6.

LISTING 9.6 Output of the show ip protocols Command

```
B1#show ip protocols
Routing Protocol is "ospf 1"
Outgoing update filter list for all interfaces is not set
Incoming update filter list for all interfaces is not set
Router ID  192.168.120.9
It is an autonomous system boundary router
Redistributing External Routes from,
```

```
connected, includes subnets in redistribution
static with metric mapped to 2, includes subnets in redistribution
Number of areas in this router is 1. 1 normal 0 stub 0 nssa
Maximum path: 6
Routing for Networks:
192.168.119.0 0.0.0.255 area 5
192.168.120.0 0.0.0.255 area 5
Passive Interface(s):
Serial3/0
Serial3/1
Routing Information Sources:
Gateway     Distance       Last Update
192.168.119.122    110     00:00:00
192.168.120.90     110     00:00:00
192.168.120.222    110     00:00:00
  Distance: (default is 110)
```

Debug Commands

The debug commands used for thorough, packet-level troubleshooting of problems in OSPF networks are listed in Table 9.2.

TABLE 9.2 Debug Commands and Descriptions

Command	Description
debug ip ospf adj	Monitors adjacencies between OSPF neighbors.
debug ip ospf database-timer	Monitors the database timer in OSPF.
debug ip ospf events	Displays the packet-level exchanges that occur in different events in the functioning of the OSPF routing process.
debug ip ospf flood	Displays events during the exchange state of adjacency development, when the entire database is being exchanged.
debug ip ospf hello	Displays when Hello packets are received and sent by the router over the various OSPF enabled interfaces in a router.
debug ip ospf lsa-generation	Displays the events that occur when an LSA is generated.

(continued)

TABLE 9.2 *(continued)*

Command	Description
debug ip ospf packet	Displays types of packets generated during all subprocesses of the OSPF routing process.
debug ip ospf retransmission	Displays the OSFP retransmission events.
debug ip ospf spf	Displays packet-level exchanges that take place during computation of the SPF algorithm to generate the best path to a destination.
debug ip ospf tree	Displays events leading to the computation of the OSPF tree structure.

The debug ip ospf packet command should be used with care, because this command can be quite resource consuming in an OSPF network with a large number of nodes.

Consider the example depicted in Figure 9.1. The output of some of the debug commands is discussed. The output for the debug ip ospf hello command is shown in Listing 9.7.

LISTING 9.7 Output of the debug ip ospf hello Command at B1

```
B1#debug ip ospf hello
OSPF hello events debugging is on
B1#
B1#
B1#
12w2d: OSPF: Rcv hello from 203.197.119.122 area 5 from Ethernet0/0
203.197.119.122
12w2d: OSPF: End of hello processing
B1#
12w2d: OSPF: Rcv hello from 203.197.120.90 area 5 from Ethernet0/0
203.197.119.4
12w2d: OSPF: End of hello processing
B1#
B1#
12w2d: OSPF: Rcv hello from 203.197.120.222 area 5 from Ethernet0/0
203.197.119.10
12w2d: OSPF: End of hello processing
B1#
```

```
12w2d: OSPF: Rcv hello from 203.197.119.122 area 5 from Ethernet0/0
203.197.119.122
B1#
12w2d: OSPF: Rcv hello from 203.197.120.90 area 5 from Ethernet0/0
203.197.119.4
12w2d: OSPF: End of hello processing
B1#undebug all
All possible debugging has been turned off
B1#
B1#
12w2d: OSPF: Rcv hello from 203.197.119.122 area 5 from Ethernet0/0
203.197.119.122
12w2d: OSPF: End of hello processing
B1#
```

Output of the debug ip ospf adj command is shown in Listing 9.8.

LISTING 9.8 Output of the debug ip ospf adj Command at B1

```
B1#debug ip ospf adj
OSPF adjacency events debugging is on
B1#
12w2d: OSPF: Interface Ethernet0/0 going Down
12w2d: OSPF:  192.168.119.1 address  192.168.119.1 on Ethernet0/0 is
  dead, state D
OWN
12w2d: OSPF: Neighbor change Event on interface Ethernet0/0
12w2d: OSPF: DR/BDR election on Ethernet0/0
12w2d: OSPF: Elect BDR  192.168.120.90
12w2d: OSPF: Elect DR   192.168.120.222
12w2d:      DR:  192.168.120.222 (Id)  BDR: 203.200.163.1  (Id)
12w2d: %OSPF-5-ADJCHG: Process 1, Nbr  192.168.119.122 on Ethernet0/0
  from 2WAY t
o DOWN, Neighbor Down: Interface down or detached
12w2d: OSPF: Neighbor change Event on interface Ethernet0/0
12w2d: OSPF: DR/BDR election on Ethernet0/0
12w2d: OSPF: Elect BDR  192.168.120.90
12w2d: OSPF: Elect DR  192.168.120.222
12w2d:  DR:  92.168.120.222 (Id)  BDR:  192.168.120.90 (Id)
12w2d: OSPF: Send DBD to  192.168.120.90 on Ethernet0/0 seq
  0x2063 opt 0x42 flag
0x7 len 32
12w2d: OSPF:  192.168.120.90 address  92.168.119.4 on Ethernet0/0
  is dead, state
```

```
DOWN
12w2d: %OSPF-5-ADJCHG: Process 1, Nbr    192.168.120.90 on Ethernet0/0
  from EXSTART
to DOWN, Neighbor Down: Interface down or detached
12w2d: OSPF: Neighbor change Event on interface Ethernet0/0
12w2d: OSPF: DR/BDR election on Ethernet0/0
12w2d: OSPF: Elect BDR 0.0.0.0
12w2d: OSPF: Elect DR  192.168.120.222
12w2d:  DR:  192.168.120.222 (Id) BDR: none
12w2d: OSPF:  192.168.120.222 address  192.168.119.10 on Ethernet0/0 is
  dead, stat
eDOWN
12w2d: %OSPF-5-ADJCHG: Process 1, Nbr    192.168.120.222 on
Ethernet0/0 from FULL t
oDOWN, Neighbor Down: Interface down or detached
B1#
```

Output of the debug ip ospf events command is shown in Listing 9.9.

LISTING 9.9 Output of the debug ip ospf events Command at B1

```
B1#debug ip ospf events
12w2d: OSPF: Rcv hello from  192.168.119.122 area 5 from Ethernet0/0
  192.168.119.122
12w2d: OSPF: End of hello processing
12w2d: OSPF: Rcv hello from  192.168.120.90 area 5 from Ethernet0/0
  192.168.119.4
12w2d: OSPF: End of hello processing
12w2d: OSPF: Rcv hello from  192.168.119.122 area 5 from Ethernet0/0
  192.168.119.122
12w2d: OSPF: Rcv hello from  192.168.120.90 area 5 from Ethernet0/0
  192.168.119.4
12w2d: OSPF: End of hello processing
```

The output for the debug ip ospf packet command is shown in Listing 9.10.

LISTING 9.10 Output of the debug ip ospf packet Command at B1

```
B1#debug ip ospf packet
12w2d: OSPF: rcv. v:2 t:1 l:100 rid:192.168.120.90
  aid:0.0.0.5 chk:97FE aut:1 auk: from Ethernet0/0
12w2d: OSPF: rcv. v:2 t:1 l:100 rid:192.168.119.122
  aid:0.0.0.5 chk:87FE aut:1 auk: from Ethernet0/0
12w2d: OSPF: rcv. v:2 t:1 l:100 rid:192.168.120.90
```

```
  aid:0.0.0.5 chk:97FE aut:1 auk: from Ethernet0/0
12w2d: OSPF: rcv. v:2 t:1 l:100 rid:192.168.119.122
  aid:0.0.0.5 chk:87FE aut:1 auk: from Ethernet0/0
```

PROBLEMS WITH ASSIGNING PRIORITY

The priority value is the primary parameter used in DR and BDR elections in the OSPF routing process. The priority values range between 1 and 255 with the default value as 1. If the priority values of two routers are same, the Router ID is used to determine the DR and BDR. This section discusses the problems that can arise while assigning priority values.

Consider the example depicted in Figure 9.2. This figure shows an OSPF network composed of routers B1, B2, and B3. A single broadcast-based multi-access segment is represented by the Ethernet segment.

In Figure 9.2, as per the default configuration, the interfaces FE0/0 of B1 and E0 of B2 and B3 have the priority 1. As a result, the DR and BDR election takes place comparing the Router IDs. In this example, B2 is the DR, and B3 becomes the BDR. This is shown in the output of the show ip ospf neighbor command, shown in Listing 9.11.

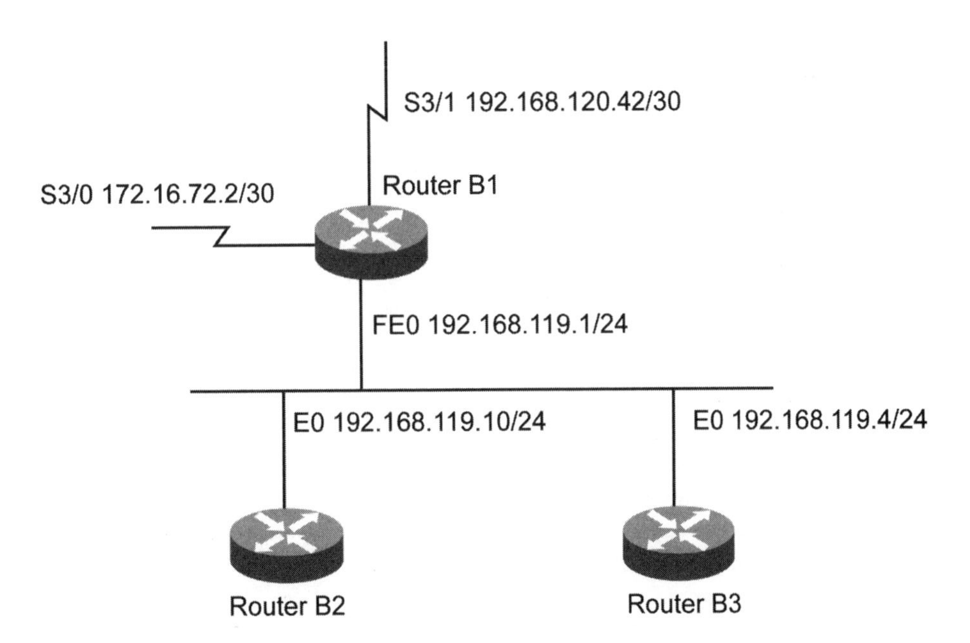

FIGURE 9.2 An OSPF network.

LISTING 9.11 Output of the show ip ospf neighbor Command at B1

```
B1#show ip ospf neighbor
Neighbor ID         Pri State         Dead Time    Address  Interface
0
192.168.119.10      1                 FULL/DR      00:00:36 192.168.119.10
FastEthernet0/
0
192.168.119.4       1                 FULL/BDR     00:00:35 192.168.119.4
FastEthernet0/
0
```

Let us discuss the problem that can arise in the given scenario. B1 is a Cisco 7206 Router while routers B2 and B3 are 2511 routers. It is more logical that B1 be the DR for the OSPF network. To ensure this, we need to set the interface priority accordingly by using the following commands:

```
B1(config)#interface FastEthernet 0/0
B1(config-if)#ip ospf priority 10
```

Cisco 7206 routers are more capable than Cisco 2511 routers. B1 should become the DR, because it needs to maintain adjacency with the remaining routers in the OSPF domain.

The output of the show ip ospf interface command confirms the priority value, as shown in Listing 9.12.

LISTING 9.12 Output of the show ip ospf interface Command at B1

```
B1#show ip ospf interface
FastEthernet0/0 is up, line protocol is up
Internet Address   192.168.119.1/25, Area 5
Process ID 1, Router ID   192.168.120.9, Network Type BROADCAST,
Cost: 1
Transmit Delay is 1 sec, State DR, Priority 10
Designated Router (ID)    192.168.119.10, Interface
address192.168.119.10
Backup Designated router (ID)     192.168.119.4, Interface address
  192.168.119.4
Timer intervals configured, Hello 10, Dead 40, Wait 40, Retransmit 5
Hello due in 00:00:02
Index 20/20, flood queue length 0
Next 0x0(0)/0x0(0)
Last flood scan length is 0, maximum is 37
```

```
Last flood scan time is 0 msec, maximum is 4 msec
Neighbor Count is 2, Adjacent neighbor count is 2
Adjacent with neighbor  192.168.119.10 (Designated Router)
Adjacent with neighbor  192.168.119.4  (Backup Designated Router)
Suppress hello for 0 neighbor(s)
  Simple password authentication enabled
```

Listing 9.12 shows a part of the command output generated. The priority value is 10. B1 is elected as the DR.

The priority should be set to 0 to elect a router as the DR or BDR.

OSPF NEIGHBOR STATES

This section discusses the methods of troubleshooting problems that arise while establishing neighbor relationships in the OSPF routing process. Table 9.3 lists some common situations encountered while troubleshooting OSPF neighbor adjacency problems. The table also shows the commands used to verify the problems.

TABLE 9.3 Causes of Neighbor Adjacency Problems and Diagnostic Commands

Cause of Neighbor Adjacency Problems	Diagnostic Command
OSPF is not configured on any of the routers in the network.	show ip ospf
OSPF is not configured on the needed interface.	show ip ospf interface
Neighbor routers are incorrectly configured.	show ip ospf show ip ospf interface
OSPF Hello packets are not processed.	show memory summary show memory processor

Consider the example depicted in Figure 9.3. This figure shows part of an OSPF network with a single area. Routers A1 and A2 belong to the same LAN, 10.0.1.0/24.

In Figure 9.3, A2 is not able to establish a neighbor relationship with A1, as ascertained with the show ip ospf neighbor command.

Run the show ip ospf neighbor command many times in A2 to ascertain that there is no change in the state of A1. This command can be used to find out the

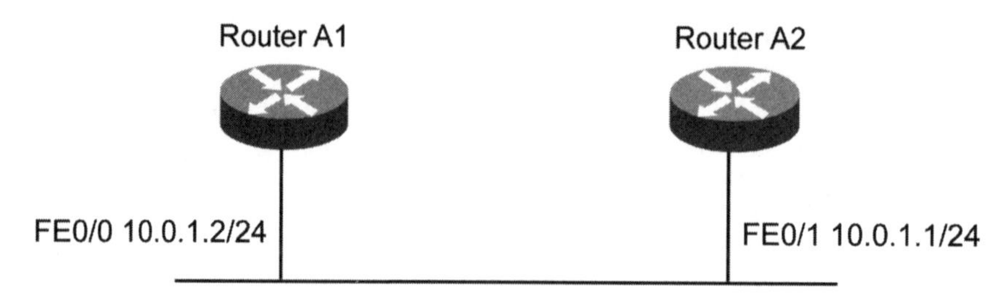

FIGURE 9.3 An OSPF network showing routers A1 and A2.

possible problem areas in the neighbor state. The different neighbor states in OSPF are:

- INIT state
- EXCHANGE/EXSTART state
- 2-way state
- Nothing at all state
- Loading state

Let us discuss the problems encountered in each of these neighbor states.

INIT State

The INIT state implies that A1 has been able to detect Hello packets from A2, but A2 has not been able to detect Hello packets from A1. If establishing the neighbor relationship is stuck at the INIT state, the output of the show ip ospf neighbor command is as shown in Listing 9.13.

LISTING 9.13 Output of show ip ospf neighbor Command in INIT State

```
A2#show ip ospf neighbor
Neighbor  ID  Pri  State  Dead Time  Address  Interface
0
10.10.1.2   1  INIT  00:00:31  10.0.1.2  FastEthernet0/1
```

To troubleshoot OSPF neighbors in INIT state:

1. Check if OSPF authentication is turned on for all OSPF routers in the LAN by using the show ip ospf interface command.
2. Check if the authentication type and authentication key match across the LAN.

3. Check and modify any access lists hampering the flow of hello packets by using the show ip interface command.
4. Check the switch configuration and physical cabling of the LAN.

EXCHANGE or EXSTART State

In the EXSTART state, bidirectional communication with the neighbor is already established. The routers try to establish the initial sequence number that would be used in the exchange of information. In the EXCHANGE state, routers exchange the complete link state database by sending database description packets.

If establishing the neighbor relationship is stuck in EXCHANGE or EXSTART state, the output of the show ip ospf neighbor command is as shown in Listing 9.14.

LISTING 9.14 Output of show ip ospf neighbor Command in EXCHANGE State

```
A2#show ip ospf neighbor
Neighbor IDPri  State   Dead Time   Address   Interface
0
10.10.1.2    1   EXCHANGE   00:00:31   0.0.1.2  FastEthernet0/1
```

To troubleshoot OSPF neighbors:

1. Check the MTU size using the show ip interface command. This parameter should be the same for potential neighbors.
2. Check the pinging of neighboring interfaces with a large-sized packet, if MTU is OK.
3. Troubleshoot the physical and software levels in Layer 2.

2-Way State

In the 2-way state, a bidirectional communication is established with a neighbor. DR and BDR election takes place at the end of the 2-way state. Full adjacencies are to be built only by the DR and BDR with the other routers, in the case of a broadcast network. If establishing neighbor relationships is stuck at 2-way state, the output of the show ip ospf neighbor command is as shown in Listing 9.15.

LISTING 9.15 Output of show ip ospf neighbor Command in 2-way State

```
A2#show ip ospf neighbor
Neighbor IDPri     State   Dead Time      Address Interface
0
10.10.1.2    1    2WAY/DROTHER      00:00:31       10.0.1.2
FastEthernet0/1
```

Remaining stuck in 2-way state is normal behavior for routers forming neighbor relationship with non-DR and non-BDR routers.

Nothing at All State

In the nothing at all state, the show ip ospf neighbor command gives no output. To troubleshoot OSPF routers in the nothing at all state:

1. Check the common interface between the router and potential neighbors by using the show ip interface FastEthernet0/1 command. The output of the show ip interface FastEthernet0/1 command is shown in Listing 9.16.

LISTING 9.16 Output of show ip interface FastEthernet0/1 Command at A2

```
A2#show ip interface FastEthernet0/1
FastEthernet0/1 is up, line protocol is down
Internet address is 10.0.1.1/30
Broadcast address is 255.255.255.255
Address determined by non-volatile memory
MTU is 1500 bytes
Helper address is not set
Directed broadcast forwarding is disabled
Multicast reserved groups joined: 224.0.0.10
Outgoing access list is not set
Inbound access list is not set
Proxy ARP is enabled
Security level is default
Split horizon is enabled
ICMP redirects are never sent
ICMP unreachables are always sent
ICMP mask replies are never sent
IP fast switching is enabled
IP fast switching on the same interface is enabled
IP Flow switching is disabled
IP Feature Fast switching turbo vector
IP multicast fast switching is enabled
IP multicast distributed fast switching is disabled
IP route-cache flags are Fast
Router Discovery is disabled
IP output packet accounting is disabled
IP access violation accounting is disabled
TCP/IP header compression is disabled
RTP/IP header compression is disabled
```

```
Probe proxy name replies are disabled
Policy routing is disabled
Network address translation is enabled, interface in domain inside
WCCP Redirect outbound is disabled
WCCP Redirect exclude is disabled
BGP Policy Mapping is disabled
```

In Listing 9.16, the code FastEthernet0/1 is up, line protocol is down shows there is a problem with the physical connectivity of the switch configuration. This problem should be rectified to establish neighbor relationships.

2. Check if the output of the show ip interface FastEthernet0/1 command is as shown.

FastEthernet0/1 is administratively down, line protocol is down implies that the FastEthernet interface has to be manually unshut using the commands:

```
A2(config)#interface FastEthernet0/1
A2(config-if)#no shut
```

3. Check if the output of the show ip interface FastEthernet0/1 command is showing the status as:

```
FastEthernet0/1 is up, line protocol is up
```

In this case, ping testing is to be performed using the A2#ping 10.10.1.2 command.

4. Check for the presence of any access lists applied to the FastEthernet interfaces in both routers, if the ping command is not successful.
5. Check the quality of the physical link if the ping command is still unsuccessful.
6. Check whether OSPF is configured for the interfaces at both routers by using the A2#show ip ospf interface fastEthernet0/1 command, if the ping command is successful.
7. Configure OSPF using the commands:

```
A2(config)#router ospf 1
A2(config-router)#network 10.0.1.0 0.0.0.255
```

If OSPF is not configured at the router interfaces.

8. Check if the routers are configured as passive interfaces by using the A2#show ip ospf interface fastEthernet0/1 command.

9. Remove passive interface configuration, because OSPF does not exchange Hello packets to form neighbor relationships over a passive interface.
10. Check if the network type is broadcast for all the members connected in a LAN. Check if the subnet mask is same for all the routers in the broadcast network segment by using the A2#show ip ospf interface fastEthernet0/1 command.
11. Check if neighboring interfaces are configured in the same area by using the A2#show ip ospf interface fastEthernet0/1 command.
12. Check if the values of dead and hello timers match between neighboring interfaces by using the A2#show ip ospf interface fastEthernet0/1 command.
13. Use the A2#debug ip ospf adjacency command to find out more information.

Load State

In a loading state, the information exchange between the OSPF routers that have formed adjacency is finalized. If a router receives an outdated or incomplete LSA during adjacency formation, it is out in the request list. These are sent out as link state requests. The requested information is sent by the requested neighbor in the form of link state updates.

```
A2#show ip ospf neighbor
Neighbor IDPri     State   Dead Time        Address Interface
0
10.10.1.2    1    LOADING    00:00:31        10.0.1.2 FastEthernet0/1
```

If the neighbor router is stuck in loading state, follow these steps:

1. Find the link state requests that are pending reply by using the A2#show ip ospf request-list command.
2. Check for the %OSPF-4-BADLSA message in the console. When the neighbor router is stuck in the loading state, it may exchange corrupt LSAs.
3. Check the compatibility of OSPF implementation in neighbor routers to detect the cause of corrupt LSAs.

OSPF ROUTING TABLE

A network may be a part of the OSPF database without being installed on the routing table. Some discrepancy is found in the database by the OSPF router that restricts it from installing the network in the routing table. The most common cause is that the router advertising the LSA is not reachable via OSPF because of misconfigurations.

Consider the example depicted in Figure 9.4. The figure shows an OSPF network in which routers A1 and A2, which are connected over a serial link, are a part of the shared LAN.

Consider in Figure 9.4 the route 10.20.1.0/24 is not installed in the routing tables of A2 and A3.

The IP address configured in the serial link between A1 and A2 is not a part of the same subnet. Adjacency is established between the two routers successfully in a point-to-point link. OSPF does not check if the neighbors belong to the same IP subnet. Even though the neighbor relationship is developed and network 10.20.1.0/24 reaches the OSPF database, it is not installed in the routing table due to the nonreachability of next-hop IP address. This problem is resolved by correctly configuring the IP address at both ends of a serial link. This problem can also occur if one of the interfaces is configured as unnumbered and the IP address is configured on the other interface.

In Figure 9.4, consider that route 10.10.1.2/24 is not established in the routing table of A3. In this case, check if any filter is applied at the incoming direction by using the show ip protocols command. We need to suitably modify the access lists to allow installation of the network in the routing table.

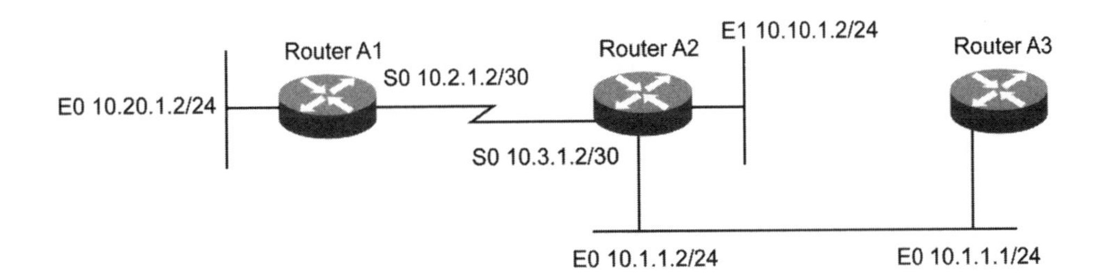

FIGURE 9.4 An OSPF network.

NBMA NETWORKS

This section discusses troubleshooting the problems in NBMA networks in OSPF.

Consider the example depicted in Figure 9.5. In the figure, a Frame Relay NBMA cloud is shown between routers A1, A2, A3, and A4. All the routers are not connected with each other via physical links.

OSPF considers an NBMA cloud to be similar to a broadcast based network. However, the exchange of Hello packets and DR and BDR elections will not take

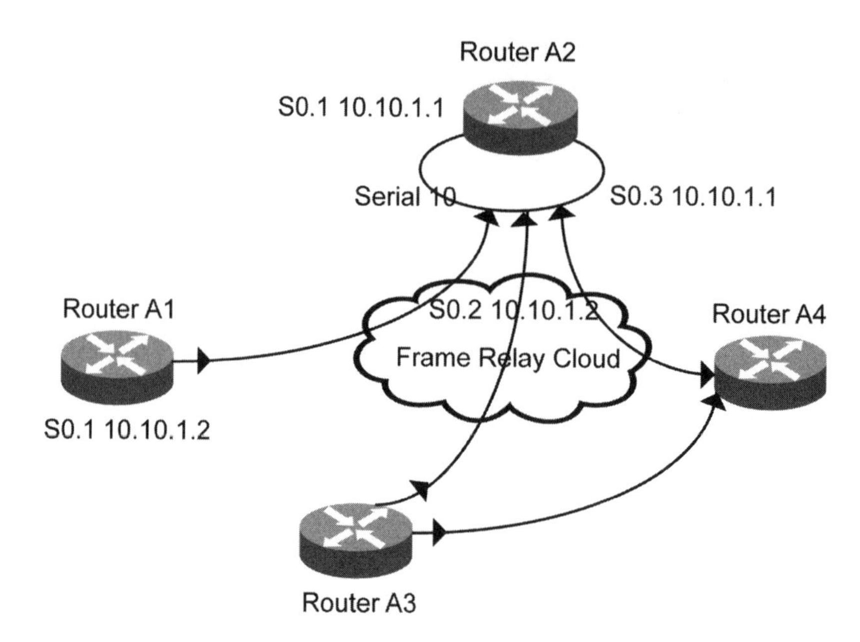

FIGURE 9.5 A Frame Relay NBMA network in OSPF.

place due to lack of any-to-any physical connectivity. This problem can be solved using any of these two methods:

- Configure subinterfaces as point-to-point links in the NBMA cloud.
- Define the OSPF network with point-to-multipoint links in the NBMA cloud.

Configure Interface as Point-to-Point Links

In this approach to solving the adjacency formation problem in an NBMA network, the subinterfaces are configured as point-to-point links in the NBMA cloud. This method solves the problem of election of DR and BDR, and adjacencies are formed. The configuration of A2 is shown in Listing 9.17.

LISTING 9.17 Configuration of A2

```
interface Serial 0
no ip address
encapsulation frame-relay
!
interface Serial0.1 point-to-point
ip address 10.10.1.1 255.255.255.252
frame-relay interface-dlci 10
```

```
!
interface Serial0.2 point-to-point
ip address 10.10.2.1 255.255.255.252
frame-relay interface-dlci 15
!
interface Serial0.3 point-to-point
ip address 10.10.3.1 255.255.255.252
frame-relay interface-dlci 20
!
```

The configuration of A1 is shown in Listing 9.18.

LISTING 9.18 Configuration of A1

```
interface Serial 0
no ip address
encapsulation frame-relay
!
interface Serial0.1 point-to-point
ip address 10.10.1.2 255.255.255.252
frame-relay interface-dlci 25
!
```

Service providers normally allocate the same subnet to all the interfaces that are part of the NBMA cloud. In the use of subinterfaces, the adjacency issue is solved, but allotting of different subnets corresponding to the subinterfaces is performed with the use of IP unnumbered on the links. However, this method is not acceptable for many network administrators as it hinders uniformity.

Define OSPF Network as Point-to-Multipoint Links

In this approach to solve the adjacency formation problem in the NBMA network, the OSPF network is defined as point-to-multipoint. The configuration of A2 is shown in Listing 9.19.

LISTING 9.19 Configuration of A2

```
!
interface Serial 0
ip address 10.10.1.1 255.255.255.0
encapsulation frame-relay
ip ospf network point-to-multipoint
!
```

The configuration of A1 is shown in Listing 9.20.

LISTING 9.20 Configuration of A1

```
!
interface Serial 0
ip address 10.10.1.2 255.255.255.0
encapsulation frame-relay
ip ospf network point-to-multipoint
!
```

In this approach, the single subnet is used for all members of the NBMA cloud. There is no issue for physical connectivity between all the routers for DR and BDR elections. Some additional information parameters are exchanged in link state updates to help in establishing connectivity with neighboring routers.

OSPF STUB AREAS

External routes, which are redistributed into OSPF from other routing protocols, are not allowed to be flooded into a stub area. Routing from a stub area to the rest of the Internet is based on a default route, which is injected by one or more of the ABRs into the area. This section discusses troubleshooting the problems encountered in OSPF stub areas.

Consider the example depicted in Figure 9.6. The figure shows part of an OSPF network with Area 0 and Area 5. Router A1 is the ABR.

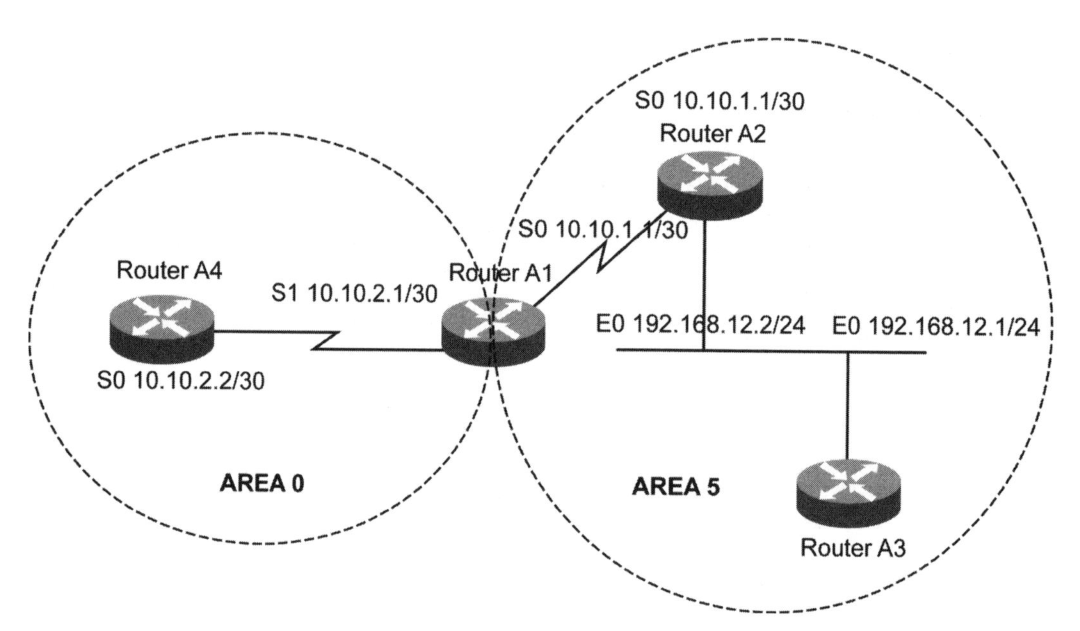

FIGURE 9.6 Part of an OSPF network showing Area 0 and Area 5.

In Figure 9.6, Area 5 is not functioning as the stub area. To troubleshoot this problem, check if all routers in Area 5 are configured as the stub area. The relevant configuration of the routers in Area 5 should be as shown in Table 9.4.

TABLE 9.4 Configuration of Routers in Stub Area

Router	Configuration
A1	router ospf 1 network 192.168.12.0 0.0.0.255 area 5 network 10.10.2.0 0.0.0.255 area 0 area 5 stub
A2	router ospf 1 network 192.168.12.0 0.0.0.255 area 5 network 10.10.1.0 0.0.0.255 area 5 area 5 stub
A3	router ospf 1 network 192.168.12.0 0.0.0.255 area 5 area 5 stub

The command area 5 stub is to be necessarily given in all the routers that are part of the stub area. This command is also executed at ABR A1 to become a neighbor to A2.

REDISTRIBUTION IN OSPF

Care should be taken to configure redistribution into OSPF and between OSPF and another IGP. This is to prevent routing loops and suboptimal routing in the network.

Consider the example shown in Figure 9.7. The figure shows a hybrid routing environment with both RIP and OSPF routing protocols. Routers A2 and A3 are running OSPF, and A4 is running RIP. A1 is redistributing RIP routes into OSPF.

In Figure 9.7, the RIP route 192.168.13.0/24 is not available in the routing table of A3. The command A3#show ip route | inc 192.168.13.0 shows no output. To troubleshoot redistribution in this scenario:

1. Check if the route is available in A1 using the A1#show ip route | inc 192.168.13.0 command. The output generated will be as shown:

```
A1#show ip route | inc 192.168.13.0
R   192.168.13.0/27 [120/1] via 10.10.2.1, 00:00:16, Serial1
```

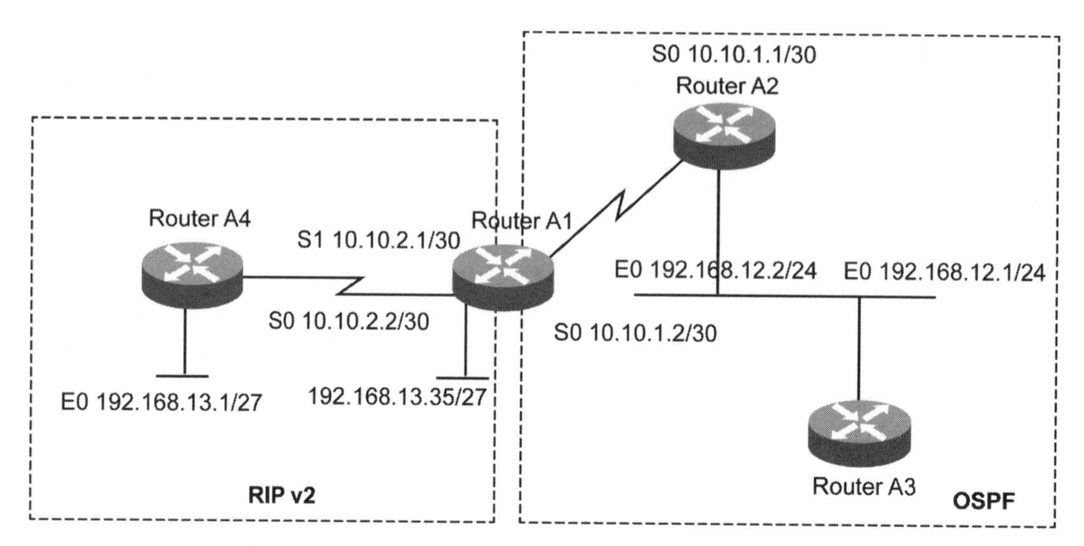

FIGURE 9.7 Example showing a network configured with OSPF and RIP routing protocols.

The output shows that the RIP route is available at A1.

2. Check if the keyword subnets is used while redistributing into OSPF. This keyword is essential, because the subject network has been subnetted. The configuration of A1 should be as shown:

```
router ospf 1
network 10.10.1.0 0.0.0.255 area 2
network 10.10.2.0 0.0.0.255 area 2
redistribute rip metric 10 subnets
```

Add the keyword if it is missing.

3. Check if the route is available in the routing table of A3, after adding the subnets keyword.
4. Check for the presence of any route map used in redistribution, if the route is unavailable. The route maps are checked as shown in Listing 9.21.

LISTING 9.21 Checking for Route Maps in A1

```
router ospf 1
network 10.10.1.0 0.0.0.255 area 2
network 10.10.2.0 0.0.0.255 area 2
redistribute rip route-map redis-rip
!
```

```
route-map redis-rip
match ip address 10
set metric 20
!
access-list 10 deny  192.168.13.0 0.0.0.31
access-list 10 permit any
!
```

Listing 9.21 shows the presence of a route map that is restricting the redistribution of network 192.168.13.0/27 into OSPF. The route map is suitably modified.

5. Check if there is any filtering of updates at A3 if there are no route maps. In the case of OSPF, we need to check for route filtering only at A3, because no outbound filtering is possible. Filtering updates are verified using the show ip protocols command. Listing 9.22 shows the output of the show ip protocols command.

LISTING 9.22 Output of the show ip protocols Command at A3

```
A3#show ip protocols
Routing Protocol is "ospf 1"
Outgoing update filter list for all interfaces is not set
Incoming update filter list for all interfaces is not set
Ethernet0 filtered by 15 (per-user), default is 15
Router ID  192.168.12.1
Number of areas in this router is 1. 1 normal 0 stub 0 nssa
Maximum path: 6
Routing for Networks:
192.168.12.0 0.0.0.255 area 5
Passive Interface(s):
Routing Information Sources:
Gateway          Distance    Last Update
192.168.12.2     110         00:00:00
  Distance: (default is 110)
```

Listing 9.22 shows that access list 15 is used to filter incoming routing updates into A3. The output of the A3#show running-config | inc access-list 15 command is as shown:

```
access-list 15 deny 192.168.13.0 0.0.0.31
access-list 15 permit any
```

Route filtering is disabled or access list 15 is suitably modified to make route 192.168.13.0/27 available in the routing table of A3.

CASE STUDY

Consider the network shown in Figure 9.8. The OSPF network consists of routers B1, B2, B3, B4, and B5. EIGRP is running among routers B5 and B6. Redistribution between OSPF and EIGRP is configured in router B5, which is running both routing protocols.

In Figure 9.8, the OSPF network faces some problems. They are:

- B2 is not visible as a neighbor of B1.
- Route 10.6.1.1/27 is not available in the routing table of B4.

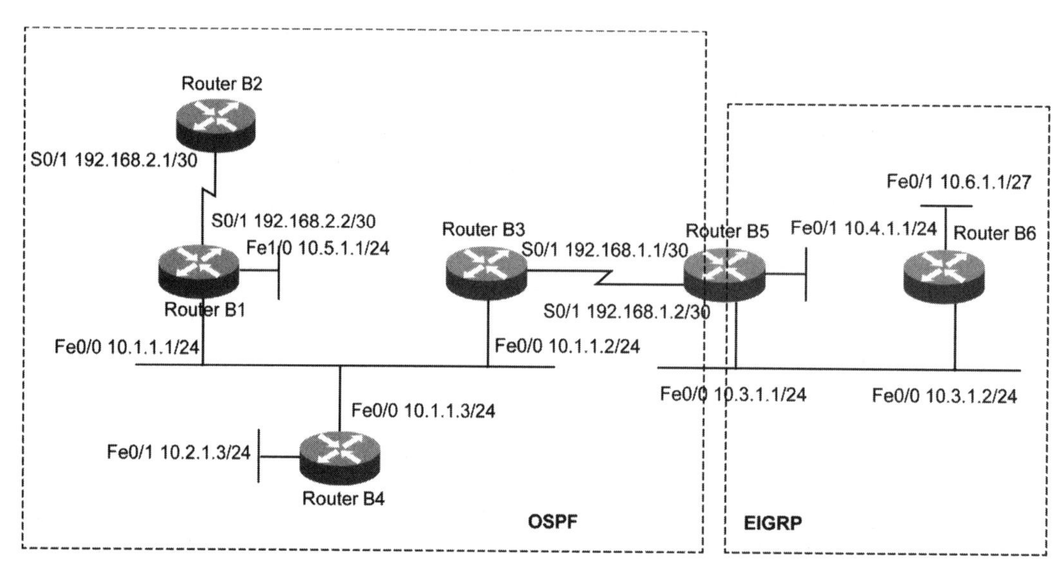

FIGURE 9.8 A network configured with OSPF and EIGRP routing protocols.

B2 Is Not Visible as a Neighbor of B1

B1 and B2 are connected via a serial link. B2 is not visible as a neighbor of B1. To troubleshoot this problem:

1. Check the status of the serial link by using the show interface Serial 0/1 command. The output of the show interface Serial 0/1 command is shown in Listing 9.23.

LISTING 9.23 Output of show interface Serial 0/1 Command

```
B1#show interface Serial 0/1
Serial0/1 is up, line protocol is up
Hardware is HD64570
Description: WAN link to router B2
Internet address is       192.168.2.2/30
MTU 1500 bytes, BW 1544 Kbit, DLY 20000 usec,
reliability 255/255, txload 1/255, rxload 1/255
Encapsulation HDLC, loopback not set
Keepalive set (10 sec)
Last input 00:00:00, output 00:00:00, output hang never
Last clearing of "show interface" counters never
Input queue: 0/75/3134350/0 (size/max/drops/flushes); Total output
  drops: 11
Queueing strategy: weighted fair
Output queue: 0/1000/64/0 (size/max total/threshold/drops)
Conversations      0/11/256 (active/max active/max total)
Reserved Conversations 0/0 (allocated/max allocated)
5 minute input rate 13000 bits/sec, 29 packets/sec
5 minute output rate 2000 bits/sec, 3 packets/sec
115001192 packets input, 4155732628 bytes, 272 no buffer
Received 425130 broadcasts, 0 runts, 0 giants, 0 throttles
228 input errors, 227 CRC, 68 frame, 0 overrun, 0 ignored, 34 abort
21650119 packets output, 3170761599 bytes, 0 underruns
0 output errors, 0 collisions, 1 interface resets
0 output buffer failures, 0 output buffers swapped out
10 carrier transitions
    DCD=up  DSR=up  DTR=up  RTS=up  CTS=up
```

Listing 9.23 shows that the interface is physically up.

2. Check the basic IP level connectivity using the B1#ping 192.168.2.1 command. The command output shows a 100% success rate to confirm that Layer 3 connectivity is working between B1 and B2. It is confirmed that there are no problems or errors in the physical link.
3. Check if OSPF is defined for the interfaces Serial0/1 in B1 and B2 by using the show ip ospf interface Serial0/1 command. The output of the show ip ospf interface Serial0/1 command at B1 is shown in Listing 9.24.

LISTING 9.24 Output of the show ip ospf interface Serial0/1 Command at B1

```
B1#show ip ospf interface Serial0/1
Serial0/1 is up, line protocol is up
```

```
Internet Address   192.168.2.2/30, Area 5
Process ID 1, Router ID   192.168.2.2, Network Type POINT_TO_POINT,
Cost: 64
Transmit Delay is 1 sec, State POINT_TO_POINT,
Timer intervals configured, Hello 10, Dead 40, Wait 40, Retransmit 5
No Hellos (Passive interface)
Index 7/7, flood queue length 0
Next 0x0(0)/0x0(0)
Last flood scan length is 0, maximum is 0
Last flood scan time is 0 msec, maximum is 0 msec
Neighbor Count is 0, Adjacent neighbor count is 0
Suppress hello for 0 neighbor(s)
  Simple password authentication enabled
```

Listing 9.24 shows that the interface Serial0/1 in B1 has been defined as a passive interface that is preventing the formation of neighbor relationship. The output of the show ip ospf interface Serial0/1 command at B2 is shown in Listing 9.25.

LISTING 9.25 Output of the show ip ospf interface Serial0/1 Command at B2

```
B2#show ip ospf interface Serial0/1
Serial0/1 is up, line protocol is up
Internet Address   192.168.2.1/30, Area 5
Process ID 1, Router ID   192.168.2.1, Network Type POINT_TO_POINT,
Cost: 64
Transmit Delay is 1 sec, State POINT_TO_POINT,
Timer intervals configured, Hello 10, Dead 40, Wait 40, Retransmit 5
 Hello due in 00:00:02
Index 7/7, flood queue length 0
Next 0x0(0)/0x0(0)
Last flood scan length is 0, maximum is 0
Last flood scan time is 0 msec, maximum is 0 msec
Neighbor Count is 0, Adjacent neighbor count is 0
Suppress hello for 0 neighbor(s)
  Simple password authentication enabled
```

Listings 9.24 and 9.25 show that OSPF is running at both the neighboring interfaces of B1 and B2. The commands to restore normalcy and establish neighbor relationship between B1 and B2 are:

```
B1(config)#router ospf 1
B1(config-router)#no passive-interface Serial 0/1
```

Route 10.6.1.1/27 Is Unavailable in Routing Table of B4

Route unavailability is caused by problems in redistribution configuration. To troubleshoot this problem:

1. Check if route 10.6.1.1/27 is available at B5 by using the #show ip route | include 10.6.1.1 command. The output is:

```
#show ip route | include 10.6.1.1
D    10.6.1.1/27 [90/1100] via 10.3.1.2, 00:00:33,
```

The command output shows that the EIGRP route is present at the routing table of B5.

2. Check if the keyword subnets is used, in case of redistribution into OSPF. The configuration in B5 should be:

```
router ospf 1
redistribute eigrp metric 10 subnets
```

The keyword subnets is mandatory so that the subnets to a major network are redistributed properly into OSPF and are visible in the routing table.

SUMMARY

In this chapter, we learned about problem isolation in OSPF routing environments. We also reminded you about resolution of the problems in OSPF routing environments. In the next chapter, we move on to troubleshooting IS-IS routing environments.

POINTS TO REMEMBER

- Open Shortest Path First (OSPF) ensures interoperability between routing devices manufactured by diverse vendors.
- Link State Advertisement (LSA) is an OSPF packet containing source, destination, and routing information, which is advertised to all OSPF routers in a hierarchical area.
- A designated router (DR) reduces the number of adjacencies formed in a broadcast network.
- A backup designated router (BDR) acts as a standby for the DR on broadcast networks by collecting routing information updated from the adjacent OSPF routers and takes the role of a DR when the DR goes down.

- Multi-access/broadcast networks are physical networks that support interconnection of more than two routers that can communicate directly.
- A single area (SA) is a logical subdivision of the greater OSPF domain, grouping routers that run OSPF with identical topological databases.
- A Totally Stubby Area (TSA) is a type of a nonstandard OSPF area that is used when few networks with limited connectivity are connected to the remaining network.
- Four types of OSPF routers are Internal Router (IR), Area Border Router (ABR), Backbone Router (BR), and Autonomous System Boundary Router (ASBR).
- The priority value is the primary parameter used in DR and BDR elections in an OSPF routing process.
- In a loading state, the information exchange between the OSPF routers that have formed adjacency is finalized.
- To prevent routing loops and sub-optimal routing in the network, configure redistribution into OSPF and between OSPF and another IGP.

10 Troubleshootir Routing Envir

IN THIS CHAPTER

- Features of IS-IS
- Problem Isolation in IS-IS
- Misconfiguration Problems in IS-IS
- L1 Router Problem
- L2 Router Problem
- Redistribution Problem

Intermediate System–to–Intermediate System (IS-IS) is designed for routing in Connectionless Network Service (CLNS) and IP-based networks. Large Internet Service Providers (ISPs) use the IS-IS routing protocol to leverage their interconnection support to large networks. ISPs need to connect both OSI and IP-based networks. As a result, IS-IS is the most appropriate routing protocol for business services similar to ISPs.

This chapter discusses various problems, their effective isolation, and the different methods to resolve problems in IS-IS routing environments.

FEATURES OF IS-IS

You should be familiar with the different protocol concepts of IS-IS to successfully isolate and troubleshoot problems. This section discusses the terminology used in the IS-IS routing protocol. The terms are:

Open Systems Interconnection (OSI): International standard protocol architecture.

Protocol Data Unit or Packet Data Unit (PDU): Single data unit of the corresponding protocol used across the network.

k State Packet data unit (LSP): Data unit of the IS-IS protocol. LSP is nalogous to the Link State Advertisement (LSA) in OSPF protocol. LSP refers to the packet containing routing information.

End System (ES): Host system in IS-IS network. ES is OSI terminology for host.

Intermediate System (IS): Router in IS-IS terminology.

Designated Intermediate System (DIS): Designated Router (DR) in IS-IS network.

Pseudo Node (PN): Broadcast link emulated as a virtual link by DIS.

SubNetwork Point of Attachment (SNPA): Data-link interface of the IS-IS network. This is the point at which a device is connected to the network.

Network Service Access Point (NSAP): OSI address. It consists of area address, system identifier, and NSEL selector byte. If NSEL for an NSAP is equal to 0, it is known as Network Entity Title (NET). NSEL for a router is always set to 0.

Network Layer Protocol ID (NLPID): Octet field that identifies a network layer protocol.

Adjacency: Nearby systems in IS-IS network.

End System–to–Intermediate System (ES-IS): Protocol similar to IS-IS that delivers the ISO defined CLNS.

IS-IS process tag: IS-IS process name given using the command router isis.

L1 router: Performs routing based on the area ID portion of the IS-IS packet.

Network Entity Title (NET): Unique address for an IS within an area. This address contains three fields: System ID, Area ID, and N-SEL.

System ID: ID used by L1 routers for locating a specific IS within an area.

Area ID: ID used by L2 routers for locating a specific area within an IS-IS network.

NSEL: Network level services offered to the IS. This value is 00 by default.

Number of manual areas: Areas configured using net command.

Number of active areas: Areas learned by adjacency relationships.

IS-IS level: Routing level of an IS-IS router.

Distance: Distance covered by the IS-IS route.

Maximum paths: Maximum number of parallel routes available to the same destination as in the routing table.

Number of SPF runs: Number of times of SPF execution for Level 1 (L1) and Level 2 (L2) routers.

The IS-IS routing domain is divided into two areas:

Level 1 (L1) area: Routing is performed between adjacent ISs within the same area.

Level 2 (L2) area: Connects two different L1 areas within the same routing domain.

There are three types of routers in IS-IS routing protocol: L1, L2, and L1-L2. A comparison of these three routers is given in Table 10.1.

TABLE 10.1 Comparing IS-IS Routers

L1 Router	L2 Router	L1-L2 Router
Nonbackbone router	Backbone router	Similar to OSPF Area Border Router
Knowledge of intra-area routes only	Knowledge of inter-area routes only	Knowledge of both intra-area and inter-area routes
Maintains L1 link state database	Maintains L2 link state database	Maintains two link state databases

The different types of data packets used in IS-IS routing protocol are shown in Table 10.2.

TABLE 10.2 Data Packets in the IS-IS Routing Protocol

IS-IS Packet	Function
IS-IS Hello Packet (IIH)	Establishes and maintains adjacency relationships between ISs.
Link State Packet (LSP)	Consists of two types of LSPs: non–psuedo node LSP and psuedo node LSP.
Complete Sequence Number Protocol data unit (CSNP)	Includes a list of LSPs known to the IS. CSNPs are generated and flooded by DIS periodically. Also explains every LSP in the link state database.
Partial Sequence Number Protocol data unit (PSNP)	Requests for missing route information after receipt of a CSNP. PSNP also acknowledges a routing update on point-to-point links.

Every IS generates and propagates a fresh non–psuedo node LSP when:

- New neighbor IS comes alive into the IS-IS network and subsequently shuts down.
- New IP prefixes are added or eliminated.
- Link cost varies.
- Update period for LSPs expires.

Every IS network generates and propagates the psuedo node LSPs when:

- New neighbor IS comes alive into the IS-IS network and subsequently shuts down.
- Update period for LSPs expires.
- New LAN is connected to IS-IS routing domain.

PROBLEM ISOLATION IN IS-IS

This section discusses the different commands available in the Cisco IOS to isolate and troubleshoot problems in IS-IS networks. Table 10.3 lists the various show command options for IS-IS networks.

TABLE 10.3 Show Commands and Descriptions

Command	Description
router#show isis	Shows if IS-IS routing protocol is enabled on a specified router. This command also displays the system level information about the IS-IS execution.
router#show clns	Provides global CLNS information such as number of interfaces, NET, type of routing (L1, L2, or L1-L2), and routing area ID.
router#show clns neighbor	Shows System ID, Interface ID, SNPA, Hold time, IS type, and protocol used.
router#show clns interface interface-name	Displays details about the CLNS interface where interface-name represents the name of the interface whose details are displayed.
router#show clns protocol	Displays details of the CLNS protocol, including the areas involving redistribution details.
router#show clns traffic	Displays details about network traffic sequence in a router.

TABLE 10.3 *(continued)*

Command	Description
router#show isis database level-1	Displays the Level 1 link-state database.
router#show isis database level-2	Displays the Level 2 link-state database.
router#show isis spf-log	Displays time and reason for the shortest path algorithm execution.

Consider the example depicted in Figure 10.1. The output of commands from Table 10.1 will be discussed with respect to this scenario.

Figure 10.1 shows part of an IS-IS network with four routers R1, R2, R3, and R4. R1 and R2 belong to Area 49.0001. R3 and R4 belong to Area 49.0002. R2 connects Area 49.0001 to Area 49.0002. R1 and R4 function as L1 routers, and R2 and R4 are L1-L2 routers.

In Figure 10.1, consider that the show commands are issued at R2. The configuration of R1 is shown in Listing 10.1.

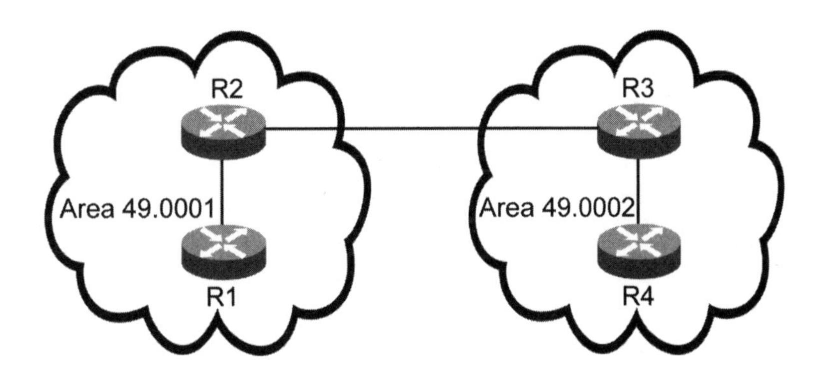

FIGURE 10.1 A network configured with an IS-IS routing protocol.

LISTING 10.1 Configuration of R1

```
Router#enable
Router#conf terminal
Router(conf)#hostname R1
R1(conf)#Interface Loopback0
R1(conf)#ip address 172.148.1.3 255.255.255.255
!
```

```
!
R1(conf)#interface FastEthernet0/0
R1(conf)#ip address 172.148.120.5 255.255.255.0
R1(conf)#ip router isis
!
!
R1(conf)#router isis
R1(conf)#is-type level-1
R1(conf)#passive-interface Loopback0
R1(conf)#net 49.0001.1721.4800.1003.00
```

The configuration of R2 is shown in Listing 10.2.

LISTING 10.2 Configuration of R2

```
Router#enable
Router#conf terminal
Router(conf)#hostname R2
R2(conf)#Interface Loopback0
R2(conf)#ip address 172.148.1.4 255.255.255.255
!
!
R2(conf)#Interface Pos2/0/0
R2(conf)#ip address 172.148.222.1 255.255.255.0
R2(conf)#ip router isis
R2(conf)#isis circuit-type level-2
!
!
R2(conf)#FastEthernet4/0/0
R2(conf)#ip address 172.148.120.10 255.255.255.0
R2(conf)#ip router isis
R2(conf)#isis circuit-type level-1
!
!
R2(conf)#router isis
R2(conf)#passive-interface Loopback0
R2(conf)#net 49.0001.1721.4800.1004.00
```

The configuration of R3 is shown in Listing 10.3.

LISTING 10.3 Configuration of R3

```
Router#enable
Router#conf terminal
```

```
Router(conf)#hostname R3
R3(conf)#Interface Loopback0
R3(conf)#ip address 172.148.2.2 255.255.255.255
!
!
R3(conf)#Interface Pos1/0/0
R3(conf)#ip address 172.148.222.2 255.255.255.0
R3(conf)#ip router isis
R3(conf)#isis circuit-type level-2
!
R3(conf)#interface Fddi3/0
R3(conf)#ip address 172.148.111.2 255.255.255.0
R3(conf)#ip router isis
R3(conf)#isis circuit-type level-1
!
R3(conf)#router isis
R3(conf)#passive-interface Loopback0
R3(conf)#net 49.0002.1721.4800.2002.00
```

The configuration of R4 is shown in Listing 10.4.

LISTING 10.4 Configuration of R4

```
Router#enable
Router#conf terminal
Router(conf)#hostname R4
R4(conf)#Interface Loopback0
R4(conf)#ip address 172.148.2.6 255.255.255.255
!
R4(conf)#interface Fddi6/0
R4(conf)#ip address 172.148.111.4 255.255.255.0
R4(conf)#ip router isis
!
R4(conf)#router isis
R4(conf)#is-type level-1
R4(conf)#passive-interface Loopback0
R4(conf)#net 49.0002.1721.4800.2006.00
```

The output of the show isis command in R2 is shown in Listing 10.5.

LISTING 10.5 Output for show isis Command in R2

```
R2#show isis
Global ISIS information
```

```
ISIS process tag: R2
System ID: 1721.4800.1004
NET: 49.0001.1721.4800.1004.00
Maximum number of areas: 2
There is 1 manual area address
49
There is 1 active area address
49
ISIS level-1
ISIS is enabled on 2 interfaces
Distance : 120
Maximum of 1 path per route
Number of SPF runs, L1: 14, L2: 5
```

The output of the router#show clns command in R2 is shown in Listing 10.6.

LISTING 10.6 Output of router#show clns Command in R2

```
R2#show clns
Global CLNS Information:
2 Interfaces Enabled for CLNS
NET: 49.0001.1721.4800.1004.00
Configuration Timer: 60, Default Holding Timer: 300, Packet
Lifetime 64
ERPDU's requested on locally generated packets
Intermediate system operation enabled (forwarding allowed)
IS-IS level-1-2 Router:
Routing for Area: 49.0001
```

The output of the router#show clns neighbor command in R2 is shown in Listing 10.7.

LISTING 10.7 Output of router#show clns neighbor Command in R2

```
R2#show clns neighbors
System Id       SNPA      Interface     State   Holdtime   Type
  Protocol
1721.4800.2002  *PPP*       PO2/0/0       Up        19        L2      IS-
  IS
1721.4800.1003 00e0.1492.2c00 Fa4/0/0    Up        11        L1      IS-
  IS
```

The output of router#show clns interface interface-name command in R2 is shown in Listing 10.8.

LISTING 10.8 Output of router#show clns interface interface-name Command in R2

```
R2#show clns interface POS2/0/0
POS2/0/0 is up, line protocol is up
Checksums enabled, MTU 4470, Encapsulation PPP
ERPDUs enabled, min. interval 10 msec.
RDPDUs enabled, min. interval 100 msec. , Addr Mask enabled
Congestion Experienced bit set at 4 packets
CLNS fast switching disabled
CLNS SSE switching disabled
DEC compatibility mode OFF for this interface
Next ESH/ISH in 47 seconds
Routing Protocol: IS-IS
Circuit Type: level-1-2
Interface number 0x0, local circuit ID 0x100
Level-1 Metric: 10, Priority: 64, Circuit ID: 1721.4800.2002.00
Number of active level-1 adjacencies: 0
Level-2 Metric: 10, Priority: 64, Circuit ID: 1721.4800.1004.00
Number of active level-2 adjacencies: 1
Next IS-IS Hello in 2 seconds
```

The output of router#show clns protocol command in R2 is shown in Listing 10.9.

LISTING 10.9 Output of router#show clns protocol Command in R2

```
R2#show clns protocol
IS-IS Router: <Null Tag>
System Id: 1721.4800.1004.00 IS-Type: level-1-2
Manual area address(es):
49.0001
Routing for area address(es):
49.0001
Interfaces supported by IS-IS:
FastEthernet4/0/0 - IP
POS2/0/0 - IP
Redistributing:
static
Distance: 120
```

The output of router#show clns traffic command in R2 is shown in Listing 10.10.

LISTING 10.10 Output of router#show clns traffic Command in R2

```
R2#show clns traffic
CLNS & ESIS Output: 14, Input: 436
```

```
CLNS Local: 0, Forward: 0
CLNS Discards:
Hdr Syntax: 0, Checksum: 0, Lifetime: 0, Output cngstn: 0
No Route: 0, Dst Unreachable 0, Encaps. Failed: 0
NLP Unknown: 0, Not an IS: 0
CLNS Options: Packets 0, total 0, bad 0, GQOS 0, cngstn exprncd 0
CLNS Segments: Segmented: 0, Failed: 0
CLNS Broadcasts: sent: 0, rcvd: 0
Echos: Rcvd 0 requests, 0 replies
Sent 0 requests, 0 replies
ESIS(sent/rcvd): ESHs: 0/0, ISHs: 14/15, RDs: 0/0, QCF: 0/0
ISO-IGRP: Querys (sent/rcvd): 0/0 Updates (sent/rcvd): 0/0
ISO-IGRP: Router Hellos: (sent/rcvd): 0/0
ISO-IGRP Syntax Errors: 0
IS-IS: Level-1 Hellos (sent/rcvd): 84/244
IS-IS: Level-2 Hellos (sent/rcvd): 0/0
IS-IS: PTP Hellos (sent/rcvd): 84/82
IS-IS: Level-1 LSPs sourced (new/refresh): 7/1
IS-IS: Level-2 LSPs sourced (new/refresh): 5/0
IS-IS: Level-1 LSPs flooded (sent/rcvd): 5/5
IS-IS: Level-2 LSPs flooded (sent/rcvd): 1/6
IS-IS: LSP Retransmissions: 0
IS-IS: Level-1 CSNPs (sent/rcvd): 0/82
IS-IS: Level-2 CSNPs (sent/rcvd): 1/1
IS-IS: Level-1 PSNPs (sent/rcvd): 1/0
IS-IS: Level-2 PSNPs (sent/rcvd): 3/1
IS-IS: Level-1 DR Elections: 2
IS-IS: Level-2 DR Elections: 1
IS-IS: Level-1 SPF Calculations: 3
IS-IS: Level-2 SPF Calculations: 3
IS-IS: Level-1 Partial Route Calculations: 0
IS-IS: Level-2 Partial Route Calculations: 0
IS-IS: LSP checksum errors received: 0
IS-IS: Update process queue depth: 0/200
IS-IS: Update process packets dropped: 0
```

The output of router#show isis database command in R2 is shown in Listing 10.11.

LISTING 10.11 Output of router#show isis database Command in R2

```
R2#show isis database
IS-IS Level-1 Link State Database
LSPID LSP Seq Num LSP Checksum LSP Holdtime ATT/P/OL
```

```
1721.4800.1004.00-00* 0x00000019 0x2783 1153 1/0/0 (11)
1721.4800.1003.00-00 0x0000000C 0x2179 905 0/0/0 (5)
1721.4800.1003.01-00 0x00000009 0x40EC 831 0/0/0 (4)
IS-IS Level-2 Link State Database
LSPID LSP Seq Num LSP Checksum LSP Holdtime ATT/P/OL
1721.4800.1004.00-00* 0x00000010 0xFC45 1153 0/0/0 (9)
1721.4800.1004.01-00* 0x00000001 0x4CB7 1137 0/0/0 (10)
1721.4800.2002.00-00 0x00000018 0x86A6 1141 0/0/0 (3)
1721.4800.2002.02-00 0x00000004 0x8558 881 0/0/0 (2)
R2> sh isis database 1721.4800.1004.00-00 detail
IS-IS Level-1 LSP 1721.4800.1004.00-00
LSPID LSP Seq Num LSP Checksum LSP Holdtime ATT/P/OL
1721.4800.1004.00-00* 0x00000006 0x4D70 991 1/0/0 (1)
Area Address: 49.0001
NLPID: 0xCC
IP Address: 172.148.120.10
Metric: 10 IP 172.148.222.0 255.255.255.0
Metric: 10 IP 172.148.120.0 255.255.255.0
Metric: 0 IP 172.148.1.4 255.255.255.255
Metric: 10 IS 1721.4800.1003.01
Metric: 0 ES 1721.4800.1004
IS-IS Level-2 LSP 1721.4800.1004.00-00
LSPID LSP Seq Num LSP Checksum LSP Holdtime ATT/P/OL
1721.4800.1004.00-00* 0x00000004 0x1539 980 0/0/0 (2)
Area Address: 49.0001
NLPID: 0xCC
IP Address: 172.148.222.1
Metric: 10 IS 1721.4800.1004.01
Metric: 10 IS 1721.4800.2002.00
Metric: 10 IP 172.148.120.0 255.255.255.0
Metric: 0 IP 172.148.1.4 255.255.255.255
Metric: 10 IP 172.148.1.3 255.255.255.255
Metric: 10 IP 172.148.222.0 255.255.255.0
```

The output of router#show isis spf-log command in R2 is shown in Listing 10.12.

LISTING 10.12 Output of router#show isis spf-log Command in R2

```
R2#show isis spf-log
Level 1 SPF log
When Duration Nodes Count Last trigger LSP Triggers
00:05:30 0 1 6 1721.4800.1004.00-00 NEWAREA RTCLEARED NEWADJ NEWLSP
TLVCONTENT
00:05:20 0 3 3 1721.4800.1003.01-00 NEWADJ TLVCONTENT
```

```
00:05:05 0 3 2 1721.4800.1004.00-00 ATTACHFLAG LSPHEADER
Level 2 SPF log
When Duration Nodes Count Last trigger LSP Triggers
00:05:31 0 1 2 1721.4800.1004.00-00 RTCLEARED NEWLSP
00:05:21 0 1 3 1721.4800.1004.00-00 NEWADJ TLVCODE TLVCONTENT
00:05:11 0 2 1 1721.4800.2002.00-00 LSPHEADER
```

Table 10.4 shows the different debug commands that can be used for the thorough troubleshooting of problems in IS-IS networks.

TABLE 10.4 debug Commands and Descriptions

Command	Description
debug isis adj-packets	Enables IS-IS adjacency packets and displays IS-IS packet information.
debug isis spf-events	Enables debugging in the Shortest Path Function (SPF) events.
debug isis snp-packets	Enables and displays CSNP/PSNP packets information.
Debug isis spf-update	Shows SPF updated information.

Consider the example depicted in Figure 10.1. The output of commands from Table 10.4 will be discussed with respect to this scenario.

The output of debug isis adj-packets command in R2 is shown in Listing 10.13.

LISTING 10.13 Output of debug isis adj-packets Command in R2

```
R2#debug isis adj-packets
ISIS-Adj: Rec L1 IIH from 00e0.1492.2c00
 (FastEthernet4/0/0), cir type 1,
cir id 1721.4800.1003.01
ISIS-Adj: Sending L1 IIH on FastEthernet4/0/0
ISIS-Adj: Rec L1 IIH from 00e0.1492.2c00
 (FastEthernet4/0/0), cir type 1,
cir id 1721.4800.1003.01
ISIS-Adj: Sending serial IIH on POS2/0/0
ISIS-Adj: Rec serial IIH from *PPP* on POS2/0/0, cir type 3, cir id 00
```

The output of the debug isis spf-events command in R2 is shown in Listing 10.14.

LISTING 10.14 Output of debug isis spf-events Command in R2

```
R2#debug isis spf-events
ISIS-SPF: Compute L1 SPT
ISIS-SPF: Move 1721.4800.1004.00-00 to PATHS, metric 0
ISIS-SPF: thru 2147483647/2147483647/2147483647, delay 0/0/0, mtu
2147483647/2147483647/2147483647, hops 0/0/0, ticks 0/0/0
ISIS-SPF: Add 1721.4800.1003.01-00 to TENT, metric 10
ISIS-SPF: Next hop local
ISIS-SPF: Move 1721.4800.1003.01-00 to PATHS, metric 10
ISIS-SPF: thru 2147483647/2147483647/2147483647, delay 0/0/0, mtu
2147483647/2147483647/2147483647, hops 0/0/0, ticks 0/0/0
ISIS-SPF: considering adj to 1721.4800.1003
  (FastEthernet4/0/0) metric 10
ISIS-SPF: (accepted)
ISIS-SPF: Add 1721.4800.1003.00-00 to TENT, metric 10
ISIS-SPF: Next hop 1721.4800.1003 (FastEthernet4/0/0)
ISIS-SPF: Move 1721.4800.1003.00-00 to PATHS, metric 10
ISIS-SPF: Add 172.148.120.0/255.255.255.0 to IP route table, metric 20
ISIS-SPF: Next hop 1721.4800.1003/172.148.120.5
  (FastEthernet4/0/0) (rejected)
ISIS-SPF: Add 172.148.1.3/255.255.255.255 to IP route table, metric 10
ISIS-SPF: Next hop 1721.4800.1003/172.148.120.5
  (FastEthernet4/0/0) (accepted)
ISIS-SPF: Add 144.254.0.0/255.255.0.0 to IP route table, metric 60
ISIS-SPF: Next hop 1721.4800.1003/172.148.120.5
  (FastEthernet4/0/0) (rejected)
```

The output of the debug isis snp-packets command in R2 is shown in Listing 10.15.

LISTING 10.15 Output of debug isis snp-packets Command in R2

```
R2#debug isis snp-packets
ISIS-SNP: Rec L1 CSNP from 1721.4800.1003
  (FastEthernet4/0/0)
ISIS-SNP: CSNP range 0000.0000.0000.00-00 to FFFF.FFFF.FFFF.FF-FF
ISIS-SNP: Same entry 1721.4800.1004.00-00, seq 92
ISIS-SNP: Same entry 1721.4800.1003.00-00, seq 79
ISIS-SNP: Same entry 1721.4800.1003.01-00, seq 77
Area 49.0001 Area 49.0002
```

The output of debug isis spf-update command in R2 is shown in Listing 10.16.

LISTING 10.16 Output of debug isis spf-update Command in R2

```
R2( config )# int fa4/0/0
R2( config -if)#isis metric 13
R2( config -if)#^Z
R2#
ISIS-SPF-TRIG: L1, new metric
ISIS-Update: Building L1 LSP
ISIS-Update: TLV contents different, code 80
ISIS-Update: TLV contents different, code 2
ISIS-SPF-TRIG: L1, 1721.4800.1004.00-00 TLV contents changed, code 2
ISIS-Update: Full SPF required
ISIS-Update: Sending L1 LSP 1721.4800.1004.00-00, seq 96, ht 1199 on
FastEthernet4/0/0
ISIS-SNP: Rec L1 CSNP from 1721.4800.1003
  (FastEthernet4/0/0)
ISIS-Stats : Compute L1 SPT
ISIS-Stats : Complete L1 SPT, Compute time 0.000, 3 nodes, 2 links on
SPT, 0 suspends
ISIS-Update: Building L2 LSP
ISIS-Update: TLV contents different, code 80
ISIS-Update: TLV contents different, code 80
ISIS-Update: Leaf routes changed
ISIS-Update: Sending L2 LSP 1721.4800.1004.00-00, seq 96, ht 1199 on
  POS2/0/0
ISIS-Update: Building L2 LSP
ISIS-Update: Rate limiting L2 LSP 1721.4800.1004.00-00, seq 97
ISIS-Update: TLV contents different, code 80
ISIS-Update: TLV contents different, code 80
ISIS-Update: Leaf routes changed
ISIS-Update: Sending L2 LSP 1721.4800.1004.00-00, seq 97, ht 1196 on
POS2/0/0
```

MISCONFIGURATION PROBLEMS IN IS-IS

Misconfiguration can be caused by manual or system errors. You need to take extreme care when configuring an IS-IS routing protocol in a router, because even a small configuration error can affect all network traffic.

Misconfiguration can occur if the NET configuration is inaccurate. As a result, the IS-IS network may not route as expected.

To discuss the misconfiguration problem, consider the example shown in Figure 10.2. In the figure, router R1 with System ID 190.65.4.1 is connected to router

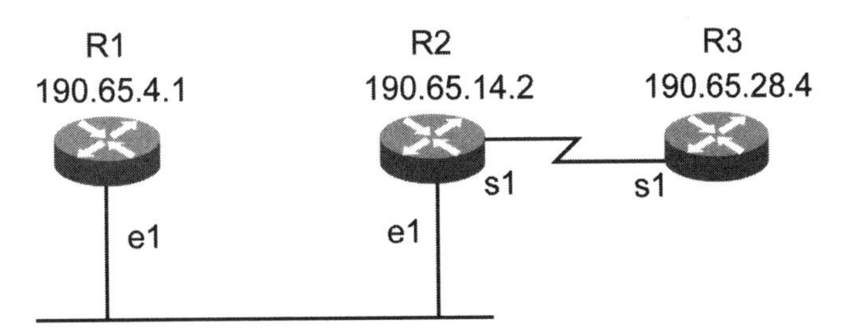

FIGURE 10.2 An IS-IS network depicting a NET misconfiguration scenario.

R2 with System ID 190.65.14.2 via an Ethernet interface, e1. R2 is connected to a third router, R3, with System ID 190.65.28.4 via a serial interface, s1.

The configuration of R1 is shown in Listing 10.17.

LISTING 10.17 Configuration of R1

```
!
hostname R1
!
interface loopback0
   ip address 190.65.4.1  255.255.255.255
!
interface Ethernet1
    ip address 190.65.24.1 255.255.255.0
    ip router isis
!
router isis
  passive-interface Loopback0
  net 49.001.1900.6500.4001.01
!
```

In Listing 10.17, NET is configured incorrectly for R1. This is the misconfiguration. The last field of NET is 01; whereas, it should be 00 for an IS-IS router. The configuration of R2 is shown in Listing 10.18.

LISTING 10.18 Configuration of R2

```
!
hostname R2
!
interface Loopback0
    ip address 190.65.14.2 255.255.255.255
```

```
!
interface Ethernet1
   ip address 190.65.24.2 255.255.255.0
   ip router isis
!
interface Serial1
  ip address 190.65.28.4 255.255.255.253
  ip router isis
!
router isis
  passive-interface Loopback0
  net 49.001.1906.5001.4002.00
!
```

The show commands are used to troubleshoot NET for R1 and R2. The output of the show isis command in R1 is shown in Listing 10.19.

LISTING 10.19 Output of show isis Command at R1

```
R1#show isis
Global ISIS information
ISIS process tag: R1
System ID: 1900.6500.4001
NET: 49.001.1900.6500.4001.01
Maximum number of areas: 2
There is 1 manual area address
49
There is 1 active area address
49
ISIS level-1
ISIS is enabled on 2 interfaces
Distance : 250
Maximum of 1 path per route
Number of SPF runs, L1: 14, L2: 0
```

The output of the show isis command in R2 is shown in Listing 10.20.

LISTING 10.20 Output of show isis Command at R2

```
R2#show isis
Global ISIS information
ISIS process tag: R2
System ID: 1906.5001.4002
NET: 49.001.1906.5001.4002.00
```

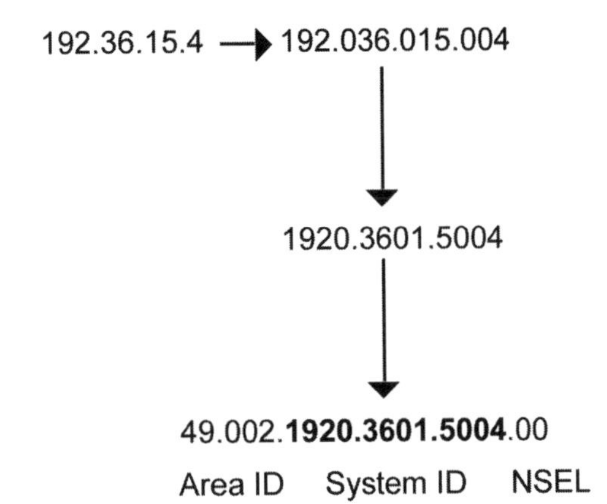

FIGURE 10.3 Calculation of NET value in IS-IS networks.

```
Maximum number of areas: 2
There is 1 manual area address
49
There is 1 active area address
49
ISIS level-1
ISIS is enabled on 2 interfaces
Distance : 250
Maximum of 1 path per route
Number of SPF runs, L1: 14, L2: 0
```

To troubleshoot the NET misconfiguration in an IS-IS network, check the NET field and calculate the NET address. For example, consider an IS with System ID 192.36.15.4 and area ID 49.002. The NET is calculated as shown in Figure 10.3.

In Figure 10.2, the NET for R1 is calculated as shown:

- Area ID: 49.001
- IP address to be configured for R1 is 190.65.4.1
- System ID => 190.065.004.001=>1900.6500.4001
- NSEL = 00 for an IS-IS router
- NET = AreaID.SystemID.NSEL

The NET for R2 is calculated as shown:

- Area ID is 49.001
- IP address to be configured for R2 is 190.65.14.2

- System ID => 190.065.014.002=>1900.6501.4002
- NSEL = 00 for an IS-IS router
- NET = AreaID.SystemID.NSEL

From the calculations, it is clear that the System ID and NSEL are configured incorrectly for R1, and the System ID is configured incorrectly for R2. The NET for R1 and R2 should be configured correctly. The correct configuration for R1 is shown in Listing 10.21.

LISTING 10.21 Correct Configuration for R1

```
!
hostname R1
!
interface loopback0
   ip address 190.65.4.1  255.255.255.255

!

interface e1
    ip address 190.65.24.1 255.255.255.0
    ip router isis
!
router isis
  passive-interface Loopback0
  net 49.001. 1900.6500.4001.00
!
```

In Listing 10.21, the change in configuration is shown in bold. The correct configuration for R2 is shown in Listing 10.22.

LISTING 10.22 Correct Configuration of R2

```
!
hostname R2
!
interface Loopback0
    ip address 190.65.14.2 255.255.255.255
!
interface e1
   ip address 190.65.24.2 255.255.255.0
   ip router isis
!
interface s1
  ip address 190.65.28.4 255.255.255.253
```

```
  ip router isis
!
router isis
  passive-interface Loopback0
  net 49.001.1900.6500.4001.00
!
```

In Listing 10.22, the change in configuration is shown in bold.

L1 ROUTER PROBLEM

L1 routers perform routing based on the area ID of the IS-IS packet. If the destination address specified in the packet is not within the same area as the L1 router, the L1 router sends these packets to the nearest L2 router.

Configuring two L1 routers within the same IS-IS area with the same System ID can cause misconfiguration problems in IS-IS. A warning "Possible Duplicate System IDs" is issued and the L1 routers treat each other's LSPs as their own. Each misconfigured L1 router with the same System ID will generate a new LSP with a higher sequence number. This process continues infinitely, resulting in a major performance bottleneck in the L1 routers.

Consider the scenario depicted in Figure 10.1. In this IS-IS network, the same System ID (1900.6500.4001) is configured for both R1 and R2. R1 and R4 function as L1 routers and R2 and R4 are L1-L2 routers. Each router will send LSPs to every other router within Area 49.001 while flooding Hello packets for establishing adjacency relationships.

Both the routers cannot distinguish from which router the current LSP has arrived. R1 and R2 will consider each other's LSPs as their own, because the routers have the same System ID. R1 and R2 increment the sequence numbers infinitely. As a result, the adjacency relationship is never established between R1 and R2. To troubleshoot this adjacency problem in L1 routers:

1. Use the command show isis in R1 and R2 and verify the NET. The output of show isis command in R1 is as shown in Listing 10.23.

LISTING 10.23 Verification of System ID on R1

```
R1#show isis
Global ISIS information
ISIS process tag: R1
System ID: 1900.6500.4001
NET: 49.001.1900.6500.4001.00
Maximum number of areas: 2
There is 1 manual area address
```

```
49
There is 1 active area address
49
ISIS level-1
ISIS is enabled on 2 interfaces
Distance : 50
Maximum of 1 path per route
Number of SPF runs, L1: 4, L2: 0
```

The output of show isis command in R2 is as shown in Listing 10.24.

LISTING 10.24 Verification of System ID on R2

```
R2#show isis
Global ISIS information
ISIS process tag: R2
System ID: 1900.6500.4001
NET: 49.001.1900.6500.4001.00
Maximum number of areas: 2
There is 1 manual area address
49
There is 1 active area address
49
ISIS level-1
ISIS is enabled on 2 interfaces
Distance : 150
Maximum of 1 path per route
Number of SPF runs, L1: 14, L2: 10
```

If the middle fields in the NET are same, the System IDs are identical to 1900.6500.4001.

2. Reconfigure one of the routers with a different System ID using the net command and enable IS-IS protocol in the newly reconfigured router. Let us reconfigure R2 as shown in Listing 10.25.

LISTING 10.25 Reconfiguration of R2

```
hostname R2
!
interface loopback0
   ip address 190.65.4.1  255.255.255.255
!
interface e1
    ip address 190.65.24.1 255.255.255.0
```

```
    ip router isis
!
router isis
  passive-interface Loopback0
  net 49.001.1900.6500.4002.00
!
```

In Listing 10.25, the code in bold shows the reconfigured System ID as 1900.6500.4002. The NET for R1 remains unchanged.

3. Verify the router configuration by using the show isis command to check whether both the routers have the same System ID. The output for the show isis command at R1 is shown in Listing 10.26.

LISTING 10.26 Output for show isis Command at R1

```
R1#show isis
Global ISIS information
ISIS process tag: R1
System ID: 1900.6500.4001
NET: 49.001.1900.6500.4001.00
Maximum number of areas: 2
There is 1 manual area address
49
There is 1 active area address
49
ISIS level-1
ISIS is enabled on 2 interfaces
Distance : 50
Maximum of 1 path per route
Number of SPF runs, L1: 4, L2: 0
```

The output for show isis command at R2 is shown in Listing 10.27.

LISTING 10.27 Output for show isis Command at R2

```
R2#show isis
Global ISIS information
ISIS process tag: R2
System ID: 1900.6500.4002
NET: 49.001.1900.6500.4002.00
Maximum number of areas: 2
There is 1 manual area address
49
```

```
There is 1 active area address
49
ISIS level-1
ISIS is enabled on 2 interfaces
Distance : 150
Maximum of 1 path per route
Number of SPF runs, L1: 14, L2: 10
```

L2 ROUTER PROBLEM

L2 routers perform routing based on the area address in the IS-IS packet and route packets toward the area ID without considering the internal structure of the area. An L2 router is always connected to another L2 router or L1-L2 router in a different area. An L2 router can have neighbors in other areas and has a Level 2 link-state database with all routing information about inter-area routing.

L2 routers form adjacency relationships with only L2 or L1-L2 routers. If an L2 router is forced to form adjacency relationship with an L1 router, it is considered a misconfiguration. Consider the example shown in Figure 10.4.

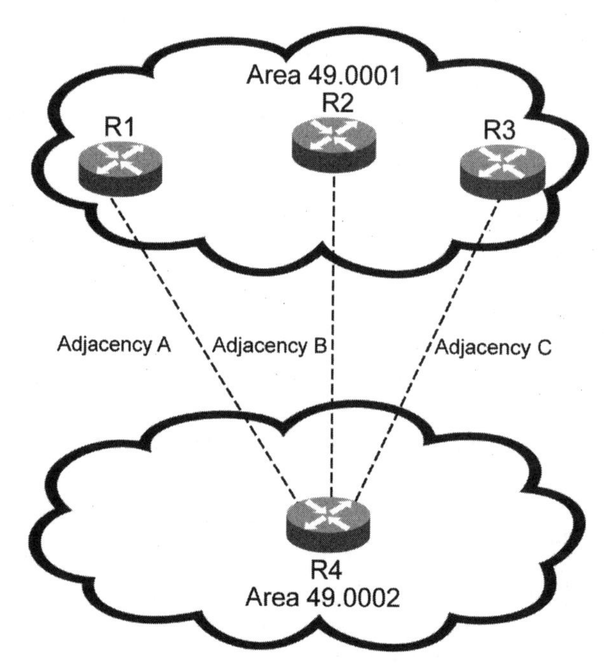

FIGURE 10.4 An IS-IS network depicting an L2 router problem.

In Figure 10.4, there are two areas with Area 49.0001 and Area 49.0002. There are four routers—R1, R2, R3, and R4—in the IS-IS network. R1 is an L1 router, R2 and R4 are L2 routers, and R3 is an L1-L2 router. Three possibilities of adjacency establishment are shown in the figure: Adjacency A between R1 and R4, Adjacency B between R2 and R4, and Adjacency C between R3 and R4. The problem is Adjacency A is illegal, whereas Adjacencies B and C are correct configurations. To troubleshoot the routers in IS-IS networks:

1. Use the show clns neighbor command to check the adjacency relationships of routers. The output of show clns neighbor command at R4 is as shown in Listing 10.28.

LISTING 10.28 Output of show clns neighbor Command at R4

```
R4#show clns neighbor
System Id Interface SNPA          State    Holdtime     Type    Protocol
R1          E1    0000.0b47.c957 Up       14           L1      ISIS
R2          E2    0000.0c34.c324 Up       10           L2      ISIS
R3          E3    0000.0c45.b445 Up        9           L1L2    ISIS
```

In Listing 10.28, Adjacency A established between R1 and R4 is incorrect.

2. Form the correct adjacency relationship with an L2 router or L1-L2 router. The correct configurations are Adjacency B or Adjacency C.
3. Verify the newly formed adjacency on R4 by using the show clns neighbor command. The output of show clns neighbor at R4 is shown in Listing 10.29.

LISTING 10.29 Output of show clns neighbor Command at R4

```
R4#show clns neighbor
System Id Interface SNPA          State    Holdtime     Type    Protocol
R2          E2    0000.0c34.c324  Up       0            L2      ISIS
R3          E3    0000.0c45.b445  Up       9            L1L2    ISIS
```

The show clns neighbor command on any L2 router should give the Type column as either type L2 or L1-L2 but not L1.

REDISTRIBUTION PROBLEM

In IS-IS, L1 and L2 routers maintain a single link-state database, which stores routing information. An L1-L2 router maintains a 2-link-state database. An L1-L2 router propagates information about an L1 area to an L2 area with the help of a 2-link-state database. As a result, full routing is achieved between different IS-IS area levels. This is known as redistribution in IS-IS. Consider the scenario depicted in Figure 10.5 to understand redistribution between IS-IS and other routing protocols.

Figure 10.5 shows part of a network that consists of an IS-IS domain and an OSPF routing domain. Routers R1 and R2 are configured in the IS-IS routing protocol and redistribute routing information to router R4, which is in the OSPF routing domain.

In Figure 10.5, incorrect redistribution configuration will result in loss of routing information, because the routing information does not reach the intended recipient properly. This is known as *route leaking*. To troubleshoot the redistribution problem in this network, configure R2 with the correct router redistribution configuration as shown in Listing 10.30.

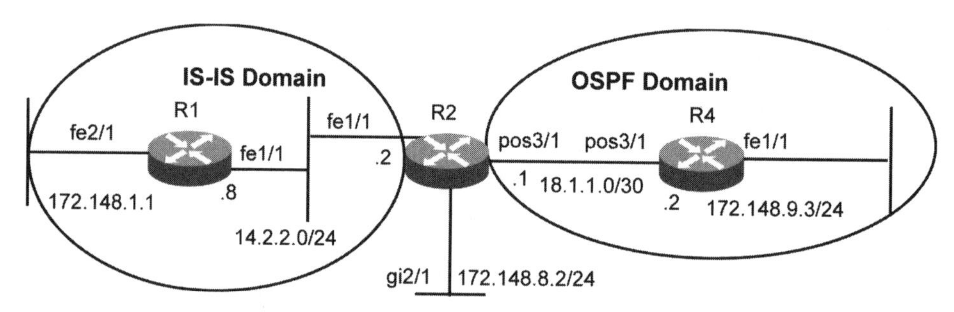

FIGURE 10.5 Redistribution between OSPF and IS-IS routing domains in a network.

LISTING 10.30 Correct Router Configuration at R2

```
interface create ip pos3/1 address-netmask 18.1.1.1/30 port so.3.1
interface create ip fe1/1 address-netmask 14.2.2.2/24 port et.1.1
interface create ip gi2/1 address-netmask 172.148.8.2/24 port gi.2.1
interface add ip lo0 address-netmask 14.2.2.2/32
!
ip-router global set router-id 14.2.2.2
!
ospf create area 0.0.0.0
ospf add interface pos3/1 to-area 0.0.0.0
ospf start
!
isis add area 49.0001
isis add interface fe1/1
isis set system-id 0000.0000.0002
isis start
!
```

```
ip-router policy redistribute from-proto direct to-proto ospf
ip-router policy redistribute from-proto isis-level-2 to-proto
  ospf
ip-router policy redistribute from-proto direct to-proto isis-
  level-2
ip-router policy redistribute from-proto ospf-ase to-proto isis-
  level-2
!
system set name R2
```

In Listing 10.30, the correct configuration for router redistribution policy is shown in bold.

CASE STUDY

Consider an IS-IS network in which there are three ISs: R1, R2, and R3. This scenario is depicted in Figure 10.6

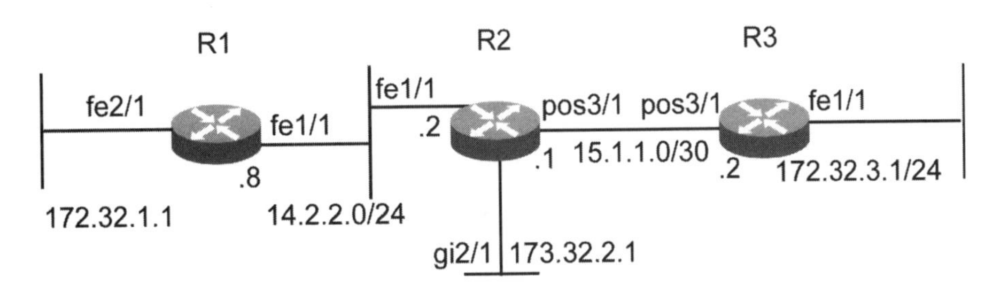

FIGURE 10.6 An IS-IS network scenario.

In Figure 10.6, R1 is connected to R2 using fast Ethernet, fe1. R2 is connected to R3 using interface pos3. The IP addresses of the interfaces are shown in the figure. The IP addresses of the three routers are listed in Table 10.5.

TABLE 10.5 IP Addresses of IS-IS Routers

Router	IP Address
R1	172.32.1.1
R2	172.32.2.1
R3	172.32.3.1

The following problems need to be solved during configuration of individual routers:

- Redistribution of routing information among R1, R2, and R3.
- R1 and R2 are configured with the same System ID.

Redistribution of Routing Information

In the configuration of the routers, the IP-redistribution policy needs to be established so that redistribution of routing information takes place correctly within the same routing domain. The command for redistribution configuration in R1 is shown in Listing 10.31.

LISTING 10.31 Redistribution Configuration in R1

```
interface create ip fe1/1 address-netmask 14.2.2.8/24 port et.1.1
interface create ip fe2/1 address-netmask 172.32.1.1/24 port gi.3.1
!
isis add area 49.0001
isis add interface fe1/1
isis add interface fe2/1
isis set system-id 0000.0000.0001
isis start
!
ip-router policy redistribute from-proto direct to-proto isis-level-1
!
system set name R1
```

In Listing 10.31, for R1, fe1 and fe2 represent the fast Ethernet interfaces that connect R1 to R2, or to any other router in the network. To configure R1 in the IS-IS network:

1. Add the R1 area to the IS-IS network after creating interfaces and assigning IP addresses.
2. Add interfaces to the area, which is added currently. In R1, interfaces fe1 and fe2 are added to Area 49.001.
3. Configure the System ID of R1 as 0000.0000.0001.
4. Enable the IS-IS routing protocol in R1 by using the isis start command.
5. Set R1 for redistribution by using the ip-router policy redistribute command.
6. Set the name R1 by using the system set name command.

The configuration of R1 is complete. R2 and R3 follow similar steps for the configuration in the IS-IS network. The command for redistribution configuration in R2 is shown in Listing 10.32.

LISTING 10.32 Redistribution Configuration in R2

```
interface create ip pos3/1 address-netmask 15.1.1.1/30 port so.3.1
interface create ip fe1/1 address-netmask 14.2.2.2/24 port et.1.1
interface create ip gi2/1 address-netmask 172.32.2.1/24 port gi.2.1
!
isis add area 49.0002
isis add interface pos3/1
isis add interface fe1/1
isis add interface gi2/1
isis set system-id 0000.0000.0002
isis start
!
ip-router policy redistribute from-proto direct to-proto isis-level-1
!
system set name R2
```

The command for redistribution configuration in R3 is shown in Listing 10.33.

LISTING 10.33 Redistribution Configuration in R3

```
interface create ip pos3/1 address-netmask 15.1.1.2/30 port so.3.1
interface create ip fe1/1 address-netmask 172.32.3.1/24 port et.1.1
!
isis add area 49.0003
isis add interface pos3/1
isis set system-id 0000.0000.0003
isis start
!
ip-router policy redistribute from-proto direct to-proto isis-level-1
!
system set name R3
```

R1 and R2 Configured with Same System ID

Consider that in Figure 10.6, the configuration of R1 is as shown in Listing 10.34.

LISTING 10.34 R1 Configuration

```
interface create ip fe1/1 address-netmask 14.2.2.8/24 port et.1.1
interface create ip fe2/1 address-netmask 172.32.1.1/24 port gi.3.1
```

```
!
isis add area 49.0001
isis add interface fe1/1
isis add interface fe2/1
isis set system-id 1720.3200.1001
isis start
!
ip-router policy redistribute from-proto direct to-proto isis-level-1
!
system set name R1
```

Consider that in Figure 10.6, the configuration of R2 is as shown in Listing 10.35.

LISTING 10.35 R2 Configuration

```
interface create ip pos3/1 address-netmask 15.1.1.1/30 port so.3.1
interface create ip fe1/1 address-netmask 14.2.2.2/24 port et.1.1
interface create ip gi2/1 address-netmask 172.32.2.1/24 port gi.2.1
!
isis add area 49.0002
isis add interface pos3/1
isis add interface fe1/1
isis add interface gi2/1
isis set system-id 1720.3200.1001
isis start
!
ip-router policy redistribute from-proto direct to-proto isis-level-1
!
system set name R2
```

The System ID for both routers is 1720.3200.1001. R1 and R2 will send LSPs to each other with the same System ID. While trying to establish adjacency, both R1 and R2 will not be able to distinguish each other's LSPs. No adjacency will be established as a result of this misconfiguration.

To troubleshoot this problem, configure R1, R2, and R3 with three distinct System IDs. The new configuration of R1 is as shown in Listing 10.36.

LISTING 10.36 New Configuration of R1

```
interface create ip fe1/1 address-netmask 14.2.2.8/24 port et.1.1
interface create ip fe2/1 address-netmask 172.32.1.1/24 port gi.3.1
!
isis add area 49.0001
```

```
isis add interface fe1/1
isis add interface fe2/1
isis set system-id 0000.0000.0001
isis start
!
ip-router policy redistribute from-proto direct to-proto isis-level-1
!
system set name R1
```

The new configuration of R2 is as shown in Listing 10.37.

LISTING 10.37 New Configuration of R2

```
interface create ip pos3/1 address-netmask 15.1.1.1/30 port so.3.1
interface create ip fe1/1 address-netmask 14.2.2.2/24 port et.1.1
interface create ip gi2/1 address-netmask 172.32.2.1/24 port gi.2.1
!
isis add area 49.0002
isis add interface pos3/1
isis add interface fe1/1
isis add interface gi2/1
isis set system-id 0000.0000.0002
isis start
!
ip-router policy redistribute from-proto direct to-proto isis-level-1
!
system set name R2
```

The new configuration of R3 is as shown in Listing 10.38.

LISTING 10.38 New Configuration of R3

```
interface create ip pos3/1 address-netmask 15.1.1.2/30 port so.3.1
interface create ip fe1/1 address-netmask 172.32.3.1/24 port et.1.1
!
isis add area 49.0003
isis add interface pos3/1
isis set system-id 0000.0000.0003
isis start
!
ip-router policy redistribute from-proto direct to-proto isis-level-1
!
system set name R3
```

SUMMARY

In this chapter, you learned about the different troubleshooting techniques used in IS-IS routing environments. In the next chapter, you will look at the troubleshooting methods employed in BGP environments.

POINTS TO REMEMBER

- The router#show isis command shows if the IS-IS routing protocol is enabled on a specified router. This command also displays the system level information about IS-IS execution.
- The router#show clns command provides global CLNS information such as the number of interfaces, NET, type of routing (L1, L2, or L1L2), and the routing area ID.
- The router#show clns neighbor command shows the System ID, Interface ID, SNPA, Hold time, IS Type, and protocol used.
- The router#show clns interface interface-name command displays details about the CLNS interface, where interface-name represents name of interface whose details are displayed.
- The router#show clns protocol command displays details of CLNS protocol, including the areas involved with redistribution details.
- The router#show clns traffic command displays details about network traffic sequence in a router.
- The router#show isis database level-1 command displays the Level 1 link-state database.
- The router#show isis database level-2 command displays the Level 2 link-state database.
- The router#show isis spf-log command displays time and reason for the shortest path algorithm execution.
- The debug isis adj-packets command enables IS-IS adjacency packets and displays IS-IS packet information.
- The debug isis spf-events command enables debugging in the Shortest Path Function (SPF) events.
- The debug isis snp-packets command enables and displays CSNP/PSNP packet information.
- The debug isis spf-update command shows SPF updated information.

11 Troubleshooting BGP for Routing Environments

IN THIS CHAPTER

- Problem Isolation in BGP
- BGP Neighbor Relationship
- BGP Route Advertisement
- Missing Routes
- Misconfiguration Problems
- Attribute Problems
- Route Dampening
- Redistribution Problems
- BGP Communities
- BGP Multihoming and Loadsharing

Border Gateway Protocol (BGP) is the standard routing protocol in the Internet. It is used for routing among different Autonomous Systems (ASs). This protocol is an interface between two different forms of administration, which may be different ISPs, companies, or educational institutions. Therefore, troubleshooting BGP problems is more complex than that of any of the Interior Gateway Protocols (IGPs). It requires maximum coordination between two ASs to solve BGP-related problems and ensure error-free routing conforming to decided policies between two ASs.

In this chapter, we'll look at the various issues with BGP and the different methods to troubleshoot these problems.

PROBLEM ISOLATION IN BGP

In the case of all routing scenarios in IP, the basic troubleshooting commands such as ping, traceroute, show interface, show ip interface, and show ip protocols are used extensively for root cause analysis of any problem in a network. There are other protocol-specific commands available in the Cisco IOS, which are used for troubleshooting at a more specific level.

In this section, you will look at the available BGP troubleshooting commands, as well as explanations for each. Table 11.1 lists some of the commonly used show commands associated with BGP.

TABLE 11.1 BGP show Commands and Descriptions

Command	Description
show ip bgp	Displays complete BGP table of a router.
show ip bgp A.B.C.D	Shows corresponding BGP table entry to network A.B.C.D.
show ip bgp cidr-only	Displays routes in BGP table with non-natural network masks.
show ip bgp community	Lists routes in a BGP table that match the named communities.
show ip bgp community-list	Displays routes in a BGP table that match the named community lists.
show ip bgp dampening	Displays detailed information about BGP dampening, if configured in the router.
show ip bgp filter-list	Lists routes matching the named filter list.
show ip bgp inconsistent-as	Displays only the routes with inconsistent origin ASs.
show ip bgp injected-paths	Lists all injected paths.
show ip bgp ipv4	Displays entries in a BGP table corresponding to the IP Version 4 address family.
show ip bgp labels	Lists labels for IPv4 NLRI specific information.
show ip bgp neighbors	Displays detailed information on BGP neighbors.
show ip bgp paths	Lists the BGP path information.
show ip bgp peer-group	Displays information on BGP peer groups.
show ip bgp prefix-list	Lists routes in a BGP table matching a named prefix list.
show ip bgp quote-regexp	Displays routes in a BGP table matching the AS path regular expression.
show ip bgp regexp	Lists routes in a BGP table matching the AS path regular expression.

TABLE 11.1 *(continued)*

Command	Description
show ip bgp replication	Displays replication status of update groups.
show ip bgp route-map	Lists BGP table route entries matching the named route map.
show ip bgp summary	Displays status summary of connections with all BGP neighbors.
show ip bgp template	Lists peer-policy or peer-session templates.
show ip bgp update-group	Displays information on update groups.
show ip bgp vpnv4	Lists VPNv4 NLRI specific information.
show ip bgp neighbor A.B.C.D advertised-routes	Displays routes that form a part of the BGP update sent to neighbor A.B.C.D. This can be either the total BGP table or selected routes satisfying any condition that might be set using an access list or a route map.
show ip bgp neighbor A.B.C.D routes	Lists routes, which are learned from BGP neighbor A.B.C.D and installed at BGP table. This can include all routes that are advertised by the neighbor. This can also include a subset of the same that satisfies certain inbound conditions applied by an access list or a route map.

Consider a BGP network as an example and see the output of the show commands. Figure 11.1 shows part of a BGP network.

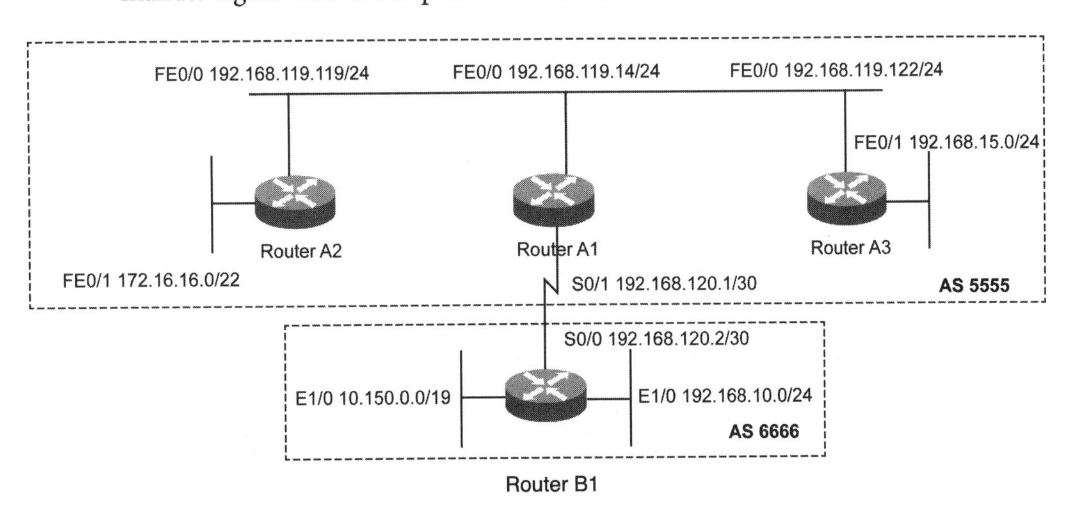

FIGURE 11.1 Part of a BGP network showing routers A1, A2, A3, and B1.

Figure 11.1 shows part of a BGP network. Routers A1, A2, and A3 belong to AS5555 and are running iBGP with each other. Router B1 belongs to AS6666, is connected to A1 via a WAN link, and is running eBGP with it. We will look at the output of some of the commonly used commands listed in Table 11.1 as applicable to the network shown in Figure 11.1. Listing 11.1 shows the output of the show ip bgp command.

LISTING 11.1 Output of show ip bgp Command

```
A1#show ip bgp 10.150.0.0
BGP routing table entry for 10.150.0.0/19, version 20439
Paths: (1 available, best #1)
Multipath: eBGP
  Advertised to update-groups:
     3
  5400, (received & used)
  192.168.120.2 from 192.168.120.2
  Origin IGP, metric 0, localpref 200, valid, internal, best
  Community:
  Originator: 192.168.120.2, Cluster list:
```

Listing 11.2 shows the output of the show ip bgp cidr-only command.

LISTING 11.2 Output of show ip bgp cidr-only Command

```
A1#show ip bgp cidr-only
BGP table version is 26948, local router ID is 192.168.120.1
Status codes: s suppressed, d damped, h history, valid, > best, i -
internal,
S Stale
Origin codes: i - IGP, e - EGP, ? - incomplete
   Network          Next Hop            Metric LocPrf Weight Path
*>i10.150.0.0/19     192.168.0.2                   200      0 6666 i
*>i172.16.16.0/22    192.168.119.119               100      0 i
```

Listing 11.3 shows the output of the show ip bgp community command.

LISTING 11.3 Output of show ip bgp community Command

```
A1#show ip bgp community 0:100
BGP table version is 26948, local router ID is 192.168.120.1
Status codes: s suppressed, d damped, h history, valid, > best, i -
internal,
S Stale
Origin codes: i - IGP, e - EGP, ? - incomplete
```

```
Network              Next Hop            Metric  LocPrf  Weight  Path
*>i192.168.15.0      192.168.119.122     0       100       0     i
```

Listing 11.4 shows the output of the show ip bgp neighbors command.

LISTING 11.4 Output of show ip bgp neighbors Command

```
A1#show ip bgp neighbors 192.168.119.122
BGP neighbor is 192.168.119.122,  remote AS 5555, internal link
  BGP version 4, remote router ID 192.168.119.122
  BGP state = Established, up for 05:24:26
  Last read 00:00:25, hold time is 180, keepalive interval is 60
seconds
  Neighbor capabilities:
    Route refresh: advertised and received(new)
    Address family IPv4 Unicast: advertised and received
  Message statistics:
    InQ depth is 0
    OutQ depth is 0
                         Sent       Rcvd
    Opens:                  9          9
    Notifications:          0          0
    Updates:           259104        259
    Keepalives:          7185       7184
    Route Refresh:          0          0
    Total:             266298       7452
  Default minimum time between advertisement runs is 5 seconds
For address family: IPv4 Unicast
  BGP table version 27219, neighbor versions 27219/0 27219/0
  Output queue sizes: 0 self, 0 replicated
  Index 8, Offset 1, Mask 0x1
  Route-Reflector Client
  Member of update-group 3
  Default weight 100
                                 Sent       Rcvd
  Prefix activity:               —          —
    Prefixes Current:            101         28 (Consumes 1344 bytes)
    Prefixes Total:              133         34
    Implicit Withdraw:            23          0
    Explicit Withdraw:             9          6
    Used as bestpath:            n/a         28
    Used as multipath:           n/a          0

                                   Outbound    Inbound
  Local Policy Denied Prefixes:    ——          ——
```

```
    Suppressed due to dampening:    2          n/a
    Total:                          2          0
  Number of NLRIs in the update sent: max 290, min 0
  Connections established 9; dropped 8
  Last reset 05:24:53, due to User reset
Connection state is ESTAB, I/O status: 1, unread input bytes: 0
Local host: 192.168.119.14, Local port: 19455
Foreign host: 192.168.119.122, Foreign port: 179
Enqueued packets for retransmit: 0, input: 0  mis-ordered: 0 (0 bytes)
Event Timers (current time is 0x19A67740):
Timer          Starts    Wakeups            Next
Retrans          4040          0             0x0
TimeWait            0          0             0x0
AckHold           334        179             0x0
SendWnd             0          0             0x0
KeepAlive           0          0             0x0
GiveUp              0          0             0x0
PmtuAger            0          0             0x0
DeadWait            0          0             0x0
iss:  370882673  snduna:  371469486  sndnxt:  371469486   sndwnd: 15007
irs: 1820013194  rcvnxt: 1820020445  rcvwnd:      14997  delrcvwnd:
  1387
SRTT: 300 ms, RTTO: 607 ms, RTV: 3 ms, KRTT: 0 ms
minRTT: 0 ms, maxRTT: 476 ms, ACK hold: 200 ms
Flags: higher precedence, nagle
Datagrams (max data segment is 1460 bytes):
Rcvd: 4234 (out of order: 0), with data: 334, total data bytes: 7250
Sent: 4413 (retransmit: 0, fastretransmit: 0), with data: 4228, total
data bytes
: 586812
```

Listing 11.5 shows the output of the show ip bgp paths command.

LISTING 11.5 Output of show ip bgp paths Command

```
A1#show ip bgp paths
Address     Hash Refcount Metric Path
0x43A20410    0        4      0 i
0x43D71FA8    0       27      0 i
0x43D73EE0    1        1      0 i
0x43D75FC8    1        1      0 6666 i
0x43D76490    2        2      0 6666 i
0x43D76FD0    3       56      0 6666 i
0x437B9378    4        1      5 i
```

Listing 11.6 shows the output of the show ip bgp summ command.

LISTING 11.6 Output of show ip bgp summ Command

```
A1#show ip bgp summ
BGP router identifier 192.168.119.14, local AS number 5555
BGP table version is 27, main routing table version 27392
7 network entries using 766 bytes of memory
5 path entries using 362 bytes of memory
6 BGP path attribute entries using 101 bytes of memory
2 BGP rrinfo entries using 1 bytes of memory
2 BGP AS-PATH entries using 37 bytes of memory
2 BGP community entries using 23 bytes of memory
1 BGP route-map cache entries using 5 bytes of memory
0 BGP filter-list cache entries using 0 bytes of memory
BGP using 135 total bytes of memory
BGP activity 135382/127865 prefixes, 184428/176882 paths, scan interval
  60 secs
Neighbor        V    AS MsgRcvd MsgSent   TblVer  InQ OutQ Up/Down
State/PfxRcd
192.168.119.122   4  5555  119977    7738    27392    0    0 03:06:26
  7428
192.168.119.119   4  5555    7138    7467    27392    0    0 02:22:55
  8

192.168.120.2     4  6666    7188    7454    27392    0    0 02:22:49
  2
```

Listing 11.7 shows the output of the show ip bgp neighbor A.B.C.D advertised-routes command.

LISTING 11.7 Output of show ip bgp neighbor A.B.C.D advertised-routes Command

```
B1#show ip bgp neighbor 192.168.120.1 advertised-routes
BGP table version is 2330382, local router ID is 192.168.120.2
Status codes: s suppressed, d damped, h history, valid, > best,
i - internal,
   r RIB-failure, S Stale
Origin codes: i - IGP, e - EGP, ? - incomplete
  Network          Next Hop            Metric LocPrf Weight Path
*> 192.168.10.0    192.168.120.2         0
32768 i
*>i 10.150.0.0/19   192.168.120.2            0           32768 i
```

Listing 11.8 shows the output of the show ip bgp neighbor A.B.C.D routes command.

LISTING 11.8 Output of show ip bgp neighbor A.B.C.D routes Command

```
B1#show ip bgp neighbor 192.168.120.1 routes
BGP table version is 2330386, local router ID is 192.168.120.2
Status codes: s suppressed, d damped, h history, valid, > best,
  i - internal,
  r RIB-failure, S Stale
Origin codes: i - IGP, e - EGP, ? - incomplete
  Network          Next Hop            Metric LocPrf Weight Path
  *> 192.168.119.0      192.168.120.1       0        100 5555 i
  *> 192.168.15.0       192.168.120.1       0        100 5555 i
  *> 172.16.16.0/22     192.168.120.1       0        100 5555 i
```

Table 11.2 shows some of the debug commands available in Cisco IOS that can be used to troubleshoot problems with BGP routing.

Consider the BGP network shown in Figure 11.2 as an example and look at output of some of the debug commands as described in Table 11.2.

TABLE 11.2 debug Commands to Troubleshoot BGP Problems

Command	Description
debug ip bgp A.B.C.D updates	Displays flow of packets corresponding to routing updates directed to a BGP neighbor address A.B.C.D.
debug ip bgp dampening	Monitors activities related to dampening of BGP routes.
debug ip bgp events	Displays all BGP-related events associated with a BGP-enabled router.
debug ip bgp in	Displays BGP exchange of information with a neighbor corresponding to routes in updates that are accepted after inbound filtering and installed in a BGP table.
debug ip bgp keepalives	Tracks BGP keepalives that are exchanged among BGP neighbors at a regular interval.
debug ip bgp out	Displays information related to BGP outbound updates.
debug ip bgp updates	Displays flow of packets corresponding to routing updates directed to all BGP peers with whom a neighbor relationship is established.

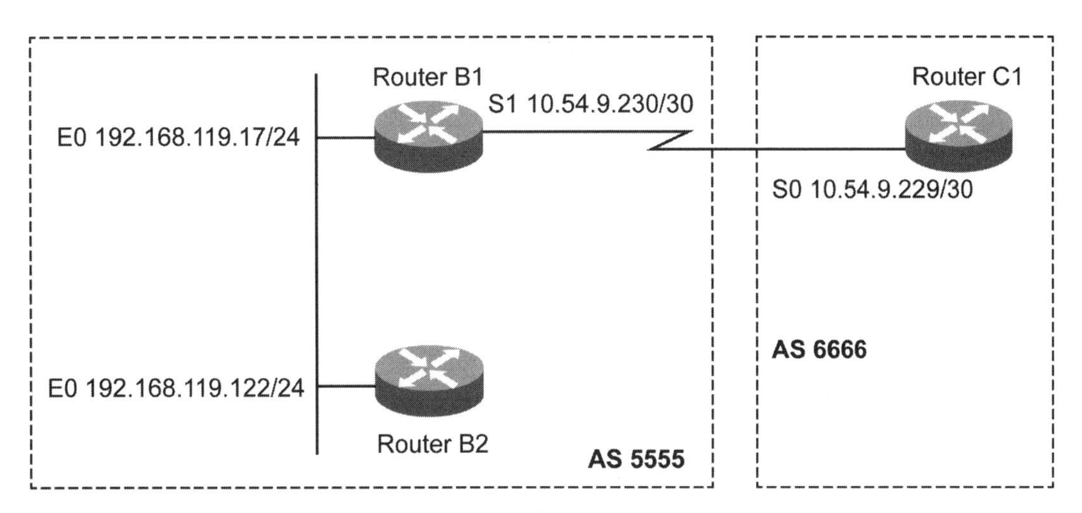

FIGURE 11.2 Part of a BGP network showing routers B1, B2, and C1.

Figure 11.2 shows part of a BGP network in which router B1 of AS5555 is running eBGP with the router C1 of AS6666 and iBGP with router B2 of AS5555. We must run certain debug commands at B1. Listing 11.9 shows output of the debug ip bgp 192.168.119.122 updates command at B1.

LISTING 11.9 Output of debug ip bgp 192.168.119.122 updates Command

```
B1#debug ip bgp 192.168.119.122 updates
BGP updates debugging is on for neighbor 192.168.119.122
B1#
Feb  3 12:23:33.425: BGP(0): 192.168.119.122 rcv UPDATE about
10.54.32.0/24 — withdrawn
Feb  3 12:23:33.525: BGP(0): 192.168.119.122 computing updates, afi 0,
neighbor version 2308872, table version 2308873, starting at 0.0.0.0
Feb  3 12:23:33.525: BGP(0): 192.168.119.122 update run completed, afi 0,
ran for 0ms, neighbor version 2308872, start version 2308873, throttled
to 2308873
B1#
B1#
Feb  3 12:23:38.841: BGP(0): 192.168.119.122 rcvd UPDATE w/ attr:
nexthop 10.54.120.192, origin i, localpref 100, metric 3, originator
192.168.158.243, clusterlist 192.168.120.8 10.54.120.10 10.54.120.192
Feb  3 12:23:38.841: BGP(0): 192.168.119.122 rcvd 10.54.32.0/24
Feb  3 12:23:38.961: BGP(0): 192.168.119.122 computing updates, afi 0,
neighbor version 2308873, table version 2308874, starting at 0.0.0.0
```

```
Feb  3 12:23:38.965: BGP(0): 192.168.119.122 update run completed, afi
0, ran for 4ms, neighbor version 2308873, start version 2308874,
throttled to 2308874
Feb  3 12:23:43.533: BGP(0): 192.168.119.122 rcv UPDATE about
10.54.32.0/24 —withdrawn
Feb  3 12:23:43.633: BGP(0): 192.168.119.122 computing updates, afi 0,
neighbor version 2308874, table version 2308875, starting at 0.0.0.0
Feb  3 12:23:43.633: BGP(0): 192.168.119.122 update run completed, afi
0, ran for 0ms, neighbor version 2308874, start version 2308875,
throttled to 2308875
Feb  3 12:23:49.046: BGP(0): 192.168.119.122 rcvd UPDATE w/ attr:
nexthop 10.54.120.192, origin i, localpref 100, metric 3, originator
192.168.158.243, cluster list 192.168.120.8 10.54.120.10 10.54.120.192
Feb  3 12:23:49.050: BGP(0): 192.168.119.122 rcvd 10.54.32.0/24
Feb  3 12:23:49.150: BGP(0): 192.168.119.122 computing updates, afi 0,
neighbor version 2308875, table version 2308876, starting at 0.0.0.0
Feb  3 12:23:49.150: BGP(0): 192.168.119.122 update run completed, afi 0,
ran for 0ms, neighbor version 2308875, start version 2308876, throttled
to 2308876
B1#
Feb  3 12:23:54.238: BGP(0): 192.168.119.122 rcv UPDATE about
10.54.32.0/24 —withdrawn
Feb  3 12:23:54.338: BGP(0): 192.168.119.122 computing updates, afi 0,
neighbor version 2308876, table version 2308877, starting at 0.0.0.0
Feb  3 12:23:54.338: BGP(0): 192.168.119.122 update run completed, afi 0,
ran for 0ms, neighbor version 2308876, start version 2308877, throttled
to 2308877
Feb  3 12:23:59.302: BGP(0): 192.168.119.122 rcvd UPDATE w/ attr:
nexthop 10.54.120.192, origin i, localpref 100, metric 3, originator
192.168.158.243, clusterlist 192.168.120.8 10.54.120.10 10.54.120.192
Feb  3 12:23:59.302: BGP(0): 192.168.119.122 rcvd 10.54.32.0/24
Feb  3 12:23:59.402: BGP(0): 192.168.119.122 computing updates, afi 0,
neighbor version 2308877, table version 2308878, starting at 0.0.0.0
Feb  3 12:23:59.402: BGP(0): 192.168.119.122 update run completed, afi 0,
ran for 0ms, neighbor version 2308877, start version 2308878, throttled
to 2308878
```

Listing 11.10 shows output of the debug ip bgp events command at B1.

LISTING 11.10 Output of debug ip bgp events Command

```
B1#debug ip bgp events
BGP events debugging is on
B1#
```

```
B1#
Feb  3 12:24:41.537: BGP: Import timer expired. Walking from 1 to 1
B1#
B1#
Feb  3 12:24:56.538: BGP: Import timer expired. Walking from 1 to 1
B1#
B1#
Feb  3 12:25:11.539: BGP: Import timer expired. Walking from 1 to 1
B1#
B1#
B1#
B1#
Feb  3 12:25:26.540: BGP: Performing BGP general scanning
Feb  3 12:25:26.540: BGP(0): scanning IPv4 Unicast routing tables
Feb  3 12:25:26.652: BGP(IPv4 Unicast): Performing BGP Nexthop scanning
for general scan
Feb  3 12:25:26.888: BGP(1): scanning IPv6 Unicast routing tables
Feb  3 12:25:26.888: BGP(IPv6 Unicast): Performing BGP Nexthop scanning
for general scan
Feb  3 12:25:26.888: BGP(2): scanning VPNv4 Unicast routing tables
Feb  3 12:25:26.888: BGP(VPNv4 Unicast): Performing BGP Nexthop
scanning for general scan
Feb  3 12:25:26.888: BGP(3): scanning IPv4 Multicast routing tables
Feb  3 12:25:27.040: BGP(IPv4 Multicast): Performing BGP Nexthop
scanning for general scan
Feb  3 12:25:42.045: BGP: Import timer expired. Walking from 1 to 1
B1#
```

Listing 11.11 shows output of the debug ip bgp keepalives command at B1.

LISTING 11.11 Output of debug ip bgp keepalives Command

```
B1#debug ip bgp keepalives
Feb  3 12:27:36.716: BGP: 192.168.119.122 received KEEPALIVE, length
(excl. header) 0
B1#
B1#
B1#
Feb  3 12:27:40.620: BGP: 192.168.119.122 sending KEEPALIVE (io)
B1#
Feb  3 12:27:43.552: BGP: 10.54.9.229 received KEEPALIVE, length (excl.
  header) 0
B1#
Feb  3 12:27:45.780: BGP: 10.54.9.229 sending KEEPALIVE (io)
```

```
B1#
B1#
Feb  3 12:28:36.715: BGP: 192.168.119.122 received KEEPALIVE, length
  (excl. header) 0
B1#
Feb  3 12:28:40.699: BGP: 192.168.119.122 sending KEEPALIVE (io)
B1#
Feb  3 12:28:43.556: BGP: 10.54.9.229 received KEEPALIVE, length (excl.
  header)0
B1#
Feb  3 12:28:45.860: BGP: 10.54.9.229 sending KEEPALIVE (io)
B1#
```

The triggered events when a BGP neighbor flaps are shown in Listing 11.12. The same events are simulated by manually resetting the BGP connection.

LISTING 11.12 Triggered Events by BGP Neighbor Flap at B1

```
B1#cle ip bgp 10.54.9.229
B1#
B1#
Feb  3 12:36:52.685: BGP: 10.54.9.229 went from Established to Idle
Feb  3 12:36:52.685: %BGP-5-ADJCHANGE: neighbor 10.54.9.229 Down User
  reset
B1#
B1#
Feb  3 12:36:52.685: BGP: 10.54.9.229 closing
Feb  3 12:36:52.685: BGPNSF state: 10.54.9.229 went from nsf_not_active
  to nsf_not_active
Feb  3 12:36:52.689: BGP: 10.54.9.229 went from Idle to Active
Feb  3 12:36:52.689: BGP: 10.54.9.229 open active, delay 17723ms
B1#
B1#
Feb  3 12:37:10.414: BGP: 10.54.9.229 open active, local address
  10.54.9.230
Feb  3 12:37:10.442: BGP: 10.54.9.229 went from Active to OpenSent
Feb  3 12:37:10.442: BGP: 10.54.9.229 sending OPEN, version 4, my as:
4755
Feb  3 12:37:10.750: BGP: 10.54.9.229 rcv message type 1, length (excl.
  header)
22
Feb  3 12:37:10.750: BGP: 10.54.9.229 rcv OPEN, version 4
Feb  3 12:37:10.750: BGP: 10.54.9.229 rcv OPEN w/ OPTION parameter
  len: 12
```

```
Feb  3 12:37:10.750: BGP: 10.54.9.229 rcvd OPEN w/ optional parameter
type 2 (Capability) len 6
Feb  3 12:37:10.750: BGP: 10.54.9.229 OPEN has CAPABILITY code: 1,
length 4
Feb  3 12:37:10.750: BGP: 10.54.9.229 OPEN has MP_EXT CAP for afi/safi:
  1/1
Feb  3 12:37:10.750: BGP: 10.54.9.229 rcvd OPEN w/ optional parameter
type 2 (Capability) len 2
Feb  3 12:37:10.750: BGP: 10.54.9.229 OPEN has CAPABILITY code: 128,
length 0
Feb  3 12:37:10.750: BGP: 10.54.9.229 OPEN has ROUTE-REFRESH
  capability(old) for all address-families
Feb  3 12:37:10.750: BGP: 10.54.9.229 went from OpenSent to OpenConfirm
Feb  3 12:37:10.786: BGP: 10.54.9.229 went from OpenConfirm to
Established
B1#
Feb  3 12:37:10.786: %BGP-5-ADJCHANGE: neighbor 10.54.9.229 Up
```

BGP NEIGHBOR RELATIONSHIP

Formation of a neighbor relationship among the BGP peer routers is the first step to set up eBGP among routers belonging to different ASs or iBGPs among routers belonging to the same AS. A neighbor relationship is formed when there is successful communication at port TCP 179 between the BGP peer routers. Look at an example as shown in Figure 11.3 to troubleshoot the problem of an unsuccessful neighbor relationship.

Figure 11.3 shows routers A1 and C1 belonging to AS5555 and AS6666, respectively, which are connected via parallel serial links. These routers are eBGP peers and require establishment of a successful neighbor relationship to exchange routing updates among the ASs.

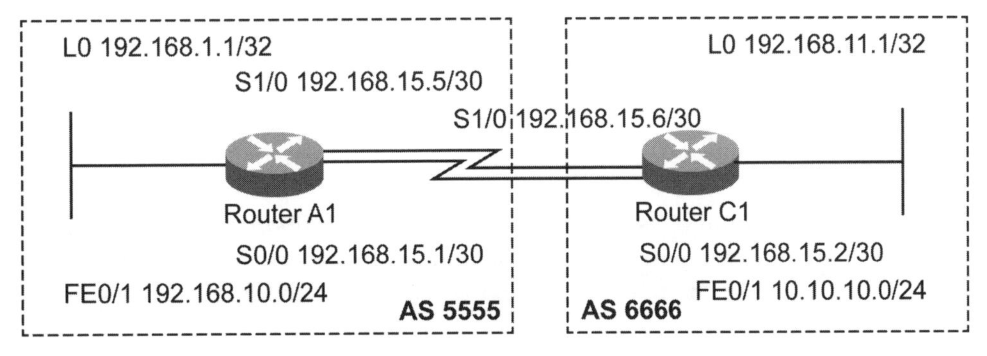

FIGURE 11.3 A BGP network depicting a neighbor relationship problem.

In Figure 11.3, the neighbor relationship is not established between A1 and C1, as ascertained from these commands:

```
A1#show ip bgp summary
A1#show ip bgp neighbor
C1#show ip bgp summary
C1#show ip bgp neighbor
```

To troubleshoot the neighbor relationship problem in BGP:

1. Check the configuration for both A1 and C1 to ascertain whether the neighbors have been defined correctly. Check both the neighbor IP addresses and AS numbers. Configure these commands on A1:

```
router bgp 5555
neighbor 192.168.11.1 remote-as 6666

Configure these command on C1:
router bgp 6666
neighbor 192.168.1.1 remote-as 6666
```

2. Load balance the two neighbors connected over parallel serial links by configuring these commands:

At A1:

```
router bgp 5555
neighbor 192.168.11.1 remote-as 6666
neighbor 192.168.11.1 ebgp-multihop 2
neighbor 192.168.11.1 update-source Loopback0
```

At C1:

```
router bgp 6666
neighbor 192.168.1.1 remote-as 6666
neighbor 192.168.1.1 ebgp-multihop

neighbor 192.168.1.1 update-source Loopback0
```

An ebgp multihop is mandatory, because the neighbors are loopback interfaces and not connected directly. Also, the update source is mandatory in these cases. These commands are not required in the case of directly connected neighbors.

3. Check the neighbor reachability if the commands are already present. Listing 11.13 shows the output of the command to check neighbor reachability.

LISTING 11.13 Checking Neighbor Reachability

```
A1#ping 192.168.11.1
Type escape sequence to abort.
Sending 5, 100-byte ICMP Echos to 202.54.9.1, timeout is 2 seconds:
....
Success rate is 0 percent (0/5)
C1#ping 192.168.11.1
Type escape sequence to abort.
Sending 5, 100-byte ICMP Echos to 202.54.9.1, timeout is 2 seconds:
....
Success rate is 0 percent (0/5)
```

As per the observation, ping is unsuccessful.

4. Check the route to BGP peers at both routers, using the commands:

```
A1#show ip route 192.168.11.1
% Network not in table
C1#show ip route 192.168.1.1
% Network not in table
```

There is no route available. In A1 and C1, configure the command:

```
A1(config)#ip route 192.168.11.1 255.255.255.255 Serial 0/0
A1(config)#ip route 192.168.11.1 255.255.255.255 Serial1/0
C1(config)#ip route 192.168.1.1 255.255.255.255 Serial 0/0
C1(config)#ip route 192.168.1.1 255.255.255.255 Serial1/0
```

5. Check the presence of any access list in the interfaces in both A1 and C1 if ping is unsuccessful. Listing 11.14 shows the output of the show ip int serial 0/0 command.

LISTING 11.14 Output of show ip int serial 0/0 Command

```
A1#show ip int serial 0/0
Serial0 is up, line protocol is up
  Internet address is 192.168.15.1/30
  Broadcast address is 255.255.255.255
  Address determined by non-volatile memory
  MTU is 1500 bytes
  Helper address is not set
  Directed broadcast forwarding is disabled
  Multicast reserved groups joined: 224.0.0.10
```

```
Outgoing access list is 101
Inbound  access list is not set
Proxy ARP is enabled
Security level is default
Split horizon is enabled
ICMP redirects are never sent
ICMP unreachables are always sent
ICMP mask replies are never sent
IP fast switching is enabled
IP fast switching on the same interface is enabled
IP Flow switching is disabled
IP Feature Fast switching turbo vector
IP multicast fast switching is enabled
IP multicast distributed fast switching is disabled
IP route-cache flags are Fast
Router Discovery is disabled
IP output packet accounting is disabled
IP access violation accounting is disabled
TCP/IP header compression is disabled
RTP/IP header compression is disabled
Probe proxy name replies are disabled
Policy routing is disabled
Network address translation is disabled
WCCP Redirect outbound is disabled
WCCP Redirect exclude is disabled
BGP Policy Mapping is disabled
```

An outbound access list applied at serial 0/0 is access list 101. Let us check the contents of the same. Listing 11.15 shows the output of the show access-list 101 command.

LISTING 11.15 Output of show access-list 101 Command

```
A1#show access-list 101
Extended IP access list 101
  0 deny tcp any any eq 179 (500 matches)
  20 deny udp any any eq 179
  30 permit ip any any (724 matches)
```

The access list has statements that deny all packets with the port number 179 to move out of serial 0/0.

A successful neighbor relationship can be established by either removing the access list from the interface or suitably modifying this list.

6. Check the physical link status and quality, and rectify the error, if there is no access list.

BGP ROUTE ADVERTISEMENT

Some networks may not be advertised properly, which results in their non-reachability from networks connected to BGP peers. Consider the example shown in Figure 11.4 to understand and troubleshoot this scenario.

In Figure 11.4, BGP is configured between AS6666 and AS7777. Routers A1 and B1 are BGP peers. A1 and A2 in AS6666 are connected by a serial link with static routing configured. Consider the scenario in which network 192.168.20.0/24 is not getting advertised to AS7777. The command A1#show ip bgp neighbor 10.1.1.1 advertised-routes | include 192.168.20 will yield no output. This confirms that the route is not getting advertised.

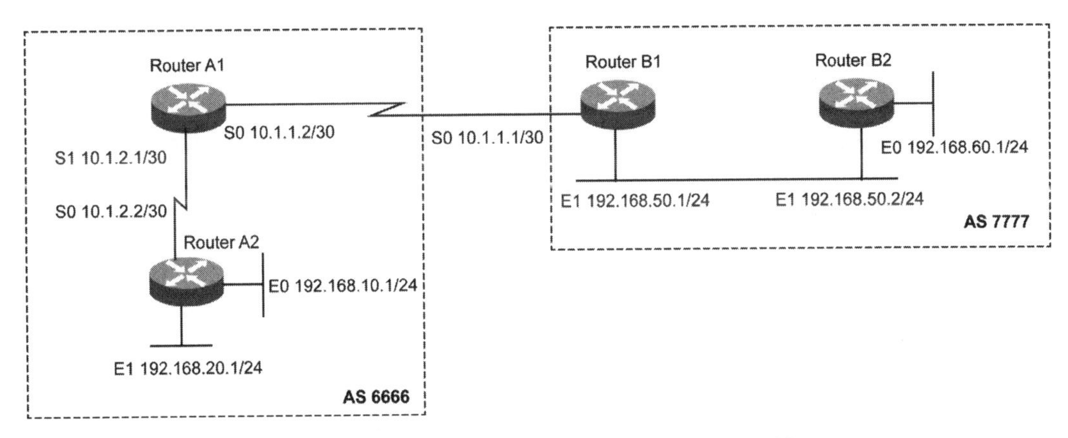

FIGURE 11.4 A BGP network depicting a BGP route advertisement problem.

To troubleshoot this router problem:

1. Check if the route is configured in BGP by using network command. The network command should be present in the configuration of A1. It will be shown as network 192.168.20.0. If the command is not available, the network will not be advertised.
2. Check for the actual presence of the route in the routing table of A1 by using the show ip route command. The output for the show ip route command is as shown in Listing 11.16.

LISTING 11.16 Output of show ip route Command

```
A1#show ip route 192.168.20.0
Routing entry for 192.168.20.0/24
```

```
Known via "static", distance 1, metric 0
Routing Descriptor Blocks:
* 10.1.2.2
    Route metric is 0, traffic share count is 1
```

In Listing 11.16, the route is shown as available. If the route is not available in the routing table, it would not be advertised.

3. Check if the route is part of the BGP table at A1 by using the show ip bgp command. The output of the show ip bgp command is shown in Listing 11.17.

LISTING 11.17 Output of show ip bgp Command

```
A1#show ip bgp 192.168.20.0
BGP routing table entry for 192.168.20.0/24, version 5703
Paths: (1 available, best #1, table Default-IP-Routing Table)
  Advertised to non peer-group peers:
  10.1.1.1 Local
    0.0.0.0 from 0.0.0.0 (10.1.2.1)
      Origin IGP, metric 0, localpref 100, weight 32768, valid,
sourced, local, best
```

In Listing 11.17, the route is found to be part of the BGP table at A1.

4. Check the BGP configuration of BGP peers for any filtering implemented. The relevant part of the BGP configuration of A1 is shown in Listing 11.18.

LISTING 11.18 Configuration of A1

```
router bgp 6666
neighbor  10.1.1.1 remote-as 7777
neighbor  10.1.1.1 route-map NEI out
!
route-map NEI permit 10
match ip address 20
set metric 100
!
access-list 20 192.168.0.0 0.0.255.255
access-list 20 permit any
```

Listing 11.18 shows that the route map NEI has been applied for updates to the BGP peer B1 at A1, which matches routes against access list 20. Access list 20 denies the network 192.168.10.0/24. This causes the nonadvertisement of the route. Suitable modification of the access list will result in the successful advertisement of the route.

MISSING ROUTES

Some destinations may become unreachable due to the absence of BGP routes from the routing table. In this section, you will have a look at a similar scenario with the help of an example and steps to overcome the problem. Figure 11.5 shows BGP running between AS5555 and AS6666.

In Figure 11.5, B1 does not have any route to network 10.10.1.0/24. This is evident from the commands shown:

```
B1#show ip route 10.10.1.1
% Network not in table
```

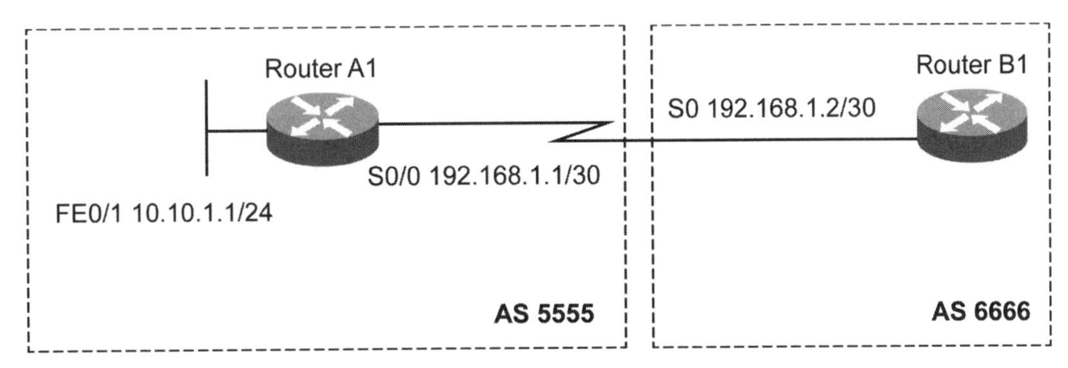

FIGURE 11.5 BGP running between A1 and B1 in AS5555 and AS6666.

To troubleshoot this missing route problem:

1. Check the existence of the path in the BGP table by using the command:

```
B1#show ip bgp 10.10.1.1
% Network not in table
```

The command output shows that there is no corresponding entry in the BGP table.

2. Check the BGP neighbor relation establishment with A1. If the neighbor relationship is not established, execute the troubleshooting steps for the same.

3. Check if route 10.10.1.1/24 is being learned from A1 after the neighbor relationship is established by using the command:

```
B1#show ip bgp neighbor 192.168.1.1 routes | inc 10.10
```

4. From the command output, it is established that the route has not been learned from A1.
5. Check for any filtering or route map present at the peer routers, which could cause nonentry of this route in the BGP table. Listing 11.19 shows the relevant part of the configuration that is checked using the command show running-config.

LISTING 11.19 Output of show running-config Command at A1

```
router bgp 5555
neighbor 192.168.1.2 remote-as 6666
neighbor 192.168.1.2 distribute-list 1 in
access-list 1 deny 10.10.1.0
access-list 1 permit any
```

Listing 11.20 shows the output of the command show running-config in B1.

LISTING 11.20 Output of show running-config Command at B1

```
router bgp 6666
neighbor 192.168.1.2 remote-as 5555
neighbor 192.168.1.2 distribute-list 1 out
access-list 1 deny 10.10.1.0
access-list 1 permit any
```

In Listings 11.19 and 11.20, access list 1 shown in bold may be present in either of the peer routers. This list is used for route filtering and can result in route 10.10.1.0/24 not being available in the BGP table of B1. You can remove or modify the access list as required to solve the problem.

6. Check the route advertisement at A1 if no routes are learned from B1. First, check the presence of the route in the BGP table of A1 by using the command:

```
A1#show ip bgp 10.10.1.0
% Network not in table
```

The command output shows that the route is not available in the BGP table. The advertisement needs to be checked. For this, execute show running-configuration and examine the relevant parts of the output. As 10.10.1.0/24 is not a classful network, the advertisement command should be in the form:

```
router bgp 5555
network 10.10.1.0 mask 255.255.255.0
```

7. Check the availability of the route in the routing table if the network command is defined properly. This is performed using the command:

```
A1#show ip route 10.10.1.0
```

MISCONFIGURATION PROBLEMS

Misconfiguration or unintentional error in configuration is a common reason for problems in BGP routing. In this section, we will discuss the common misconfiguration issues and ways to detect and correct them. Figure 11.6 shows the scenario illustrating a misconfiguration problem.

In Figure 11.6, misconfiguration can be caused because:

- Numbers may be incorrectly defined.
- Networks may not be declared.

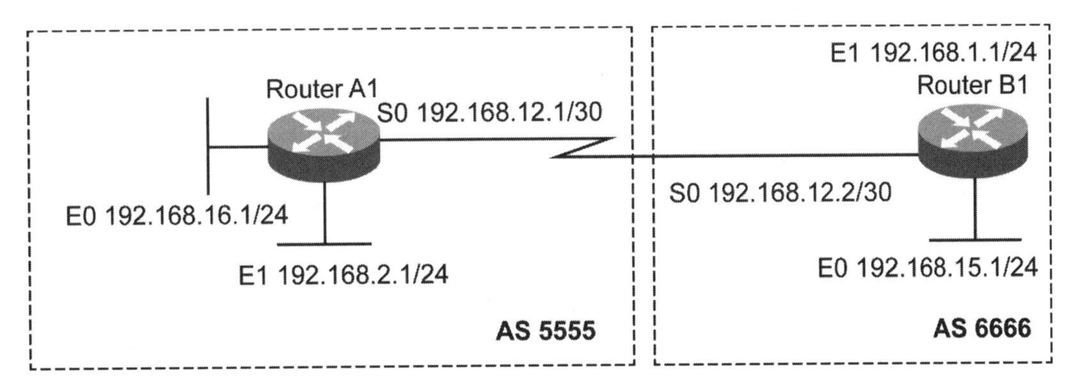

FIGURE 11.6 A BGP network showing misconfiguration problems between AS5555 and AS6666.

Incorrectly Defined Numbers

In Figure 11.6, consider the configurations for A1 and B1. The configuration of A1 is:

```
router bgp 5555
neighbor 192.168.12.2 remote-as 5555
```

The configuration for B1 is:

```
router bgp 6666
neighbor 192.168.12.1 remote-as 5555
```

In this case, the neighbor relationship will not be established between the routers. The corrected configuration for A1 is:

```
router bgp 5555
neighbor 192.168.12.2 remote-as 6666
```

The correct configuration for B1 is:

```
router bgp 6666
neighbor 192.168.12.1 remote-as 5555
```

Undeclared Networks

In Figure 11.6, let's say the configuration of A1 is:

```
router bgp 5555
neighbor 192.168.12.2 remote-as 6666
network 192.168.2.0
```

The configuration of B1 is:

```
router bgp 6666
neighbor 192.168.12.2 remote-as 5555
network 192.168.2.0
network 192.168.16.0
```

In this case, not all the directly available networks in the routing table of A1 have been included in the BGP table. As a result, the networks that are not declared would not be reachable via BGP from any direct BGP peer or any other BGP-compatible router that is a part of the Internet. The corrected configuration for A1 is:

```
router bgp 5555
neighbor 192.168.12.2 remote-as 6666
network 192.168.2.0
network 192.168.16.0
```

The correct configuration for B1 is:

```
router bgp 6666
neighbor 192.168.12.2 remote-as 5555
```

```
network 192.168.2.0
network 192.168.16.0
```

ATTRIBUTE PROBLEMS

BGP routes are characterized by a number of attributes, based upon which the best path to a destination is selected by using the BGP path selection algorithm. The BGP attributes are AS_path, Origin, Next Hop, Weight, Local Preference, Multi Exit Discriminator, and Community.

An incorrect setting of the attribute values can lead to nonoptimal path selection in BGP. Figure 11.7 illustrates an example to show the effect of problems related to BGP attributes and ways to detect and correct them. The figure shows a BGP network with three different ASs, AS5555, AS6666, and AS7777. AS5555-AS6666, AS5555-AS7777, and AS6666-AS7777 are running on eBGP. Routers B1 and B2 in AS6666 are running on iBGP.

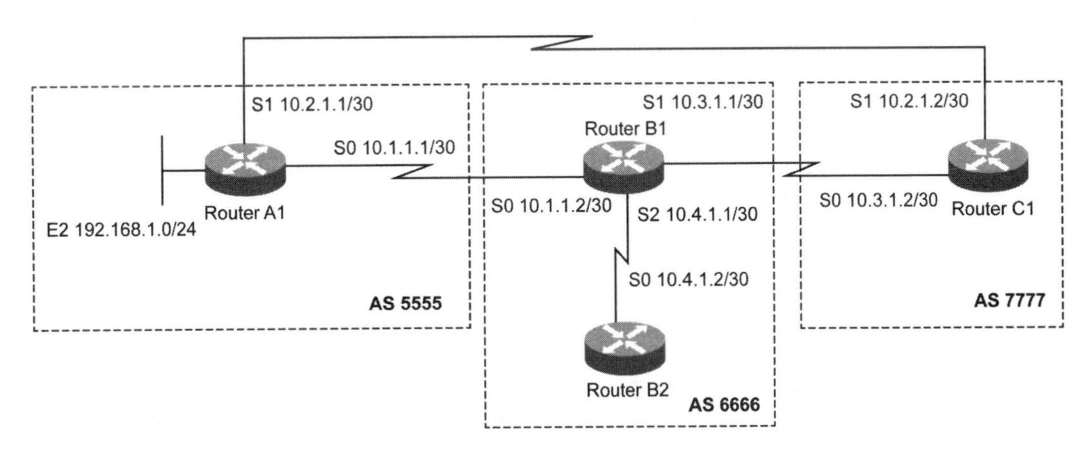

FIGURE 11.7 A BGP network depicting problems related to attributes.

In Figure 11.7, the route to the network 192.168.1.0/24 of AS5555 at C1 in AS7777 is via B1 in AS6666 and not via the direct connectivity to AS5555, as shown in Listing 11.21.

LISTING 11.21 Output of show ip route 192.168.1.0 Command

```
C1#show ip route 192.168.1.0
Routing entry for 192.168.1.0/24
  Known via "bgp 7777", distance 200, metric 0
  Tag 4657, type internal
  Last update from 10.3.1.1 01:12:49 ago
```

```
Routing Descriptor Blocks:
* 10.3.1.1, from 10.3.1.1, 01:12:49 ago
    Route metric is 0, traffic share count is 1
    AS Hops 1
```

The optimal path should be directly through AS5555. To obtain the optimal path:

1. Check if the neighbor relationship is established with A1 of AS5555.
2. Check if route 192.168.1.0/24 has been learned from A1 by using the command

```
C1#show ip bgp neighbor 10.2.1.1 routes | include 192.168.1.0
```

If the route has not been learned, execute proper troubleshooting steps as provided in a previous section. If the route has been learned from both B1 and A1, check the attributes.

3. Check the BGP configuration at C1 corresponding to neighbors B1 and A1 for any weight attribute that might be set. You can do this by using the show running-config command. The relevant part of the configuration is shown in Listing 11.22.

LISTING 11.22 Configuration at C1

```
router bgp 7777
neighbor 10.3.1.1 remote-as 6666
neighbor 10.3.1.1 weight 200
neighbor 10.2.1.1 remote-as 5555
neighbor 10.2.1.1 weight 100
```

In Listing 11.22, the bold code lines show that the weight for routes from neighbor 10.3.1.1 is set to a higher value of 200 and the routes from neighbor 10.2.1.1 are set to 100. As a result, the path via B1 is preferred for all routes.

4. Check the other attribute setting if the weight attribute is set to the same value for both neighbors or the default value is maintained.
5. Check the local preference attribute by using the show running-config command and checking the relevant parts, as shown in Listing 11.23.

LISTING 11.23 Relevant Configuration at C1

```
router bgp 7777
neighbor 10.3.1.1 remote-as 6666
neighbor 10.3.1.1 route-map LOCAL1 in
```

```
neighbor 10.2.1.1 remote-as 5555
neighbor 10.2.1.1 route-map LOCAL2 in
route-map LOCAL1 permit 10
match ip address 1
set local-preference 500
route-map LOCAL2 permit 10
match ip address 1
set local-preference 200
access-list 1 permit any
```

In Listing 11.23, a higher local preference value of 500 is set to routes learned from B1, and therefore, B1 is preferred.

6. Check the AS path attribute if there is no change in the local preference attribute. The configuration of A1 and B1 is to be checked to find if the path is being advertised with any AS path prepend. The relevant part of the configuration for A1 is shown in Listing 11.24.

LISTING 11.24 Configuration at A1

```
router bgp 5555
neighbor 10.2.1.2 remote-as 7777
neighbor 10.2.1.2 route-map AS1 out

route-map AS2 permit 10
match ip address 1
set as-path prepend 5555 5555 5555
```

The relevant part of the configuration for B1 is shown in Listing 11.25.

LISTING 11.25 Configuration at B1

```
router bgp 6666
neighbor 10.3.1.2 remote-as 7777
```

In Listings 11.24 and 11.25, the AS path corresponding to route 192.168.1.0/24 learned from the B1 is 6666 5555. This path is shorter than that learned from A1, which is 5555 5555 5555 5555. Therefore, the path via Router B1 is preferred.

ROUTE DAMPENING

The stability of a BGP route is measured in terms of the number of times it flaps in a defined time interval. The higher the number of flaps, the less stable a route is. To prevent instability in the global routing tables due to flapping of any particular route, BGP suppresses advertisement of less stable routes until they become more stable. The method of route-flap dampening takes care of this problem. This is illustrated in Figure 11.8.

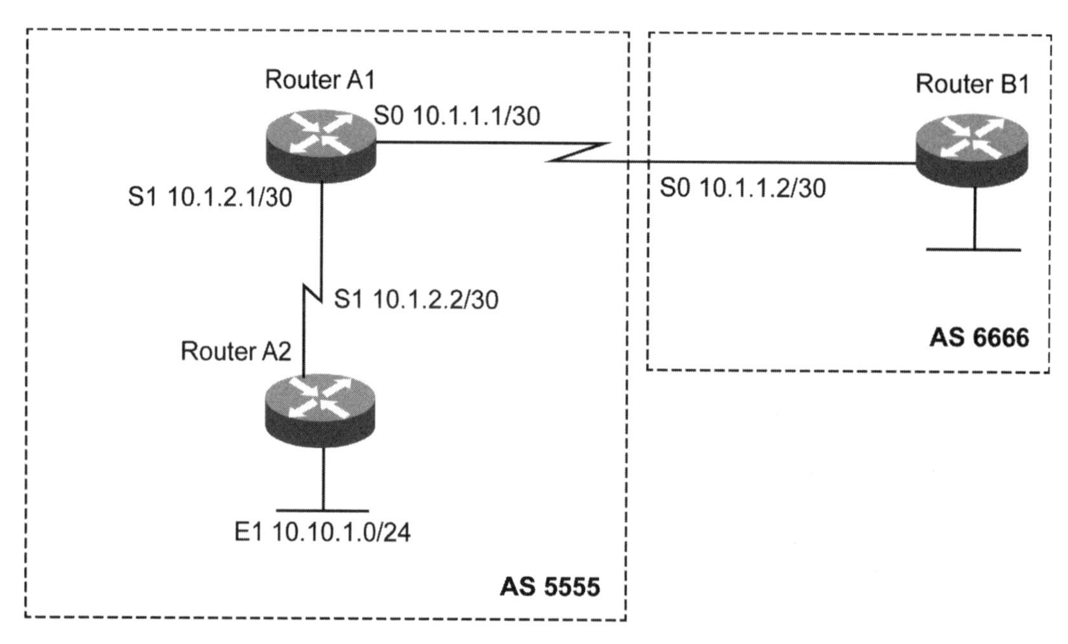

FIGURE 11.8 BGP route dampening occurs between AS5555 and AS6666.

In Figure 11.8, BGP is running between AS5555 and AS6666. AS5555 has routers A1 and A2, which are connected via a serial link with static routing configured. Route 10.10.1.0/24 is flapping due to flapping of the physical link between A1 and A2, as shown in these commands:

```
A1#show ip route | include, 00:00
S       172.16.1.0 [1/0] via 10.1.2.2, 00:00:35
```

The flapping of route 10.10.1.0/24 at A1 causes flapping in the routing table of B1 and also for any other AS to which AS6666 may be connected. To prevent this flapping, check BGP route flap dampening at A1 by using the command:

```
A1#show ip bgp dampening parameters
% dampening not enabled.
```

This command output shows that BGP dampening is not enabled at A1. Enable route flap dampening by using the commands:

```
A1(config)#router bgp 5555
A1(config-router)#bgp dampening
```

The route starts accumulating a penalty of 1000 with each flap, as shown in Listing 11.26.

LISTING 11.26 Output of show ip bgp 10.10.1.0/24 Command

```
A1#show ip bgp 10.10.1.0/24
BGP routing table entry for 10.10.1.0/24, version 76
Paths: (1 available, no best  path)
5555(history entry)
from 10.1.2.2(10.1.2.2)
Origin IGP, metric 0, external
Dampinfo: penalty 1000, flapped 1 times in 0:03:23
```

When the penalty exceeds the suppress limit after a few flaps, the route is suppressed as shown in Listing 11.27.

LISTING 11.27 Output of show ip bgp 10.10.1.0/24 Command

```
A1#show ip bgp 10.10.1.0/24
BGP routing table entry for 10.10.1.0/24, version 82
Paths: (1 available, no best  path)
5555(suppressed due to dampening)
from 10.1.2.2(10.1.2.2)
Origin IGP, metric 0, external
Dampinfo: penalty 2500, flapped 3 times in 15:20:02, reuse in 00:40:30
```

The route is suppressed due to frequent flaps and would be re-advertised after the penalty value is less than the reuse limit. The penalty value is reduced by half in each subsequent half-life period if there are no flaps, and then can fall below the reuse limit.

REDISTRIBUTION PROBLEMS

In this section, we will discuss the redistribution problems encountered in BGP. Consider the example depicted in Figure 11.9.

In Figure 11.9, BGP is running between router A1 of AS5555 and router B1 of AS6666. Routers A1 and A2 have static routes configured between them. Redistribution of static routes is configured at A1. Suppose route 192.168.1.0/24 is not available in the routing table of B1 of AS6666.

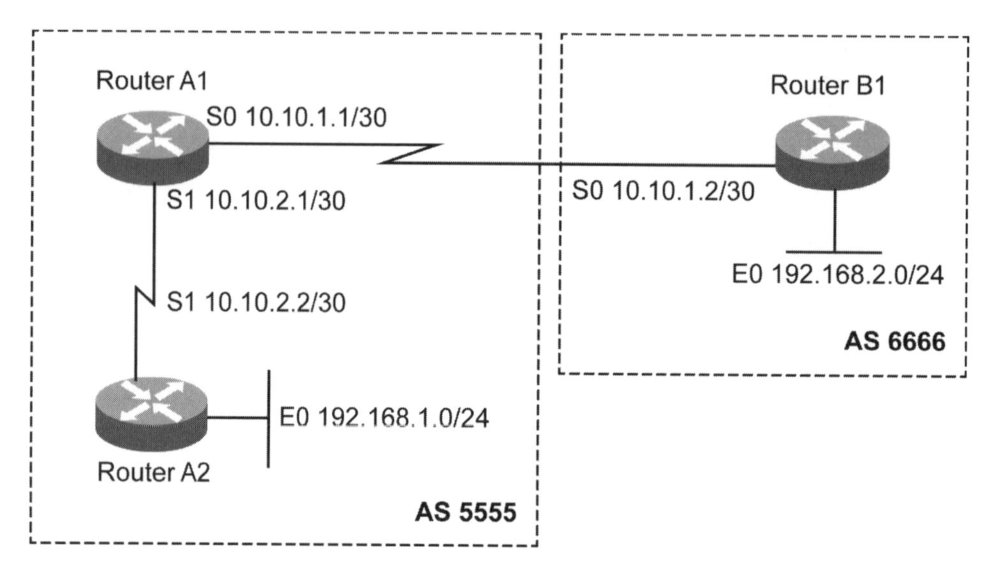

FIGURE 11.9 A BGP network showing a redistribution problem.

To troubleshoot this problem:

1. Check the availability of route 192.168.1.0/24 at A1 if any static route is configured. This is performed by using the show ip route command. The output of show ip route is shown in Listing 11.28.

LISTING 11.28 Output of show ip route Command at A1

```
A1#show ip route 192.168.1.0
Routing entry for 192.168.1.0/24
  Known via “static”, distance 1, metric 0
  Redistributing via bgp 5555
  Routing Descriptor Blocks:
  10.10.2.2
      Route metric is 0, traffic share count is 1
```

2. Check the presence of the network in the BGP table of A1 using the command

```
A1#show ip bgp 192.168.1.0
% Network not in table
```

The network is not available in the BGP table, which indicates that it is not declared in BGP.

3. Check the redistribution, because static routes are used to advertise the networks and not the network command. The relevant part of the configuration of A1 is shown in Listing 11.29.

LISTING 11.29 Configuration of A1

```
router bgp 5555
neighbor 10.10.1.2 remote-as 6666
redistribute static route-map redis-stat
!
route-map redis-stat permit 10
match ip address 15
set metric 10
!
access-list 15 permit 10.10.0.0 0.0.0.255
```

Listing 11.29 shows that redistribution of static routes into BGP is controlled by using the route map redis-stat, which allows routes matching access list 15. Network 192.168.1.0/24 does not match access list 15 and is not redistributed.

The modified access list configuration, which would allow the route 192.168.1.0/24 to be redistributed into BGP, is as shown:

```
access-list 15 permit 10.10.0.0 0.0.0.255
access-list 15 permit 192.168.1. 0.0.0.255
```

BGP COMMUNITIES

To discuss the troubleshooting scenario in BGP communities, consider the example shown in Figure 11.10. In the figure, part of a BGP-enabled network is shown.

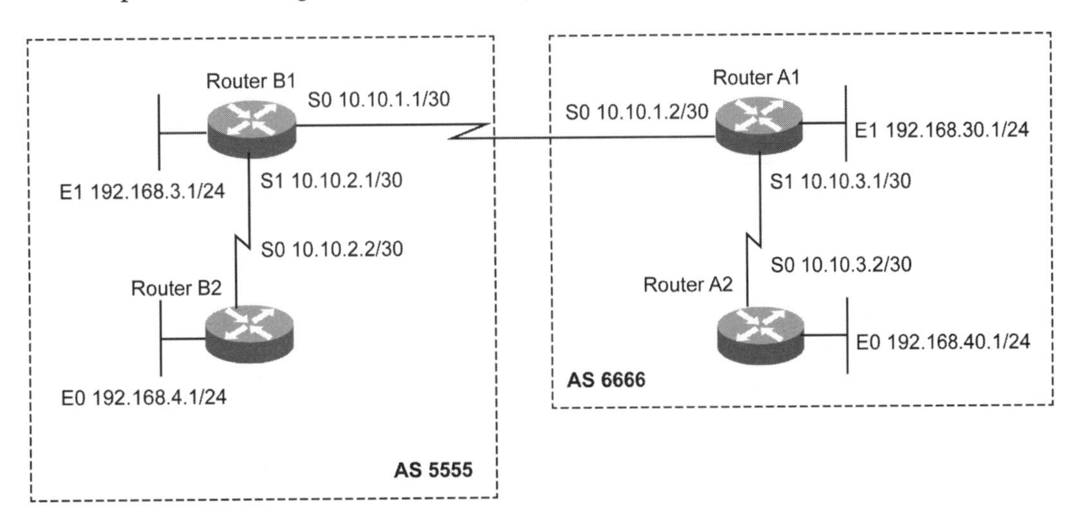

FIGURE 11.10 A BGP network depicting the BGP communities' problem.

BGP is running between router B1 of AS5555 and router A1 of AS6666. iBGP is running within B1 and B2 of AS5555, and A1 and A2 of AS6666.

In Figure 11.10, the network is configured with the help of communities. Consider a scenario in which network 192.168.4.0/24 is not to be advertised to AS6666. It is found that network 192.168.4.0/24, which was to be restricted from being advertised to AS6666, is available in the BGP table of A1 and A2.

To troubleshoot this problem, check if the proper community has been associated with network 192.168.4.0/24. The relevant part of the configuration of B2 is shown in Listing 11.30.

LISTING 11.30 Configuration of B1

```
router bgp 6666
neighbor 10.10.2.1 remote-as 6666
neighbor 10.10.2.1 route-map COMM out
!
route-map COMM permit 10
match ip address 15
set community no-export
!
access-list 15 permit 192.168.4.0
```

The community attribute of the route 192.168.4.0/24 needs to be set as no-export to prevent the advertisement to eBGP peers, which is A1 in this case.

The community attribute is not carried by default in BGP updates. In order for it to be carried, the neighbor 10.10.2.1 send-community command must be present in the BGP configuration.

BGP MULTIHOMING AND LOADSHARING

To discuss the troubleshooting of BGP multihoming and loadsharing scenarios, consider the example depicted in Figure 11.11. In the figure, AS6666 is multihomed to AS7777 and AS8888. Router A1 is connected via serial links to routers B1 and C1. Both C1 and A1 are advertising the default route to A1.

In Figure 11.11, you see that load balancing between the two outbound paths is not happening. In outbound load balancing:

- If the default route is advertised by both B1 and C1, the weight attribute of updates received from both the ISPs can be set to the same value so that load balancing occurs.

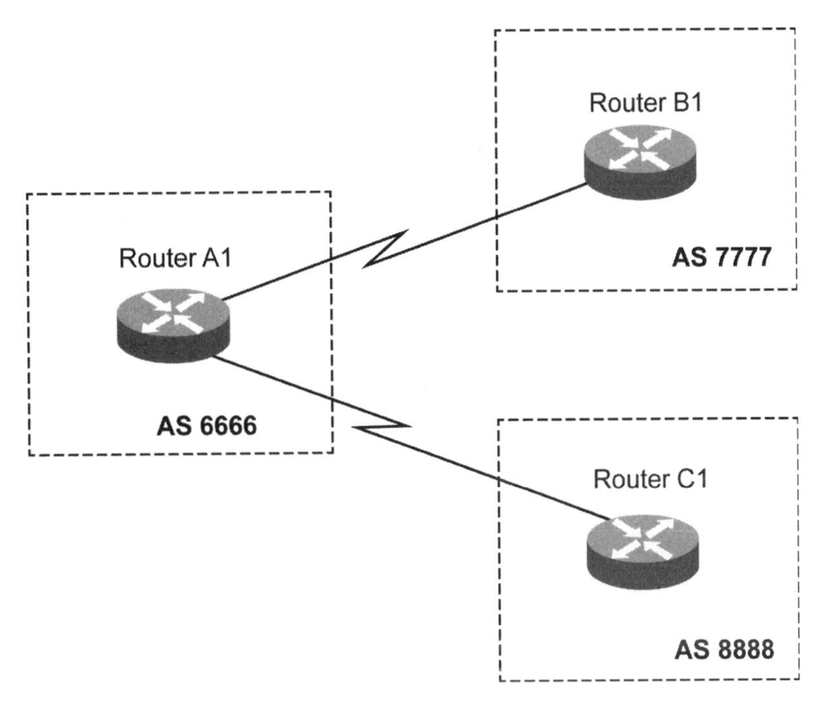

FIGURE 11.11 A BGP network depicting multihoming and loadsharing problems.

- If, in AS6666, the links to the ISPs corresponding to AS7777 and AS8888 are terminated in different routers and not in the same router, the local preference attribute instead of the weight attribute can be used to achieve loadsharing.

In order to achieve inbound load balancing:

- Different prefixes are advertised to the BGP peers belonging to different ASs to which the subject AS is multihomed. In this case, mutual backup can be configured by use of AS path prepend.
- If there are multiple links to the same ISP, terminate them in the same router on both sides. Configure peering between the loopback addresses on both sides. The route to the loopback address on either side is statically configured pointing to the physical links. The type of load balancing in this case depends on the type of switching enabled in the router. In the case of process switching, per packet load balancing is achieved, fast switching results in per destination load balancing. In case of CEF, loadsharing can be achieved per source/destination pair. In this case, load balancing can be achieved in both the inbound and outbound direction.

CASE STUDY

Consider the BGP network shown in Figure 11.12. There are three ASs, AS5555, AS6666, and AS7777. Router B1 of AS6666 is running BGP with router A1 of AS5555 and router C1 of AS7777. Routers A2 of AS5555 and C2 of AS7777 are also running eBGP among themselves.

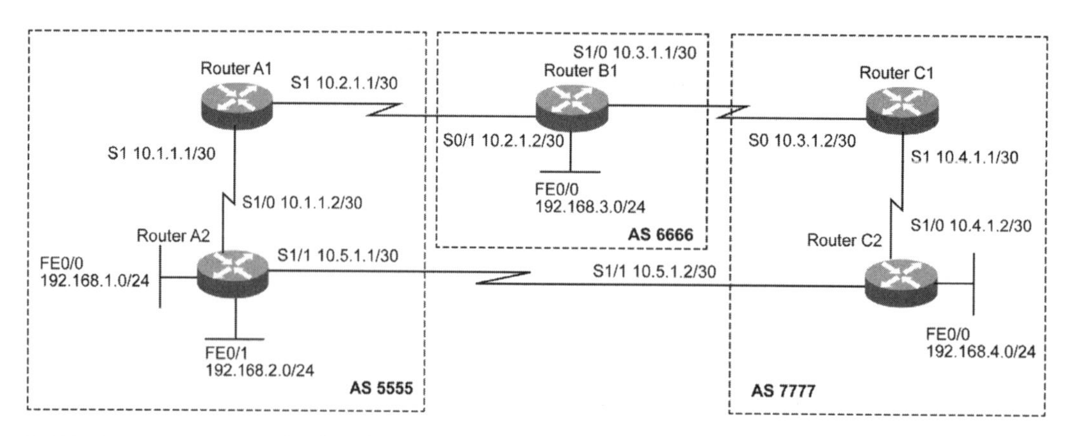

FIGURE 11.12 A scenario showing a complex BGP network.

In Figure 11.12:

1. Route 192.168.4.0/24 is not available in the routing table of B1. This is evident with the command:

```
B1#show ip route 192.168.4.0
% Network not in table
```

2. The route for network 192.168.1.0/24 from B1 is not via A1, which is the optimum path, but via C1. Listing 11.31 shows the output of the show ip route 192.168.1.0 command.

LISTING 11.31 Output of show ip route 192.168.1.0 Command

```
B1#show ip route 192.168.1.0
Routing entry for 192.168.1.0/24
  Known via "bgp 4755", distance 200, metric 0
  Tag 4657, type internal
  Last update from 10.5.1.1 01:12:49 ago
  Routing Descriptor Blocks:
  10.5.1.1, from 10.3.1.2, 01:12:49 ago
      Route metric is 0, traffic share count is 1
      AS Hops 1
```

To troubleshoot these problems, the neighbor relationship at B1 is established and checked. Listing 11.32 shows the output of the show ip bgp neighbor command.

LISTING 11.32 Output of show ip bgp neighbor Command

```
B1#show ip bgp neighbor
BGP neighbor is 10.2.1.1,  remote AS 5555, external link
  BGP version 4, remote router ID 10.2.1.1
  BGP state = Established, up for 05:24:26
  Last read 00:00:25, hold time is 180, keepalive interval is 60 seconds
    Neighbor capabilities:
  Route refresh: advertised and received(new)
  Address family IPv4 Unicast: advertised and received
  Message statistics:
    InQ depth is 0
    OutQ depth is 0
                         Sent       Rcvd
  Opens:                    9          9
  Notifications:            0          0
  Updates:             259104        259
  Keepalives:            7185       7184
  Route Refresh:            0          0
  Total:               266298       7452
  Default minimum time between advertisement runs is 5 seconds
 For address family: IPv4 Unicast
  BGP table version 27219, neighbor versions 27219/0 27219/0
  Output queue sizes: 0 self, 0 replicated
  Index 8, Offset 1, Mask 0x1
  Route-Reflector Client
  Member of update-group 3
  Default weight 100
                          Sent      Rcvd
 Prefix activity:         —         —
  Prefixes Current:       101       28 (Consumes 1344 bytes)
  Prefixes Total:         133       34
  Implicit Withdraw:      23        0
  Explicit Withdraw:      9         6
  Used as bestpath:       n/a       28
  Used as multipath:      n/a       0

                                   Outbound     Inbound
 Local Policy Denied Prefixes:     ————         ————
    Suppressed due to dampening:   2            n/a
    Total:                         2            0
  Number of NLRIs in the update sent: max 290, min 0
```

```
  Connections established 9; dropped 8
  Last reset 05:24:53, due to User reset
Connection state is ESTAB, I/O status: 1, unread input bytes: 0
Local host: 10.2.1.2, Local port: 19455
Foreign host: 10.2.1.1, Foreign port: 179
Enqueued packets for retransmit: 0, input: 0  mis-ordered: 0 (0 bytes)
Event Timers (current time is 0x19A67740):
Timer          Starts    Wakeups            Next
Retrans          4040          0             0x0
TimeWait            0          0             0x0
AckHold           334        179             0x0
SendWnd             0          0             0x0
KeepAlive           0          0             0x0
GiveUp              0          0             0x0
PmtuAger            0          0             0x0
DeadWait            0          0             0x0
iss:  370882673  snduna:  371469486  sndnxt:  371469486   sndwnd: 15007
irs: 1820013194  rcvnxt: 1820020445  rcvwnd:      14997  delrcvwnd:
  1387
SRTT: 300 ms, RTTO: 607 ms, RTV: 3 ms, KRTT: 0 ms
minRTT: 0 ms, maxRTT: 476 ms, ACK hold: 200 ms
Flags: higher precedence, nagle
Datagrams (max data segment is 1460 bytes):
Rcvd: 4234 (out of order: 0), with data: 334, total data bytes: 7250
Sent: 4413 (retransmit: 0, fastretransmit: 0), with data: 4228, total
  data bytes
: 586812
BGP neighbor is 10.3.1.2,  remote AS 7777, external link
   BGP version 4, remote router ID 10.3.1.2
          BGP state = Established, up for 05:24:26
  Last read 00:00:25, hold time is 180, keepalive interval is
  60 seconds
        Neighbor capabilities:
    Route refresh: advertised and received(new)
    Address family IPv4 Unicast: advertised and received
  Message statistics:
  InQ depth is 0
  OutQ depth is 0
                      Sent       Rcvd
  Opens:                 9          9
  Notifications:         0          0
```

```
  Updates:            259104        259
  Keepalives:           7185       7184
  Route Refresh:           0          0
  Total:              266298       7452
  Default minimum time between advertisement runs is 5 seconds
 For address family: IPv4 Unicast
  BGP table version 27219, neighbor versions 27219/0 27219/0
  Output queue sizes: 0 self, 0 replicated
  Index 8, Offset 1, Mask 0x1
  Route-Reflector Client
  Member of update-group 3
  Default weight 100
                                 Sent       Rcvd
Prefix activity:                 —          —
  Prefixes Current:               101         28 (Consumes 1344 bytes)
  Prefixes Total:                 133         34
  Implicit Withdraw:               23          0
  Explicit Withdraw:                9          6
  Used as bestpath:               n/a         28
  Used as multipath:              n/a          0
                                     Outbound    Inbound
Local Policy Denied Prefixes:        ————        ————
  Suppressed due to dampening:       2           n/a
  Total:                             2           0
  Number of NLRIs in the update sent: max 290, min 0
  Connections established 9; dropped 8
  Last reset 05:24:53, due to User reset
Connection state is ESTAB, I/O status: 1, unread input bytes: 0
Local host: 10.3.1.1, Local port: 19455
Foreign host: 10.3.1.2, Foreign port: 179
Enqueued packets for retransmit: 0, input: 0  mis-ordered: 0 (0 bytes)
Event Timers (current time is 0x19A67740):
Timer          Starts    Wakeups            Next
Retrans          4040          0             0x0
TimeWait            0          0             0x0
AckHold           334        179             0x0
SendWnd             0          0             0x0
KeepAlive           0          0             0x0
GiveUp              0          0             0x0
PmtuAger            0          0             0x0
DeadWait            0          0             0x0

iss:  370882673  snduna:  371469486  sndnxt:  371469486     sndwnd:
  15007
```

```
irs: 1820013194  rcvnxt: 1820020445  rcvwnd:      14997  delrcvwnd:
  1387
SRTT: 300 ms, RTTO: 607 ms, RTV: 3 ms, KRTT: 0 ms
minRTT: 0 ms, maxRTT: 476 ms, ACK hold: 200 ms
Flags: higher precedence, nagle
Datagrams (max data segment is 1460 bytes):
Rcvd: 4234 (out of order: 0), with data: 334, total data bytes: 7250
Sent: 4413 (retransmit: 0, fastretransmit: 0), with data: 4228, total
  data bytes
: 586812
```

In Listing 11.32, the command in bold shows the successful establishment of a neighbor relationship of B1 with both A1 and C1. The neighbor relationship between iBGP neighbors C1 and C2 is checked. Listing 11.33 shows the output of the show ip bgp neighbor 10.4.1.2 command.

LISTING 11.33 Output of show ip bgp neighbor 10.4.1.2 Command

```
C1#show ip bgp neighbor 10.4.1.2
BGP neighbor is 10.4.1.2,  remote AS 7777, internal link
BGP version 4, remote router ID 10.2.1.1
  BGP state = Established, up for 03:24:26
Last read 00:00:25, hold time is 180, keepalive interval is 60 seconds
Neighbor capabilities:
    Route refresh: advertised and received(new)
    Address family IPv4 Unicast: advertised and received
  Message statistics:
    InQ depth is 0
    OutQ depth is 0
                         Sent       Rcvd
    Opens:                  9          9
    Notifications:          0          0
    Updates:           259104        259
    Keepalives:          7185       7184
    Route Refresh:          0          0
    Total:             266298       7452
  Default minimum time between advertisement runs is 5 seconds
  For address family: IPv4 Unicast
  BGP table version 27219, neighbor versions 27219/0 27219/0
  Output queue sizes: 0 self, 0 replicated
  Index 8, Offset 1, Mask 0x1
  Route-Reflector Client
  Member of update-group 3
  Default weight 100
```

```
                                 Sent       Rcvd
  Prefix activity:               ----       ----
    Prefixes Current:             101         28 (Consumes 1344 bytes)
    Prefixes Total:               133         34
    Implicit Withdraw:             23          0
    Explicit Withdraw:              9          6
    Used as bestpath:             n/a         28
    Used as multipath:            n/a          0
                                   Outbound    Inbound
  Local Policy Denied Prefixes:    --------    -------
    Suppressed due to dampening:          2        n/a
    Total:                                2          0
  Number of NLRIs in the update sent: max 290, min 0
Connections established 9; dropped 8
  Last reset 05:24:53, due to User reset
Connection state is ESTAB, I/O status: 1, unread input bytes: 0
Local host: 10.4.1.2, Local port: 19455
Foreign host: 10.4.1.1, Foreign port: 179
Enqueued packets for retransmit: 0, input: 0  mis-ordered: 0 (0 bytes)
Event Timers (current time is 0x19A67740):
Timer          Starts    Wakeups            Next
Retrans          4040          0             0x0
TimeWait            0          0             0x0
AckHold           334        179             0x0
SendWnd             0          0             0x0
KeepAlive           0          0             0x0
GiveUp              0          0             0x0
PmtuAger            0          0             0x0
DeadWait            0          0             0x0
iss:  370882673  snduna:  371469486  sndnxt:  371469486   sndwnd: 15007
irs: 1820013194  rcvnxt: 1820020445  rcvwnd:      14997  delrcvwnd:
  1387
SRTT: 300 ms, RTTO: 607 ms, RTV: 3 ms, KRTT: 0 ms
minRTT: 0 ms, maxRTT: 476 ms, ACK hold: 200 ms
Flags: higher precedence, nagle
Datagrams (max data segment is 1460 bytes):
Rcvd: 4234 (out of order: 0), with data: 334, total data bytes: 7250
Sent: 4413 (retransmit: 0, fastretransmit: 0), with data: 4228, total
data bytes
: 586812
```

In Listing 11.33, the establishment of a neighbor relationship between C1 and C2 is confirmed. At C1, check if network 192.168.4.0/24 is declared at B1. This is performed by using the command:

```
C1#show ip bgp neighbor 10.3.1.1 advertised-routes | include
  192.168.4.0
```

There is no output for this command confirming that the route is not declared at B1. The presence of any filters or route map in the BGP configuration of C1 needs to be checked. The relevant part of the configuration of C1, obtained by executing the command show running-configuration, is shown in Listing 11.34.

LISTING 11.34 Output of show running-configuration Command at C1

```
router bgp 7777
neighbor 10.3.1.1 remote-as 6666
neighbor 10.3.1.1 distribute-list 8 out
neighbor 10.4.1.1 remote-as 7777
!
access-list 8 deny 192.168.4.0 0.0.0.255

access-list 8 permit any
```

Access list 8 is used to filter outgoing BGP updates from C1 to A1. Network 192.168.4.0/24 is matched by access list 8. As a result, the update of the network is not available from C1.

The update should be received by B1 from AS5555 as well, which is also connected to AS7777 (via the WAN link between A2 and C2). Check any filtering of the outgoing routing updates from A2 to C2. The relevant part of the configuration of C2, obtained by executing the command show running-configuration, is shown in Listing 11.35.

LISTING 11.35 Output of show running-configuration Command

```
router bgp 7777
neighbor 10.5.1.1 remote-as 5555
neighbor 10.5.1.1 distribute-list 9 out
neighbor 10.4.1.2 remote-as 7777
!
access-list 9 deny 192.168.4.0 0.0.7.255
access-list 9 permit any
```

In Listing 11.35, propagation of the update corresponding to network 192.168.4.0/24 is restricted by access list 9.

Modify the route filtering at either C1 or C2—or both—so that the route for network 192.168.4.0/24 is available in the BGP table and in the routing table of B1. Confirm establishment of a neighbor relationship of B1 with A1.

Updates for network 192.168.1.0/24 should be available at B1 from both A1 and C1. Check if any weight has been applied to the updates. The relevant part of the configuration of B1 obtained by running the command show running-config in B1 is shown in Listing 11.36.

LISTING 11.36 Output of show running-config Command

```
router bgp 6666
neighbor 10.2.1.1 remote-as 5555
neighbor 10.2.1.1 route-map A1IN in
neighbor 10.3.1.2 remote-as 7777
neighbor 10.3.1.2 route-map C1IN in
!
route-map A1IN permit 10
match ip address 10
set weight 100
!
route-map C1IN permit 10
match ip address 10
set weight 200
!
access-list 10 permit any
```

In Listing 11.36, the weight of all routes from C1 is set to a higher value of 200 than the routes from A1, which are set to 100. As a result, C1 is the preferred path for all routes, which are learned both from A1 and C1. The order can be reversed by suitably modifying the weight attribute.

SUMMARY

In this chapter, we learned about problem isolation in BGP environments. We also reminded you about resolution of the problems in BGP environments. In the next chapter, we move on to troubleshooting redistribution routing environments.

POINTS TO REMEMBER

- Border Gateway Protocol (BGP) is the standard routing protocol in the Internet, which is used for routing among different Autonomous Systems.
- The show ip bgp command displays the complete BGP table of a router.
- The show ip bgp A.B.C.D command displays the complete BGP table of a router.

- The show ip bgp cidr-only command displays routes in the BGP table with non-natural network masks.
- The show ip bgp community command lists routes in a BGP table that match the named communities.
- The show ip bgp community-lists command displays routes in the BGP table that match the named community lists.
- The show ip bgp dampening command displays detailed information about BGP dampening, if configured in the router.
- The show ip bgp filter-list command lists routes matching the named filter list.
- The show ip bgp inconsistent-as command displays only the routes with inconsistent origin ASs.
- The show ip bgp injected-paths command lists all injected paths.
- The show ip bgp ipv4 command displays entries in a BGP table corresponding to the IP Version 4 address family.
- The show ip bgp labels command lists labels for IPv4 NLRI specific information.
- The show ip bgp neighbors command displays detailed information on BGP neighbors.
- The show ip bgp paths command lists the BGP path information.
- The show ip bgp peer-group command displays information on BGP peer groups.
- The show ip bgp prefix-list command lists routes in a BGP table matching the named prefix list.
- The show ip bgp quote-regexp command displays routes in a BGP table matching the AS path regular expression.
- The show ip bgp replication command displays replication status of update groups.
- The show ip bgp route-map command lists BGP table route entries matching the named route map.
- The show ip bgp summary command displays the status summary of connections with all BGP neighbors.
- The show ip bgp template command lists peer-policy/peer-session templates.
- The show ip bgp update-group command displays information on update groups.
- The show ip bgp vpnv4 command lists VPNv4 NLRI specific information.
- The show ip bgp neighbor A.B.C.D advertised-routes command displays routes that form a part of the BGP update sent to the neighbor A.B.C.D.

12 Troubleshooting Redistribution Routing Environments

IN THIS CHAPTER

- EIGRP and OSPF Redistribution Environment
- RIP and OSPF Redistribution Environment
- Static Routing and the OSPF Redistribution Environment
- Static Routing and the RIP Redistribution Environment
- IS-IS and OSPF Redistribution Environment

Redistribution among routing protocols is common in the current networking environment. You need to configure redistribution in scenarios such as mergers and acquisition of companies, which might require combining diverse networks, expanding step by step, or incorporating equipment from multiple vendors.

Redistribution leads to sharing routing information among various routing protocols, a phenomenon that does not happen by default. The routing tables become more complex after incorporating redistribution. As a result, redistribution requires detailed study of the network, careful planning, and proper decision on the points of redistribution and the routes to be redistributed. The common problems due to redistribution are:

Probable routing loops: Arise due to inconsistencies in the routing states of different routers present in an interconnected network. These loops result in packet loss and delays in packet transmission.

Nonoptimal path selection: In the case of an optimal routing system, the path through which data traverses is considered the best available path. This path is selected by routers to achieve faster data packet delivery. In the case of nonoptimal path selection, the data delivery time will be longer, resulting in delay and less network performance.

To prevent problems related to redistribution, use route maps and distribution lists in order. There are no generic Cisco IOS commands to troubleshoot all types of redistribution scenarios. For troubleshooting redistribution related issues, a thorough understanding of the network is a primary requirement. The commands used in troubleshooting redistribution problems are generic IP routing-related commands, such as show ip route, show ip interface, show interface, show ip protocols, show running-configuration, ping, and traceroute.

Later in the chapter we will discuss different redistribution scenarios, problems that can arise in them, and ways to troubleshoot the problems.

EIGRP AND OSPF REDISTRIBUTION ENVIRONMENT

Consider Figure 12.1, which shows a redistribution environment. Here, part of a network is shown that has both EIGRP and OSPF routing protocols. Routers B2, B3, and B4 are running EIGRP exclusively. Routers B5, B6, and B7 are running OSPF only. Router A1 is running both routing protocols, OSPF and EIGRP, and is redistributing between these protocols.

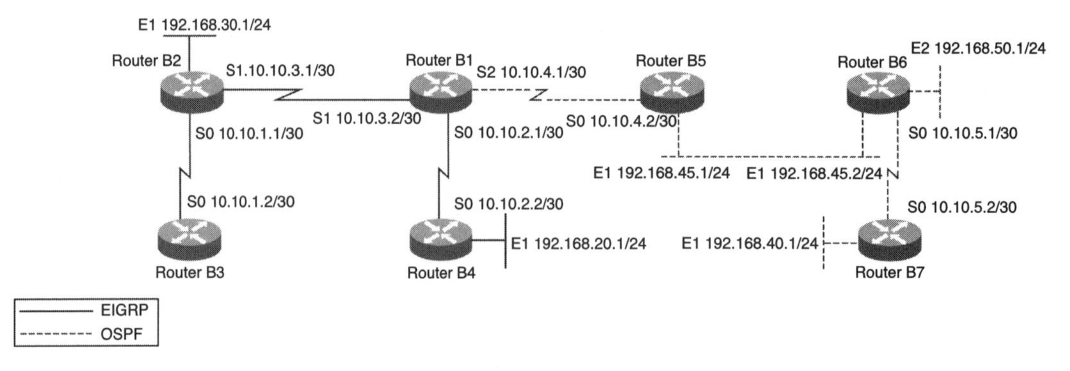

FIGURE 12.1 Redistribution scenario with routing protocols EIGRP and OSPF.

Table 12.1 lists the configuration of all seven routers B1, B2, B3, B4, B5, B6, and B7.

TABLE 12.1 Router Configurations

Router	Relevant Part of Configuration
B1	interface Serial0 ip address 10.10.2.1 255.255.255.252 ! interface Serial1 ip address 10.10.3.2 255.255.255.252 ! interface Serial2

TABLE 12.1 *(continued)*

Router	Relevant Part of Configuration
	ip address 10.10.4.1 255.255.255.252
	!
	router eigrp 100
	network 10.10.2.0
	network 10.10.3.0
	redistribute ospf 1 route-map redis-1
	!
	router ospf 1
	network 10.10.4.0 0.0.0.255 area 0
	redistribute eigrp 100 metric 1000
	!
	route-map redis 10 permit 10
	match ip address 15
	set metric 10000 100 255 1 1000
	!
	access-list 15 permit 192.168.40.0 0.0.7.255
	access-list 15 permit 10.10.0.0 0.0.0.255
	!
B2	interface Serial0
	ip address 10.10.1.1 255.255.255.252
	!
	interface Serial1
	ip address 10.10.3.1 255.255.255.252
	!
	interface Ethernet1
	ip address 192.168.30.1 255.255.255.224
	!
	router eigrp 100
	network 10.10.1.0
	network 10.10.3.0
	network 192.168.30.0
	!
B3	interface Serial0
	ip address 10.10.1.2 255.255.255.252
	!
	router eigrp 100
	network 10.10.1.0
	!

(continued)

TABLE 12.1 *(continued)*

Router	Relevant Part of Configuration
B4	interface Serial0 ip address 10.10.2.2 255.255.255.252 ! interface Ethernet1 ip address 192.168.20.1 255.255.255.252 ! router eigrp 100 network 10.10.2.0 network 192.168.20.0 !
B5	interface Serial0 ip address 10.10.4.2 255.255.255.252 ! interface Ethernet1 ip address 192.168.45.1 255.255.255.0 ! router ospf 1 network 10.10.4.0 0.0.0.255 area 0 network 192.168.45.0 0.0.0.255 area 0 !
B6	interface Serial0 ip address 10.10.5.1 255.255.255.252 ! interface Ethernet1 ip address 192.168.45.2 255.255.255.0 ! interface Ethernet2 ip address 192.168.50.1 255.255.255.0 ! router ospf 1 network 10.10.5.0 0.0.0.255 area 0 network 192.168.45.0 0.0.0.255 area 0 network 192.168.50.0 0.0.0.255 area 0 !
B7	interface Serial0 ip address 10.10.5.2 255.255.255.252 ! interface Ethernet1

TABLE 12.1 *(continued)*

Router	Relevant Part of Configuration
	ip address 192.168.40.1 255.255.255.252 ! router ospf 1 network 10.10.5.0 0.0.0.255 area 0 network 192.168.40.0 0.0.0.255 area 0 !

In Figure 12.1, consider these problems:

1. Network 192.168.30.0/27 is unavailable in the routing table of B7, as shown in the output.

```
B7#show ip route 192.168.30.0
%Network not in table
```

2. Network 192.168.50.0/24 is unavailable in the routing table of B3, as shown in the output.

```
B3#show ip route 192.168.50.0
%Network not in table
```

Network 192.168.30.0/27 Unavailable at B7

To isolate and resolve this problem:

1. Check B2 to see if interface 192.168.30.1/27 is up, because this is the only interface corresponding to network 192.168.30.0/27. Listing 12.1 displays output of the show ip interface brief command for B2.

LISTING 12.1 Output of the show ip interface brief Command

```
B2#show ip interface brief
Interface        IP-Address       OK? Method     Status       Protocol
Ethernet1        192.168.30.1     YES NVRAM      up           up
Serial0          10.10.1.1        YES NVRAM      up           up
Serial1          10.10.3.1        YES NVRAM      up           up
```

Listing 12.1 Shows that the interface is up.

2. Check if routing for this interface is taking place in EIGRP. Listing 12.2 displays output of the show ip protocols command for B2.

LISTING 12.2 Output of the show ip protocols Command

```
B2#show ip protocols
Routing Protocol is "eigrp 100"
  Outgoing update filter list for all interfaces is not set
  Incoming update filter list for all interfaces is not set
  Default networks flagged in outgoing updates
  Default networks accepted from incoming updates
  EIGRP metric weight K1=1, K2=0, K3=1, K4=0, K5=0
  EIGRP maximum hopcount 100
  EIGRP maximum metric variance 1
  Redistributing:
  Automatic network summarization is in effect
  Maximum path: 4
  Routing for Networks:
   10.10.1.0
   10.10.3.0
   192.168.30.0
  Passive Interface(s):
    Routing Information Sources:
    Gateway          Distance      Last Update
    10.10.3.2            90         00:05:35
    10.10.1.2            90         00:05:20
   Distance: internal 90 external 170
```

The output shows that routing is happening for the EIGRP network.

3. Check if route 192.168.30.0/27 is available in the routing table of B1, where redistribution is occurring. Listing 12.3 displays output of the show ip route command for B1.

LISTING 12.3 Output of the show ip route Command

```
B1#show ip route 192.168.30.0/27
Routing entry for 192.168.30.0/27
Known via "eigrp 100", distance 90, metric 41075200, type internal
Last update from 10.10.3.1 on Serial0, 01:01:05 ago
Routing Descriptor Blocks:
*10.10.3.1, from 10.10.3.1, 01:01:05 ago, via Ethernet1
    Route metric is 41075200, traffic share count is 1
    Total delay is 42000 microseconds, minimum bandwidth is 64 Kbit
```

```
Reliability 255/255, minimum MTU 1500 bytes
Loading 3/255, Hops 3
```

Listing 12.3 shows that the route is available in the routing table of B1. The route is being propagated all over the EIGRP domain and is unavailable only in the OSPF domain.

4. Check the redistribution configuration of EIGRP routes into OSPF at B1. The relevant part of the configuration is shown in Listing 12.4.

LISTING 12.4 Part of Configuration to Check Redistribution

```
!
router ospf 1
network 10.10.4.0 0.0.0.255 area 0
redistribute eigrp 100 metric 1000
!
```

In Listing 12.4, no mention is made of the subnets of the primary classful networks. The address 192.168.30.0/27 is a subnet of 192.168.30.0/24 and is not getting redistributed, because the subnets keyword is not mentioned in redistribution. The corrected configuration commands are shown in Listing 12.5.

LISTING 12.5 Corrected Configuration for Redistribution of 192.168.30.0/27

```
!
router ospf 1
network 10.10.4.0 0.0.0.255 area 0
redistribute eigrp 100 metric 1000 subnets
!
```

Now, the route will be available as shown in Listing 12.6.

LISTING 12.6 Output of the show ip route Command for B7

```
B7#show ip route 192.168.30.0
Routing entry for 192.168.30.0/27
Known via "ospf 1", distance 110, metric 3, type extern 1
Last update from 10.10.5.1 on Serial0, 01:12:50 ago
* Routing Descriptor Blocks:
  10.10.5.1, from 192.168.45.1, 01:12:50 ago, via Ethernet1
   Route metric is 3, traffic share count is 1
```

Network 192.168.50.0/24 Unavailable at B3

To isolate and resolve this problem:

1. Check B6 to see if interface 192.168.50.1/24 is up, which is the only interface corresponding to network 192.168.50.0/24. Listing 12.7 displays output of the show ip interface brief command for B6.

LISTING 12.7 Output of the show ip interface brief Command

```
B6#show ip interface brief | inc Ethernet2
Ethernet2              192.168.50.1      YES NVRAM  up                up
```

This output shows that the interface is up.

2. Check if routing for this interface is taking place in OSPF. Listing 12.8 displays output of the show ip protocols command for B6.

LISTING 12.8 Output of the show ip protocols Command

```
B6#show ip protocols
Routing Protocol is "ospf 1"
Outgoing update filter list for all interfaces is not set
Incoming update filter list for all interfaces is not set
Router ID 192.168.50.1
Number of areas in this router is 1. 1 normal 0 stub 0 nssa
Maximum path: 6
Routing for Networks:
  192.168.50.0 0.0.0.255 area 0
  192.168.45.0 0.0.0.255 area 0
  10.10.5.0 0.0.0.255 area 0
  Passive Interface(s):
  Routing Information Sources:
  Gateway          Distance       Last Update
  192.168.45.1          110        00:00:00
  10.10.5.2             110        00:00:00
Distance: (default is 110)
```

Listing 12.8 shows that routing is happening for the OSPF network.

3. Check if the route is available in the routing table of B1, where redistribution is happening. Listing 12.9 displays output of the show ip route command for B1.

LISTING 12.9 Output of the show ip route Command

```
B1#show ip route 192.168.50.1
Routing entry for 192.168.50.0/24
  Known via "ospf 1", distance 110, metric 3, type internal
  Last update from 192.168.45.2 on Ethernet1, 01:12:50 ago
  Routing Descriptor Blocks:
    192.168.45.2, from 192.168.45.2, 01:12:50 ago, via Ethernet1
     *Route metric is 3, traffic share count is 1
```

The output shows that the route is available in the routing table of R1. The route is being propagated all over the OSPF domain and is unavailable only in the EIGRP domain.

4. Check the redistribution configuration of OSPF routes into EIGRP configured at B1. The relevant part of the configuration is displayed in Listing 12.10.

LISTING 12.10 Redistribution Configuration

```
router eigrp 100
network 10.10.2.0
network 10.10.3.0
redistribute ospf 1 route-map redis-1
!
route-map redis 10 permit 10
match ip address 15
set metric 10000 100 255 1 1000
!
access-list 15 permit 192.168.40.0 0.0.7.255
access-list 15 permit 10.10.0.0 0.0.255.255
```

In Listing 12.10, we see that some restriction has been applied to redistribution of routes from OSPF to EIGRP using route-map redis-1. The route map matches routes against access list 15. The access list 15 is not redistributed, because it does not match route 192.168.50.0/24.

5. Modify the access list using these commands:

```
access-list 15 permit 192.168.0.0 0.0.255.255
access-list 15 permit 10.10.0.0 0.0.255.255
```

This will ensure redistribution of all available OSPF routes into EIGRP. The route will be available as shown in Listing 12.11.

LISTING 12.11 Available Route After Access List Modification

```
B3#show ip route 192.168.50.0
Routing entry for 192.168.50.0/24
Known via "eigrp 100", distance 90, metric 41075200, type internal
Last update from 10.10.1.1 on Serial0, 01:01:05 ago
Routing Descriptor Blocks:
 *10.10.1.1, from 10.10.3.2, 01:01:05 ago, via Serial0
      Route metric is 41075200, traffic share count is 1
      Total delay is 42000 microseconds, minimum bandwidth is 64 Kbit
      Reliability 255/255, minimum MTU 1500 bytes
      Loading 3/255, Hops 3
```

RIP AND OSPF REDISTRIBUTION ENVIRONMENT

Consider Figure 12.2, which shows a redistribution scenario in which the routing protocols in use are RIP and OSPF. Routers B3 and B4 are running OSPF, router B1 is running RIP, and router B2 is running both routing protocols, RIP and OSPF, and is redistributing between the two.

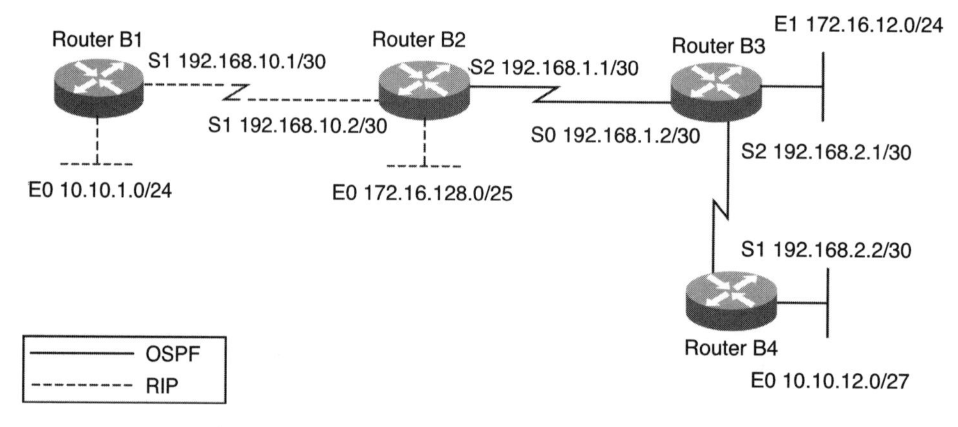

FIGURE 12.2 Redistribution scenario running OSPF and RIP.

Table 12.2 lists the configuration of all four routers B1, B2, B3, and B4.

TABLE 12.2 Router Configuration

Router	Relevant Part of Configuration
B1	interface Serial1 ip address 192.168.10.1 255.255.255.252

TABLE 12.2 *(continued)*

Router	Relevant Part of Configuration
	! interface Ethernet0 ip address 10.10.1.1 255.255.255.0 ! router rip network 192.168.10.0 network 10.10.1.0 !
B2	interface Serial1 ip address 192.168.10.2 255.255.255.252 ! interface Serial2 ip address 192.168.1.1 255.255.255.252 ! interface Ethernet0 ip address 172.16.128.1 255.255.255.128 ! router rip network 192.168.10.0 network 172.16.128.0 redistribute ospf 1 metric 1 ! router ospf 1 network 192.168.1.0 0.0.0.255 area0 redistrbute rip metric 1000 !
B3	! interface Serial0 ip address 192.168.1.2 255.255.255.252 ! interface Ethernet1 ip address 172.16.12.1 255.255.255.0 ! interface Serial2 ip address 192.168.2.1 255.255.255.252 ! router ospf 1 network 192.168.1.0 0.0.0.255 area0 network 192.168.2.0 0.0.0.255 area0

(continued)

TABLE 12.2 *(continued)*

Router	Relevant Part of Configuration
	network 172.16.12.0 0.0.0.255 area0 !
B4	! interface Ethernet0 ip address 10.10.12.1 255.255.255.224 ! interface Serial1 ip address 192.168.2.2 255.255.255.252 ! router ospf 1 network 192.168.2.0 0.0.0.255 area0 network 10.10.12.0 0.0.0.255 area0 !

In Figure 12.2:

- Network 10.10.12.0/27 is unreachable from the RIP domain.
- Network 172.16.12.0/24 is unreachable from the RIP domain.

Network 10.10.12.0/27 Unreachable from RIP Domain

To troubleshoot this problem:

1. Check if the corresponding interface in B4 is up by using the show ip interface brief command. The output of the command shows that the interface is up and running.
2. Check if routing is enabled for this network at B4 by using the show ip protocols command.
3. Check availability of route 10.10.12.0/27 in the OSPF routing domain by using the show ip route 10.10.12.0/27 command. The output of this command shows that the routing for network 10.10.12.0/27 is enabled and is propagated in the OSPF routing domain.
4. Check for the presence of route maps or filters in the redistribution configuration at B2 by using the show running-config command. The output of the command shows that there are no such restrictions on redistribution.
5. Check if there is any subnet of the major network 10.0.0.0/8 in the RIP domain. There could be a subnet mismatch, because this is a case of

redistribution between a classful routing protocol, RIP, and a classless routing protocol, OSPF. Use the show ip route | inc 10 command to check this. The output of the command shows that there is a network 10.10.1.0/24 in the RIP domain and network 10.10.12.0/27, which is redistributed into RIP from OSPF, is not propagated in the RIP domain.

The modified configuration to ensure that network 10.10.12.0/27 is available in RIP domain is given by the command:

```
B2(config)#ip route 10.10.12.0 255.255.255.0 Null0
B2(config)#router rip
B2(config-router)#redistribute static metric 1
```

Using this configuration, solve the current redistribution problem. RIP will propagate network 10.10.12.0/24 with a next-hop of B2, which has a more specific route to the subnet 10.10.12.0/27.

There can be issues if any other subnet of 10.10.12.0/24 is located outside the OSPF domain. In that case, there are two approaches to troubleshoot the problem. The IP addressing scheme can be modified to ensure that all subnets of 10.10.12.0/24 are located within the OSPF domain. Otherwise, static routes need to be defined for the other subnets that are not part of the OSPF domain. As a result, the /24 network will be pointed towards the OSPF domain, but there will be routers with longer netmasks available for a better match, pointed elsewhere, either within the RIP domain or if any other routing protocol environment is in use in the network.

Network 172.16.12. 0/24 Unreachable from RIP Domain

As discussed in the previous section, follow the steps to check and ascertain that route 172.16.12.0/24 is propagated within the OSPF domain but is not propagated within the RIP domain despite being redistributed. There could be a subnet mismatch. In that case:

1. Check if there is any subnet of the major network 172.16.0.0/16 in the RIP domain by using the show ip route | inc 172.16 command. The output of the command proves the presence of a network 172.16.128.0/25 in the RIP domain. The network 172.16.12.0/24 redistributed into RIP from OSPF is not propagated in the RIP domain.
2. Break down the OSPF network into subnets of the same netmask and redistribute into OSPF. The commands used for this are:

```
B2(config)#ip route 172.16.12.0 255.255.255.128 Serial2
B2(config)#ip route 172.16.12.128 255.255.255.128 Serial2
```

```
B2(config)#router rip
B2(config-router)#redistribute static metric 1
```

After this modification, the route for network 172.1.6.12.0/24 will be available in the RIP domain but as a part of two different subnets and not a single subnet.

STATIC ROUTING AND THE OSPF REDISTRIBUTION ENVIRONMENT

Consider Figure 12.3, which shows a routing environment with static routes being redistributed into OSPF. Routers B2, B3, and B4 are a part of the OSPF routing domain. Static routing is configured between routers B1 and B2. Redistribution is configured in router B1.

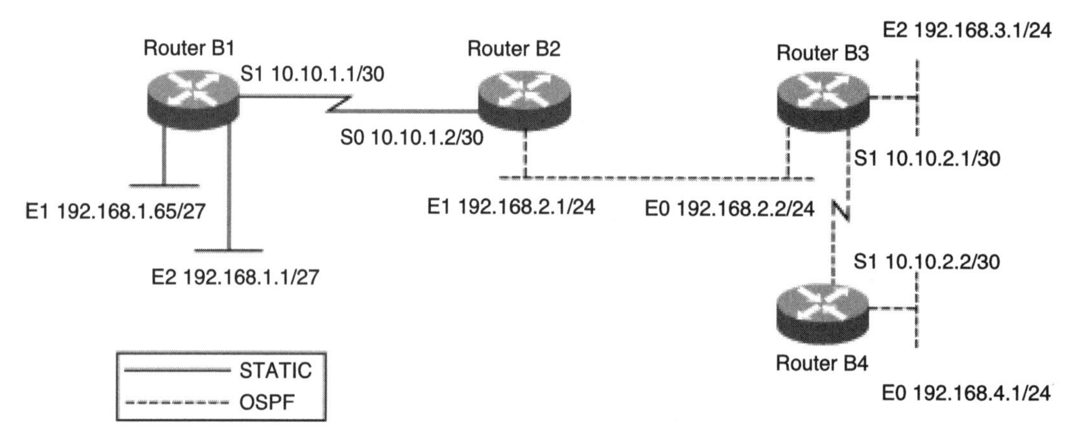

FIGURE 12.3 Redistribution scenario with static routing between B1 and B2.

Table 12.3 lists configuration of four routers B1, B2, B3, and B4.

TABLE 12.3 Router Configuration

Router	Relevant Part of Configuration
B1	interface Serial0 ip address 10.10.1.1 255.255.255.252 ! interface Ethernet1 ip address 192.168.1.65 255.255.255.224 ! interface Ethernet2

TABLE 12.3 *(continued)*

Router	Relevant Part of Configuration
	ip address 192.168.1.1 255.255.255.224 !
B2	interface Serial0 ip address 10.10.1.2 255.255.255.252 ! interface Ethernet1 ip address 192.168.2.1 255.255.255.0 ! ip route 192.168.1.0 255.255.255.224 10.10.1.1 ip route 192.168.1.64 255.255.255.224 10.10.1.1 ! router ospf 1 network 10.10.1.0 0.0.0.255 area 0 network 192.168.2.0 0.0.0.255 area 0 redistribute static metric 100 !
B3	interface Serial1 ip address 10.10.2.1 255.255.255.252 ! interface Ethernet0 ip address 192.168.2.2 255.255.255.0 ! interface Ethernet2 ip address 192.168.3.1 255.255.255.0 ! router ospf 1 network 192.168.2.0 0.0.0.255 area 0 network 192.168.3.0 0.0.0.255 area 0 network 10.10.2.0 0.0.0.255 area 0 !
B4	interface Serial1 ip address 10.10.2.2 255.255.255.252 ! interface Ethernet0 ip address 192.168.4.1 255.255.255.0 ! router ospf 1 network 192.168.4.0 0.0.0.255 area 0 network 10.10.2.0 0.0.0.255 area 0 !

In Figure 12.3, subnet 192.168.1.65/27 is inaccessible from network 192.168.3.0/24, as shown in the command:

```
B3#show ip route 192.168.1.66
% Network not in table
```

To isolate and resolve this problem:

1. Check availability of the route in B2 where redistribution is happening. The command and output to check this are shown in Listing 12.12.

LISTING 12.12 Output of the show ip route 192.168.1.66 Command

```
B2#show ip route 192.168.1.66
Routing entry for 192.168.1.64/27
  Known via "static", distance 1, metric 0
  Redistributing via ospf 1
  Advertised by ospf 1 metric 100
  Routing Descriptor Blocks:
  *10.10.1.1
      Route metric is 0, traffic share count is 1
```

Listing 12.12 shows that a static route pointing to B1 is available.

2. Check for filtering or a route map applied to redistribution from static to OSPF. No such filtering is present. This is confirmed by checking the configuration of the router by issuing the show running-config command.
3. Check the redistribution configuration to ensure that the subnets have been allowed in redistribution. The relevant part of the configuration is shown in Listing 12.13.

LISTING 12.13 Configuration to Allow Subnets in Redistribution

```
router ospf 1
network 10.10.1.0 0.0.0.255 area 0
network 192.168.2.0 0.0.0.255 area 0
redistribute static metric 100
!
```

Here, the keyword subnets is missing. For that, perform the configuration:

```
B2(config)#router ospf 1
B2(config)# redistribute static metric 100 subnets
```

STATIC ROUTING AND THE RIP REDISTRIBUTION ENVIRONMENT

Consider Figure 12.4, which depicts a routing environment in which some static routing and RIP routing is involved. Routers B1 and B2 and routers B1 and B3 are connected by WAN links over which no dynamic routing protocol is enabled. Routing for networks connected to routers B2 and B3, respectively, at router B1 is configured statically. B1, B4, and B5 are part of the RIP routing domain. B1 redistributes static routes to RIP for these routes to be visible throughout the RIP domain.

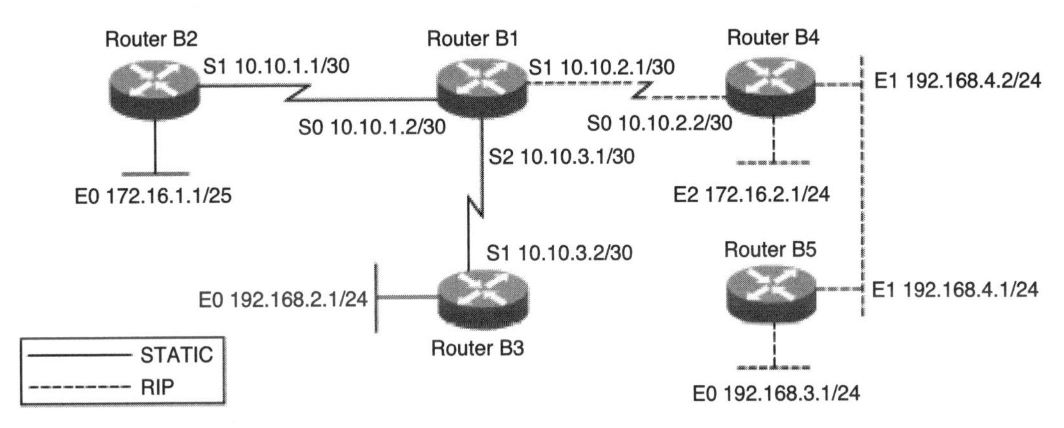

FIGURE 12.4 Routing environment with no dynamic routing protocol enabled.

Table 12.4 lists the configuration of all five routers B1, B2, B3, B4, and B5.

TABLE 12.4 Router Configuration

Router	Relevant Part of Configuration
B1	interface Serial0 ip address 10.10.1.2 255.255.255.252 ! interface Serial1 ip address 10.10.2.1 255.255.255.252 ! interface Serial2 ip address 10.10.3.1 255.255.255.252 ! ip route 192.168.2.0 255.255.255.0 10.10.3.2 ip route 172.16.1.0 255.255.255.128 10.10.1.1 !

(continued)

TABLE 12.4 *(continued)*

Router	Relevant Part of Configuration
	router rip network 10.10.1.0 network 10.10.2.0 network 10.10.3.0 redistribute static metric 1 distribute-list 3 out Serial1 ! access-list 3 deny 172.16.1.0 0.0.0.255 access-list 3 deny 10.10.1.0 0.0.0.255 access-list 3 permit any
B2	interface Serial1 ip address 10.10.1.1 255.255.255.252 ! interface Ethernet0 ip address 172.16.1.1 255.255.255.128 ! ip route 0.0.0.0 0.0.0.0 Serial1
B3	interface Serial1 ip address 10.10.3.2 255.255.255.252 ! interface Ethernet0 ip address 192.168.2.1 255.255.255.0 ! ip route 0.0.0.0 0.0.0.0 Serial1
B4	interface Serial0 ip address 10.10.2.2 255.255.255.252 ! interface Ethernet1 ip address 192.168.4.2 255.255.255.0 ! interface Ethernet2 ip address 172.16.2.1 255.255.255.0 ! router rip network 10.10.2.0 network 192.168.4.0 network 172.16.2.1 !

TABLE 12.4 *(continued)*

Router	Relevant Part of Configuration
B5	interface Ethernet0 ip address 192.168.3.1 255.255.255.0 ! interface Ethernet1 ip address 192.168.4.1 255.255.255.0 ! router rip network 192.168.4.0 network 192.168.3.0 !

In Figure 12.4, the users of the Accounts department at location A, which is connected to B5, are not able to connect to the ERP server at location B, which is connected to B2. The Accounts department LAN is 192.168.3.0/24 and the ERP server LAN is 172.16.1.0/25. To isolate and resolve this problem:

1. Check if Layer-3 connectivity exists between two network segments by using the ping command, which confirms that connectivity is absent. This is shown in the output as:

```
B5#ping 172.16.1.1
Type escape sequence to abort.
Sending 5, 100-byte ICMP Echos to 172.16.1.2, timeout is 2 seconds:
.....
Success rate is 0 percent (0/5)
```

2. Check if the respective interfaces at B2 and B5 are up. The commands to check this are:

```
B2#show ip interface brief
B5#show ip interface brief
```

This command output shows that both the interfaces are up.

3. Check the presence of route 172.16.1.0/25 in the routing table of B5. The route is not available. This confirms the absence of the route in the RIP domain as shown:

```
B5#show ip route 172.16.1.1
% Network not in table
```

4. Check the presence of the route in B1 where redistribution between static routes into RIP is configured. The commands for this are shown in Listing 12.14.

LISTING 12.14 Output of the show ip route 172.16.1.1 Command

```
B1#show ip route 172.16.1.1
Routing entry for 203.200.177.96/28
  Known via "static", distance 1, metric 0
  Redistributing via rip
  Routing Descriptor Blocks:
  10.10.1.1
      Route metric is 0, traffic share count is 1
```

Listing 12.14 proves the existence of a static route pointing to B2.

5. Check the redistribution configuration for the presence of any route maps or filters, which might be preventing availability of the mentioned route in the RIP domain. The relevant part of the configuration is shown in Listing 12.15.

LISTING 12.15 Part of the Configuration to Check Route Maps

```
router rip
network 10.10.1.0
network 10.10.2.0
network 10.10.3.0
redistribute static metric 1
distribute-list 3 out Serial1
!
access-list 3 deny 172.16.1.0 0.0.0.255
access-list 3 deny 10.10.1.0 0.0.0.255
access-list 3 permit any
```

In Listing 12.15, the bold code shows that there is an access list-based filter that is preventing redistribution of the static route corresponding to 172.16.1.0/25. Suitable modification of access list 3 should solve the problem. The access list is modified by issuing the command:

```
B1(config)#no access-list 3
B1(config)#access-list 3 deny 10.10.1.0 0.0.0.255
B1(config)#access-list 3 permit any
```

The matching of subnet masks of major networks can also be a cause of the problem, because this is a case of redistribution into RIP, which is a classful routing protocol. The network that is not advertised in an RIP domain after getting redistributed is originally a Class B network.

6. Check if RIP knows any other subnet mask of this network by using the show ip route | inc 172.16 command. The output of the command shows that there is another network available in the major network 172.16.0.0/16, that is, 172.16.2.0/24. This information is available with RIP. Therefore, it is not recognizing the network 171.16.1.0/25, which has a different subnet mask.
7. Configure a static route of the network 172.16.1.0/24 at B1 instead of 172.16.1.0/25. The route will be propagated in the RIP domain. The commands issued are:

```
B1(config)#no ip route 172.16.1.0 255.255.255.128 10.10.1.1
B1(config)#ip route 172.16.1.0 255.255.255.0 10.10.1.1
```

IS-IS AND OSPF REDISTRIBUTION ENVIRONMENT

Figure 12.5 shows an IS-IS and OSPF routing environment. Router A2 is running IS-IS, and routers B1 and B2 are running OSPF. A1 is running both IS-IS and OSPF and is redistributing between the two protocols.

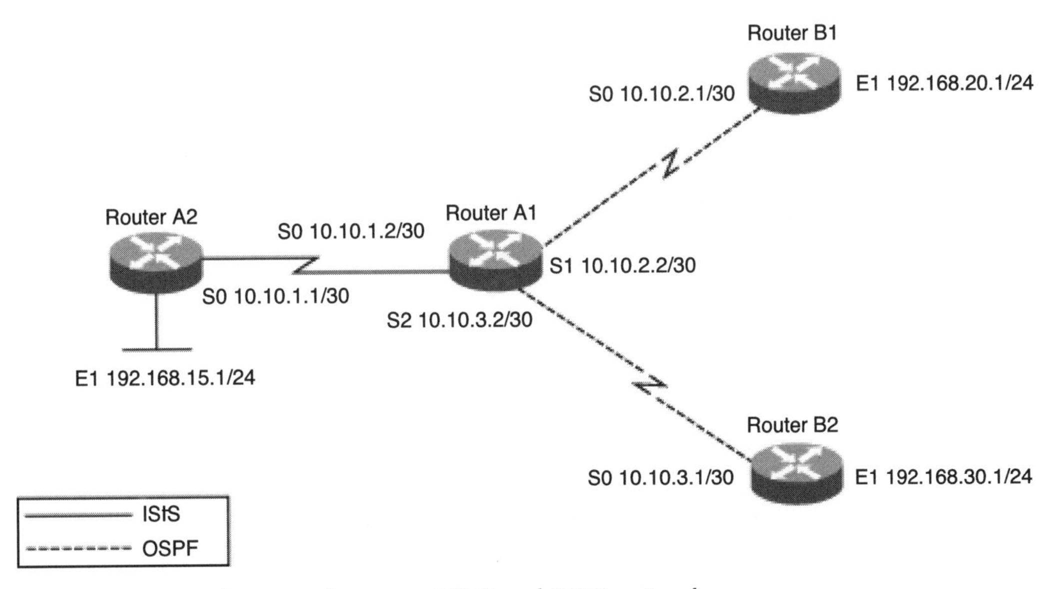

FIGURE 12.5 Routing scenario amongst IS-IS and OSPF protocols.

Table 12.5 lists the configuration of all four routers A1, A2, B1, and B2.

TABLE 12.5 Router Configuration

Router	Relevant Part of Configuration
A1	interface Serial0 ip address 10.10.1.2 255.255.255.252 ip router isis ! interface Serial1 ip address 10.10.2.2 255.255.255.252 ip router isis ! interface Serial2 ip address 10.10.3.2 255.255.255.252 ip router isis ! router ospf 1 network 10.10.1.0 0.0.0.255 area 0 network 10.10.2.0 0.0.0.255 area 0 network 10.10.3.0 0.0.0.255 area 0 passive-interface Serial0 redistribute isis route-map redis1 ! router isis net 25.0001.0000.0000.000a.00 redistribute ospf 1 metric 30 ! route-map redis1 permit 10 match ip address 20 set metric 100 ! access-list 20 permit 10.10.0.0 0.0.0.255 access-list 20 permit 192.168.0.0 0.0.248.255 !
A2	interface Serial0 ip address 10.10.1.1 255.255.255.252 ip router isis ! interface Ethernet1 ip address 192.168.15.1 255.255.255.0 ip router isis

TABLE 12.5 *(continued)*

Router	Relevant Part of Configuration
	! router isis net 25.0001.0000.0000.000a.00
B1	interface Serial0 ip address 10.10.2.1 255.255.255.252 ! interface Ethernet1 ip address 192.168.20.1 255.255.255.0 ! router ospf 1 network 10.10.2.0 0.0.0.255 area 0 network 192.168.20.0 0.0.0.255 area 0 !
B2	interface Serial0 ip address 10.10.3.1 255.255.255.252 ! interface Ethernet1 ip address 192.168.30.1 255.255.255.0 ! router ospf 1 network 10.10.3.0 0.0.0.255 area 0 network 192.168.30.0 0.0.0.255 area 0 !

In Figure 12.5, the network mail server in the IS-IS routing domain with IP address 192.168.15.2 is inaccessible from Sales LAN 192.168.30.0/24. To troubleshoot this problem:

1. Check the basic IP connectivity by using the ping command at B2. The output of the command is shown in Listing 12.16.

LISTING 12.16 Output of Ping Command

```
B2#ping 192.168.15.2
Type escape sequence to abort.
Sending 5, 100-byte ICMP Echos to 192.168.15.2, timeout is 2 seconds:
.....
Success rate is 0 percent (0/5)
```

In Listing 12.16, there is no Layer-3 connectivity existing between the mail server network and the Sales LAN.

2. Check the presence of the route at B2 by using the command:

```
#show ip route 192.168.15.0
% Network not in table
```

The command output confirms the unavailability of the route in the OSPF domain.

3. Check if any route is available at A1 by using the command:

```
A1#show ip route | inc 192.168.15.0
i  92.168.15.0/24 [115/1] via 10.10.1.1, 00:00:16, Serial0
```

The command output shows the presence of an IS-IS route with a next hop of 10.10.1.1. Route 192.168.15.0/24 is available in the IS-IS domain but is unavailable in the OSPF domain. This confirms that there is some problem in redistribution.

4. Obtain the relevant part of the configuration at A1 by issuing the command show running-config, as shown in Listing 12.17.

LISTING 12.17 Output of the show running-config Command

```
router ospf 1
network 10.10.1.0 0.0.0.255 area 0
network 10.10.2.0 0.0.0.255 area 0
network 10.10.3.0 0.0.0.255 area 0
passive-interface Serial0
redistribute isis route-map redis1
!
route-map redis1 permit 10
match ip address 20
set metric 100
!
access-list 20 permit 10.10.0.0 0.0.0.255
access-list 20 permit 192.168.0.0 0.0.248.255
!
```

Listing 12.17 shows that a route map redis1 is used to restrict redistribution from IS-IS to RIP. Only the routes matching access list 20 are redistributed. Network 192.168.15.0/24 does not match access list 20. This causes the nonredistribution of the route into OSPF. Access list 20 can be suitably modified as shown in Listing 12.18.

LISTING 12.18 Modifying Access List 20

```
A1(config)#no access-list 20
A1(config)access-list 20 permit 10.10.0.0 0.0.0.255
A1(config)access-list 20 permit 192.168.0.0 0.0.240.255
```

CASE STUDY

Figure 12.6 depicts a complex routing scenario, which has a combination of both static routing and dynamic routing protocols, including RIP and EIGRP. The places where different routing protocols have been used are marked in the figure.

In Figure 12.6, routers B1 and B2 are exclusively running EIGRP, and routers B6 and B7 are exclusively running RIP. Static routing is configured between routers

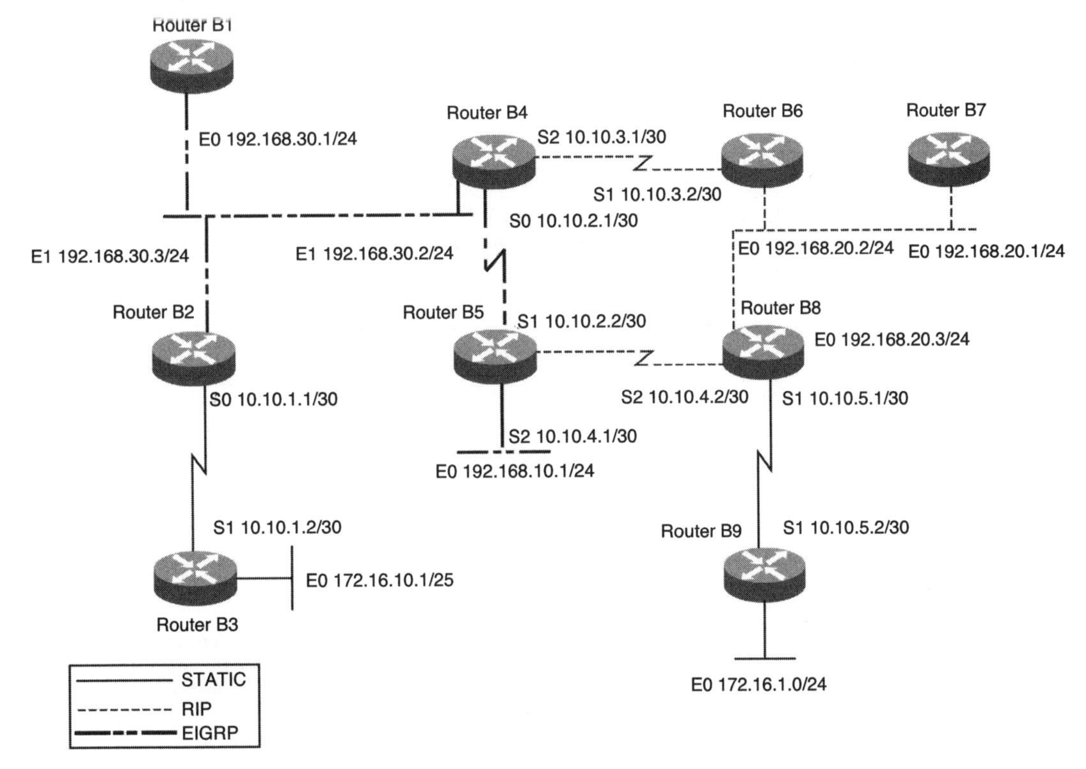

FIGURE 12.6 A complex routing scenario running RIP and EIGRP.

B2 and B3, and the static routes are redistributed to EIGRP in router B2. Static routing is also configured between routers B8 and B9, and these routes are being redistributed into RIP at router B8. Redistribution from RIP to EIGRP is configured at router B4, and that from EIGRP to RIP is configured in router B5. Table 12.6 lists configuration of all nine routers B1, B2, B3, B4, B5, B6, B7, B8, and B9.

TABLE 12.6 Router Configuration

Router	Relevant Part of Configuration
B1	interface Ethernet0 ip address 192.168.30.1 255.255.255.0 ! router eigrp 100 network 192.168.30.0 !
B2	interface Ethernet1 ip address 192.168.30.3 255.255.255.0 ! interface Serial0 ip address 10.10.1.1 255.255.255.252 ! router eigrp 100 network 192.168.30.0 network 10.10.1.0 redistribute static metric 10000 100 250 3 1500 !
B3	ip route 172.16.10.0 255.255.255.128 10.10.1.2 interface Ethernet0 ip address 172.16.10.1 255.255.255.128 ! interface Serial1 ip address 10.10.1.2 255.255.255.252 ! ip route 0.0.0.0 0.0.0.0 10.10.1.1
B4	interface Ethernet1 ip address 192.168.30.2 255.255.255.0 ! interface Serial0 ip address 10.10.2.1 255.255.255.252 ! interface Serial2

TABLE 12.6 *(continued)*

Router	Relevant Part of Configuration
	ip address 10.10.3.1 255.255.255.252 ! router eigrp 100 network 192.168.30.0 network 10.10.2.0 network 10.10.3.0 redistribute rip metric 10000 100 250 3 1500 passive-interface Serial2 ! router rip network 192.168.30.0 network 10.10.2.0 network 10.10.3.0 passive-interface Serial0 passive-interface Ethernet1 !
B5	interface Ethernet0 ip address 192.168.10.1 255.255.255.0 ! interface Serial1 ip address 10.10.2.2 255.255.255.252 ! interface Serial2 ip address 10.10.4.1 255.255.255.252 ! router eigrp 100 network 192.168.10.0 network 10.10.2.0 network 10.10.4.0 passive-interface Serial2 ! router rip network 192.168.10.0 network 10.10.2.0 network 10.10.4.0 passive-interface Serial1 passive-interface Ethernet0 redistribute eigrp 100 metric 1 !
B6	interface Ethernet0

(continued)

TABLE 12.6 *(continued)*

Router	Relevant Part of Configuration
	ip address 192.168.20.2 255.255.255.0 ! interface Serial1 ip address 10.10.3.2 255.255.255.252 ! router rip network 192.168.20.0 network 10.10.3.0 !
B7	interface Ethernet0 ip address 192.168.20.1 255.255.255.0 ! router rip network 192.168.20.0 !
B8	interface Ethernet0 ip address 192.168.20.3 255.255.255.0 ! interface Serial1 ip address 10.10.5.1 255.255.255.252 ! interface Serial2 ip address 10.10.4.2 255.255.255.252 ! router rip network 192.168.20.0 network 10.10.5.0 network 10.10.4.0 redistribute static metric 1 ! ip route 172.16.1.0 255.255.255.0 10.10.5.2 !
B9	interface Ethernet0 ip address 172.16.1.0 255.255.255.0 ! interface Serial1 ip address 10.10.5.2 255.255.255.252 ! ip route 0.0.0.0 0.0.0.0 10.10.5.1

In Figure 12.6:

- The path from network 192.168.10.0/24 to network 172.16.1.0/24 via B4 is nonoptimal.
- There is no route of network 172.16.10.0/25 from the RIP routing domain.

Nonoptimal Path from Network 192.168.10.0/24 to Network 172.16.1.0/24

To isolate and resolve this problem:

1. Redistribute static route to 172.16.1.0/24 into RIP at B8, which in turn is redistributed into EIGRP at B4. B5 learns route 172.16.1.0/24 via both EIGRP from B4 and RIP from B8. The EIGRP route is preferred, because it has a lower administrative distance of 90 as compared to 120 of RIP. This results in a nonoptimal route taken from network 192.168.10.0/24 to reach 172.16.1.0/24. The route taken via EIGRP is B5-B4-B6-B8-B9. The optimal route would be via RIP, which is B5-B8-B9.
2. Implement necessary filters to allow for selection of an optimal RIP route to send EIGRP updates from B4 to B5. The relevant part of the configuration of Router B4 is shown in Listing 12.19.

LISTING 12.19 Configuration for Optimal RIP Route Selection

```
router eigrp 100
network 192.168.30.0
network 10.10.2.0
network 10.10.3.0
redistribute rip metric 10000 100 250 3 1500
passive-interface Serial2
distribute-list 2 out Serial0
!
access-list 2 deny 172.16.1.0 0.0.0.255
access-list 2 permit any
```

The additional configuration commands are shown in bold.

3. Ensure optimal routing since there will be only a single RIP learned route available from network 192.168.10.0/24 to 172.16.1.0/24.
4. Ensure reachability in case of failures by providing a backup path by modifying the administrative distance. The relevant part of the configuration of B5 is shown in Listing 12.20.

LISTING 12.20 Part of the Backup Path Configuration

```
router eigrp 100
network 192.168.10.0
network 10.10.2.0
network 10.10.4.0
passive-interface Serial2
distance 130 0.0.0.0 255.255.255.255 10
!
access-list 10 permit 172.16.1.0 0.0.0.255
```

In Listing 12.20, some commands have been added, shown in bold. The administrative distance for route 172.16.1.0/24, learned from any EIGRP source, is set at 130, which is higher than that of an RIP route. As a result, by default, the RIP route will be selected, but in case of unavailability of the RIP route, the EIGRP route would be available for routing in the routing table.

Route Unavailable for Network 172.16.10.0/25 from RIP

To isolate and resolve this problem:

1. Redistribute static route to 172.16.10.0/24 into EIGRP at B2, which in turn, is redistributed into RIP at B5.
2. Check for the presence of the EIGRP route in B5. The command to check this is shown in Listing 12.21.

LISTING 12.21 Output of the show ip route 172.16.10.1 Command

```
B5#show ip route 172.16.10.1
Routing entry for 172.16.10.0/25
Known via "eigrp 300", distance 90, metric 41075200, type internal
Redistributing via eigrp 300
Last update from 10.10.2.1 on Serial0, 01:01:05 ago
Routing Descriptor Blocks:
10.10.2.1, from 192.168.30.3, 01:01:05 ago, via Serial0
     Route metric is 41075200, traffic share count is 1
 Total delay is 42000 microseconds, minimum bandwidth is 64 Kbit
     Reliability 255/255, minimum MTU 1500 bytes
     Loading 3/255, Hops 3
```

3. Check for the presence of any outbound filter or route map at B4 for any restriction related to network 172.16.1.0/25. This is checked by using the show running-config command. None is found.

4. Check for the presence of any RIP subnetwork of the major network 172.16.0.0/16 in the routing table of B5. You'll find that a RIP route exists for network 172.16.10/24. Because RIP is a classful routing protocol, it does not understand VLSM and understands only one subnet mask per major network.
5. Propagate any routing information related to network 172.16.0.0/16 in the /24 netmask to resolve this problem. Use this configuration at B2 to achieve this. Use IP route 172.16.1.0 255.255.255.0 10.10.1.2 instead of IP route 172.16.1.0 255.255.255.128 10.10.1.2.

Now, the /24 network will be redistributed into EIGRP at B2 and from EIGRP to RIP at B5, instead of the /25 network. This network with the same netmask as 172.16.10/24 would be advertised in the RIP domain after redistribution and would be reachable from the RIP domain.

SUMMARY

In this chapter, we learned about troubleshooting different redistribution environments such as EIGRP and OSPF, RIP and OSPF, and static routing and OSPF. We also reminded you about troubleshooting static routing environments and RIP, IS-IS, and OSPF redistribution environments.

POINTS TO REMEMBER

- Redistribution leads to sharing of routing information among various routing protocols.
- Redistribution requires detailed study of the network and careful planning, because the routing tables become more complex after incorporating redistribution.
- The common problems due to redistribution are probable routing loops and nonoptimal path selection.
- Routing loops arise due to inconsistencies in the routing states of different routers present in an interconnected network.
- In the case of nonoptimal path selection, the data delivery time is longer, resulting in delay and less network performance.
- To prevent problems related to redistribution, use route maps and distribution lists in order.
- There are no generic Cisco IOS commands with which to troubleshoot all types of redistribution scenarios.

Appendix About the CD-ROM

The CD-ROM included with Cisco IP Routing Protocols: Troubleshooting Techniques includes all the code listings, tables, and images from the various examples found in the book. In addition, this CD-ROM includes an exhaustive Question Bank for the readers to test knowledge acquired after reading the book.

CD-ROM FOLDERS

Listing: Contains all the code listings from examples covered in each chapter.

Images: Contains all the images from each chapter.

Tables: Contains all the tables from each chapter.

QuestionBank: Contains multiple choice questions to test the learner's knowledge on Cisco Troubleshooting concepts.

OVERALL SYSTEM REQUIREMENTS: RECOMMENDED

- Windows® 98, Windows NT, Windows 2000, or Windows XP Professional
- Pentium III processor or greater
- CD-ROM drive
- Hard drive
- 128 MB of RAM, 256 MB recommended
- 20 MB of hard drive space for the code examples.

INSTRUCTIONS FOR USE OF THE QUESTIONBANK

1. Insert the CD-ROM in the CD-ROM drive
2. Double-click the QuestionBank folder
3. Double-click the Troubleshooting-Setup.zip file
4. Extract the setup files in a folder (example, create a folder called TroubleshootingQB in your C, D, or any other hard disk drive).
5. Double-click the TroubleshootingQB folder
6. Double-click the Troubleshooting-Setup folder
7. Double-click the setup.exe file
8. Follow the setup instructions
9. Once the setup is over, you can access the Question Bank from the Start → Programs → Troubleshooting with Cisco

Index

A

Action Plan Definition, 22
Action Plan Implementation, 22
Addressing at the Network Layer, 44
Application Layer, 11,14
ATM, 29
Attribute Problems, 351

B

B2 Is Not Visible as a Neighbor of B1, 294
Basic Protocol Behavior, 27
  Acknowledgment number, 52,58
  Address family identifier, 54
  AppleTalk, 56
  ASN, 58
  ATM, 30
  bandwidth, 54,56
  Big internetworks, 45
  broadcasts, 54
  Checksum, 52,53,58
  CIDR, 53
  Class B, 46
  Class C, 46
  Command, 54
  CONS, 29
  DARPA, 35
  Data, 44,52
  data frames, 36
  Data offset, 52
  data packet, 31
  Destination Address, 43
  Destination Port, 52
  DUAL, 56
  dynamic routing, 50
  EIGRP, 56,57
  Exterior routes, 55
  Flag, 52
  Flags, 43,58
  FR, 27
  Fragment Offset, 43
  Framing Flag, 36
  FST, 29
  full-duplex, 51
  HDLC, 36,40
  Header Checksum, 43
  heterogeneous, 34
  hierarchical IP addressing, 46
  host, 44,47
  host address, 44
  hosts, 45
  ICMP, 49,50
  Identification, 43
  IEEE 802.2, 29
  IGRP, 54
  IHL, 43
  Interior routes, 55
  InterNIC, 45
  IP, 35,41,56
  IP address, 44,45,57
  IP addresses, 36,45,48
  IP addressing, 54
  IP protocol, 43
  IP protocols, 35
  IP routing table, 49,50
  IPX, 56
  IRDP, 49,51
  Layer 3, 28,44
  LCP, 36
  Length, 52
  LLC, 29
  LLC2, 29
  Medium internetworks, 46
  Metric, 54
  MTU, 54,57
  multicast, 48
  NCP, 36
  NetWare protocols, 35
  Network layer, 49
  nodes, 28,39,49
  octet, 46
  octets, 44,45,47
  Opcode, 57
  Opcode types, 57
  Options, 43,52
  OSI layers, 35
  OSI model, 27,28,29,51,53
  OSI reference model, 34
  OSPF, 28
  packets, 28
  PPP, 34,36
  Protocol, 43,50,52,54,57
  protocols, 27,28,29,32,53
  reliability, 54
  reliable protocols, 27
  Reserved, 52
  RFC, 32,54
  RIP, 53
  routing protocol, 41
  routing protocols, 49
  routing table, 54,55
  routing tables, 53
  RTP, 58
  RTT, 50
  SDLC, 36,39,40
  secondary nodes, 39
  Sequence, 58
  Sequence Number, 52
  seven layers, 34
  Small internetworks, 46
  Source Address, 43
  Source Port, 52
  subnet, 50
  subnet mask, 44,45,54
  subnets, 44,51,54,55
  subnetworks, 46,50,55
  System routes, 55
  TCP, 28,30,35,51,52
  TCP/IP, 29,35,44
  Time-to-Live, 43
  TLVs, 58
  Total Length, 43
  Type and Length Value, 58
  Type of Service, 43
  UDP, 51
  upper layers, 44
  Urgent pointer, 52
  VC, 29
  Version, 42,54
  VLSM, 53,56
  Window, 52
  X.25, 29
BGP Communities, 357
BGP Multihoming and Loadsharing, 358
BGP Neighbor Relationship, 341
BGP Route Advertisement, 345

C

Case Study, 23,262,294,323,360,393
  Adjacency, 300
  Area ID, 300,317
  CLNS, 299,300
  Designated Intermediate System, 300
  Distance, 300
  DR, 300
  End System, 300
  End System–to–Intermediate System, 300
  Hello packets, 317
  Intermediate System, 300
  IP, 299
  IP addresses, 324
  IS-IS, 299,312
  IS-IS domain, 322

Case Study *(cont.)*
IS-IS level, 300
IS-IS process tag, 300
ISO, 300
ISPs, 299
L1 router, 300
L1 routers, 317
L2 router, 317
L2 routers, 320
Level 1, 301
Level 2, 301
Link State Packet data unit, 300
LSA, 300
LSP, 300,302,317
LSPs, 302,317,326
Maximum paths, 300
N-SEL, 300
NET, 300,312,318
Network Entity Title, 300
network layer protocol, 300
Network Layer Protocol ID, 300
Network Service Access Point, 300
NSAP, 300
NSEL, 300,316
Number of active areas, 300
Number of manual areas, 300
Number of SPF runs, 300
Octet, 300
Open Systems Interconnection, 299
OSI, 299
OSI terminology, 300
OSPF routing domain, 322
Protocol Data Unit or Packet Data Unit, 299
Pseudo Node, 300
route leaking, 322
routing, 321,324
routing domain, 324
routing protocol, 299,301,312,324
routing protocols, 322
SubNetwork Point of Attachment, 300
System ID, 300
Challenges and Issues of Complex Networks, 1
access layer, 5,6
ACLs, 6
Application, 9
Application layer, 9,19
application-presentation interface, 9
bandwidth, 6,16
BGP, 7
core layer, 6
CRC, 17
data packets, 13,19
Data-link layer, 10,17
DES, 19
desktop layer, 5
distribution layer, 6
domains, 6
encryption, 19
encrypts, 11
fiber optic, 4
firewalls, 4
frames, 10
FTP, 11
full-duplex, 11
gateways, 10
half-duplex, 11
hierarchical, 5,6
hosts, 22
HTTP, 11
IGRP, 1
Internet Protocol, 14
IP, 14
IP addresses, 13
LANs, 1
layer, 6
layered protocol, 9
layered troubleshooting approach, 7
layers, 5
MAC-layer filtering, 6
micro segmentation, 6
multiprotocol, 91
Network Access layer, 14
Network layer, 10,19
nodes, 6
OSI, 7
OSI model, 7,9,11,13,14,15,18
OSPF, 1,7
packets, 6,10,14,17
Physical layer, 9,10,16
Presentation layer, 9,11
Presentation layers, 9
protocol, 13,14,17,18
protocols, 1,3,4,7,11,15
QoS, 7
repeaters, 16
review problem solving cycle, 23
routing, 6,7
routing algorithm, 18
RSA, 19
seven layers, 7
Simplex Stop-and-Wait Protocol, 17
SMTP, 11
subnet topology, 17,18
subnets, 5,18
subnetworks, 5,6
TCP, 13,14
TCP/IP, 1,7,13,14,15
telemedicine, 4
Token Ring, 6
topology, 10,18
Transport layer, 10,14,18
UDP, 14
WANs, 1
workgroup layer, 6
bandwidth, 16
hierarchical, 5
host, 16
OSI model, 15
Cisco Diagnostic Commands, 68
Cisco Discovery Protocol, 87
Cisco Hierarchical Approach, 5
Cisco IOS Diagnostic Commands, 109
Cisco Network Management Tools, 61
AAMP, 65
Account management, 66
Address Error, 76
AEP, 80
AppleTalk, 64,80
ATM, 87
AVVID, 61
bandwidth, 66
Banyan VINES, 64
bottlenecks, 66,90
broadcast, 65
Bus Error, 76
CDP, 68,88
CIP, 72
CiscoWorks Blue, 62
CiscoWorks2000, 62
clear cdp counters, 89
Configuration management, 63,66
Configuration snap-in management, 63
CWM, 65,66
data packets, 83
Data-link layer, 87
DECnet, 64
Device monitoring, 63
Distributed polling and threshold configuration, 64
DRAM, 72
Emulator Trap, 76
fast switching, 83
Fault management, 66
Frame Relay, 87
FTP, 90
GUI, 62,65
heterogeneous, 87

host addresses, 77
HP OpenView™, 62
IBM SNA, 62
ICMP, 79
Informix Reporting Application, 66
Initial, 74
Inventory management, 63
IP, 62,64
IP address, 81
IP addresses, 88
IP packet, 85
IPX, 80
ISO, 64
LAN Service Manager, 68
Max Free, 73
MIB, 63
Min Free, 73,74
NetSys, 67
NetSys SLM, 67
NVRAM, 70
OSI model, 68
Parity Error, 75
Performance management, 66
Performance Manager, 68
Permanent, 72,73
Planning, modeling, and analysis, 66
protocol, 80,84,90
Protocol analysis, 64
RCP, 90
RMON, 63,64
SCM, 66
Security management, 63,66
SES, 65
Seven-layer traffic analysis, 64
show cdp, 89
show cdp entry, 89
show cdp interface, 89
show commands, 63
SNA, 64
SNMP, 62,65,67,87
SPAN, 64
SRAM, 72
Statistics Collection Manager, 66
Sun Solaris(tm), 65
SunNet, 62
TFTP, 90
Tivoli TME 10™, 65
topology, 67,68
VISTA, 67
VLAN, 61
VLAN Monitoring, 64
VLANs, 64,65
WAN Service Manager, 68
Watchdog Timeout, 76
XNS protocol, 64
Connectivity Manager, 67
CWM, 67
NFS, 64
CiscoView, 62
CiscoWorks, 62
CiscoWorks QoS Policy Manager, 107
Classes of IP Addressing, 45
Classless Routing, 171
Compatibility Between RIPv1 and RIPv2, 181
Configuration Problems, 161
Configure Interface as Point-to-point Links, 288
Connection-Oriented Behavior, 30
Connection-Oriented Behavior Versus Connectionless Behavior, 32
Connectionless Behavior, 30
Connectionless Network Protocol, 29
Connectivity and Cable Testers, 94
10Base, 94
Accounting Management, 103
ACLs, 106
Apollo, 116
AppleTalk, 116
ARP, 116
ASCII, 98
Auto Update Server, 107
AVVID, 104,107
bandwidth, 96
Cisco IDS Host Sensor and Console, 107
Cisco nGenius Real-Time Monitor, 105
Cisco Secure Policy Manager, 107
CiscoView, 105
CiscoWorks ACL Manager, 106
CiscoWorks Campus Manage, 105
CiscoWorks Device Fault Manager, 105
CiscoWorks Internetworks Performance Monitor, 106
CiscoWorks Resource Manager Essential, 105
CLI, 111
CLNS, 116
coaxial, 94
Configuration management, 99,103
CSU/DSU, 96
CTS, 96
CWM, 102
DCE, 96
DECnet, 116
DFM, 102
DTE, 96
DTR, 96
EMS, 103
EMSs, 103
Fault detection, 99
Fault management, 103
FDDI, 96
Fiber optic, 94
firewall, 107
General purpose analyzers, 98
Get Bulk Request, 102
Get Next, 102
Get Request, 102
GUI, 98
heterogeneous, 103
High-end analyzers, 98
show running-config, 110
ICMP, 116
IDS, 106
Inform Request, 102
IP, 116
IP route, 105
LMS, 105
MAC layer, 94
Management Center for IDS Sensors, 106
Management Center for PIX Firewalls, 106
Management Center for VPN Routers, 106
Management information model, 102
Management protocol, 102
Management server, 102
MIB, 102
MIBs, 99
MIB, 104
Monitoring Center for Security, 106
multiprotocol network topologies, 108
NetSys, 108
network topology, 97
NEXT, 94
NIDS, 109
NMS, 102,104
NVRAM, 110
NVS, 109
OMNI, 95
on-the-reel testing, 94
OSI layer, 93
OTDRs, 94,95
PDU, 102
Performance Management, 103
Port unreachable, 117

Connection-Oriented Behavior (*cont.*)
- protocol, 97,98
- protocol layers, 97
- protocols, 103,107,108
- QoS, 99,103,107
- QPM, 107
- Resource management, 99
- Response, 102
- RME, 106
- RMON, 99,105
- routing algorithms, 107,108
- Security Management, 103
- Set, 102
- show buffers, 110
- show cdp neighbors, 110
- show controllers, 110
- show debugging, 110
- show flash, 110
- show interfaces, 110
- show memory summary, 110
- show process cpu, 110
- show stack, 110
- show startup-config, 110
- show version, 110
- SLAs, 103
- SNMP, 100
- Software-based analyzers, 98
- STP, 94
- TCP, 98
- TCP/IP, 100
- TDR, 94
- TDRs, 94,95
- Time exceeded, 117
- Token Ring, 96
- topologies, 108
- TRAP, 102,105
- TTL, 117
- twinax, 94
- UTP, 94
- VINES, 116
- VISTA, 108
- VLANs, 105
- VMS, 106
- VPN, 106
- VPN/Security, 106
- VPNs, 103
- XNS network, 116
- Performance management, 99

Consideration of Possibilities, 22
Core Dumps, 90

D

Data-Link Layer, 10
debug Commands, 137,275
Define OSPF Network as Point-to-Multipoint Links, 289
Device management, 63
Digital Interface Testers, 96
domain, 21

E

EIGRP and OSPF Redistribution Environment, 370
- classful routing protocol, 381,399
- classless routing protocol, 381
- dynamic routing protocols, 393
- EIGRP, 370,375,377,393,398
- EIGRP domain, 377
- IP addressing scheme, 381
- IP routing, 370
- IS-IS, 389
- IS-IS routing domain, 391
- ISIS domain, 392
- netmask, 399
- netmasks, 381
- Nonoptimal path selection, 369
- optimal routing, 369,397
- OSPF, 370,375,377,378,380,381,384,389
- OSPF domain, 375,377,381,392
- OSPF routing domain, 380
- Probable routing loops, 369
- protocols, 389
- RIP, 378,381,389,392,393,398
- RIP domain, 380,381,382,387,388,399
- RIP domain, 381
- RIP route, 397
- RIP routing, 385
- routing, 369,374,376,380
- routing domain, 380,382,385,397
- routing protocol, 381,385,389
- routing protocols, 369,370,378,393
- routing table, 374,376,377,387,398
- routing tables, 369
- show ip protocols, 374
- SPF, 377
- static route, 384,397,398
- static routes, 381,382
- static routing, 385,393
- subnet, 375,380,381
- subnet mask, 389
- subnet masks, 389
- subnets, 375,381
- subnets, 384
- subnetwork, 398
- VLSM, 399

EIGRP Metric Problem, 254
EIGRP Neighbor Formation Problem, 248
EIGRP Redistribution Problem, 264
EIGRP Route Problem, 252,263
Enhanced Interior Gateway Routing Protocol, 56
Enterprise Management Systems, 103
EXCHANGE, 283
EXCHANGE or EXSTART State, 283

F

Fast Sequenced Transport, 29
Features of EIGRP, 229
- access list, 264
- Ack, 234
- Acknowledgment, 230
- Adjacency, 230
- Advertised Distance, 231
- ASN, 247
- bandwidth, 254,256,257
- Data traffic, 255
- debug eigrp packets ack, 234
- debug eigrp packets hello, 234
- debug eigrp packets ipxsap, 234
- debug eigrp packets probe, 234
- debug eigrp packets query, 234
- debug eigrp packets reply, 234
- debug eigrp packets request, 234
- debug eigrp packets retry, 234
- debug eigrp packets SIAquery, 234
- debug eigrp packets SIAreply, 234
- debug eigrp packets stub, 234
- debug eigrp packets terse, 234
- debug eigrp packets update, 234
- debug eigrp packets verbose, 234
- DNS, 241
- EIGRP domain, 242,261
- EIGRP routes, 253
- EIGRP routing domain, 245
- EIGRP routing protocol, 240
- EIGRP topology, 253
- EIGRP traffic, 258
- Feasible Distance, 230
- Feasible Successor, 231
- Hello, 230,233
- Hello packets, 248,263
- how ip protocols, 243
- IP address, 261
- IP level, 263
- Ipxsap, 234
- MTU, 249,254
- multicast, 230
- network segment, 264
- network topology, 230
- network traffic, 240
- Observation, 241,245,248
- Probe, 234

Problem Isolation, 242,249
Queries, 230
Query, 233
Replies, 230
Reply, 233
RID, 253
RIP domain, 261
RIP route, 261
RIP routes, 264
routing packets, 257
routing process, 230
routing protocol, 250,251
routing protocols, 257,260,263
routing table, 252,255,263
Routing Updates, 230
RTP, 230
SIA, 258,259,260
SIAquery, 234
SIAreply, 234
Stub, 234
Successor, 231
TCP/IP, 231
topology, 253,258,260
topology table, 248,255,258,263
traceroute, 261
unicast, 230
Update, 233
Problem Isolation, 245
Request, 233

Features of IGRP, 199
ASN, 200
Bandwidth, 200,216
broadcast, 208
broadcasts, 211
Delay, 200
Delay in Convergence, 211
Exterior route, 200
flush timer, 223
IGRP, 199,200,201,211,216
IGRP protocol, 221
IGRP Routes Missing from Routing Table, 208
IGRP routing process, 219
IGRP routing updates, 212
Interior route, 200
IP addres, 208
IP address, 208
IP route, 225
Load, 200
MTU, 200
Passive Interface, 211
Reliability, 200
routing packets, 211
routing protocol, 201,219
routing table, 208,213,218,222,223
routing tables, 211
routing updates, 223
System route, 200
Unable to Ping IP Address, 208
routing tables, 221

Features of IS-IS, 299

Features of RIP, 149
BGP, 165
broadcast, 183,195
broadcasts, 190
Classful routing protocol, 149
Configuration of Passive Interface, 165
counting to infinity, 191
distance vector protocol, 187
flush timer, 150,186
hold-down timer, 150
hop count, 184
IGRP, 165
invalid timer, 150
IP address, 162,165
ip protocols, 157
ip rip receive version, 166
ip rip send version, 166
IP routing table, 162
loop-free routing, 193
Missing Network Command, 162
network topologies, 190
network topology, 154,187,190
OSPF, 165
peer router, 182
ping, 165
RIP, 149,154,161,162,184,187
RIP routing protocol, 165,183, 187
RIP routing update, 165
RIP routing updates, 167
RIPv1, 183
RIPv2, 183
RIPv2 protocol, 181
routing, 162
routing by rumor, 187
routing loop, 190
routing loops, 191
routing protocol, 167,171
routing protocols, 149,151,171,181
routing table, 150,159,161,162,165,167,181,186,187,191,194,196
routing tables, 176,190
routing update, 163,165
routing updates, 154,182
show ip route, 165
Simple Split Horizon, 191
Split Horizon with Poisoned Reverse, 193
subnet mask, 149,171,181
subnets, 150,177
topology, 181,187,190,195
triggered updates, 195
UCP, 150
update timer, 150
version, 166
VLSM, 150
VLSM subnets, 154

H

High Spectrum Cable Testers, 95
High-level Data Link Control, 40
Hold-down Timers, 195
Hop Count, 191
hybrid routing, 291

I

IGRP Does Not Install All Possible Equal Cost Paths, 213
INIT State, 282
Interior Gateway Routing Protocol, 54
Internal and external security mechanisms, 4
Internet Layer, 14
IP, 29
IS-IS and OSPF Redistribution Environment, 389
Isolation of Error, 23

L

L1 Router Problem, 317
L2 Router Problem, 320
Load State, 286
Looping, 187
Low Spectrum Cable Testers, 94

M

Management Solutions for LANs, 105
Management Solutions for VPN/Security, 106
Management Solutions for WANs, 106
Mapping the TCP/IP and OSI Models, 15
Misconfiguration, 154
Misconfiguration in EIGRP, 240
Misconfiguration in IGRP, 207
Misconfiguration Problems, 349
Misconfiguration Problems in IS-IS, 312
Misconfigured ASN, 216
Missing Routes, 347
Modeling and Simulation Tools, 107

Modeling and Simulation Tools *(cont.)*

N

NBMA Networks, 287
NetSys Baseliner 4.0, 67
NetSys Network Management Suite, 67
NetSys SLM Suite, 67
Network 10.10.12.0/27 Unreachable from RIP Domain, 380
Network 172.16.12. 0/24 Unreachable from RIP Domain, 381
Network 192.168.30.0/27 Unavailable at B7, 373
Network 192.168.50.0/24 Unavailable at B3, 376
Network Access Layer, 13
Network and Protocol Analyzers, 97
Network Layer, 10
Network Monitoring and Management System, 98
Network Not Declared in EIGRP, 241
Nonoptimal Path from Network 192.168.10.0/24 to Network 172.16.1.0/24, 397
Nothing at All State, 284

O

Observation, 252,255,261
Observe and Review Findings, 22
OSPF domain, 268
OSPF Neighbor States, 281
OSPF routing protocol, 267
OSPF Routing Table, 286
OSPF Stub Areas, 290
OSPF Terminology, 267
Other show Commands, 77

P

Physical Layer, 10
ping, 122
  ARP, 138
  debug arp, 138
  debug ip icmp, 138
  debug ip packet, 138
  debug ip rip, 137
  DNS, 145
  DNS server, 139
  domain, 145
  gateway address, 139
  ICMP, 138
  IGRP, 145
  IP address, 139,145
  IP addresses, 139
  IP protocol, 135
  local host, 139,146
  Pinging a Remote IP Address, 146
  Pinging Loopback and Local IP Addresses, 145
  Pinging the Router, 145
  protocol, 137,145
  remote host, 139,146
  repeater, 144
  RIP, 137,145
  routing loops, 145
  routing protocol, 145
  routing table, 137
  subnet mask, 141
  subnet mask address, 139
  TCP/IP, 137,138,139,141,142,143,144,145
  WINS server, 146
  Pinging DNS, Default Gateway, and WINS Servers, 145
  Using the Tracert Tool, 146
Point-to-Point Protocol, 36
Presentation Layer, 11
Problem Definition, 21
Problem Isolation, 253,255,259,261
Problem Isolation in BGP, 329
  ASs, 329,341,351
  BGP, 329,330,332,337,345,347,351,352,355,357,358,360,366
  BGP peers, 345,346
  BGP route, 353
  BGP routes, 351
  BGP routing, 336,349
  BGP table, 346,348,350,358
  CEF, 359
  eBGP peers, 358
  ISP, 359
  ISPs, 329,358,359
  loopback address, 359
  Multi Exit Discriminator, 351
  NEI, 347
  Next Hop, 351
  peer routers, 348
  Route Dampening, 353
  routing protocol, 329
  routing table, 347,350,354,355,360,366
  routing updates, 366
  static routes, 355,356
  static routing, 345
Problem Isolation in EIGRP, 231
Problem Isolation in IGRP, 201
Problem Isolation in IS-IS, 302
Problem Isolation in TCP/IP Networks, 139
Problem Resolution Model Approach, 19
Problems with Assigning Priority, 279
Protocol Characteristics, 34
Protocols of the Data-Link Layer, 36
Protocols of the Network Layer, 41
Protocols of the Session, Presentation, and Application Layers, 53
Protocols of the Transport Layer, 51

R

R1 and R2 Configured with Same System ID, 325
Redistribution in OSPF, 291
Redistribution of Routing Information, 324
Redistribution Problem, 260,321
Redistribution Problems, 355
Resolution of Problems in OSPF, 268
Resolve the Problem, 23
Review Problem Solving Cycle, 23
RIP and OSPF Redistribution Environment, 378
RIP Is Not Installing All Possible Equal-Cost Paths, 167
RIP Routes Missing from Routing Table, 162
Route 10.6.1.1/27 Is Unavailable in Routing Table of B4, 297
Route Poisoning, 194
Route Unavailable for Network 172.16.10.0/25 from RIP, 398
routing, 268
Routing at the Network Layer, 49
Routing Information Protocol, 53

S

Server, 4
Session Layer, 11
show Commands, 128,268
Split Horizon, 191
Static Routing and the OSPF Redistribution Environment, 382
Static Routing and the RIP Redistribution Environment, 385
Stuck in Active State, 258
Synchronous Data Link Control, 39

T

TCP, 29,51
TCP/IP Problems and Symptoms, 141
TCP/IP Router Diagnostic Tools, 121
  Address, 123
  ARP, 134
  ARPA, 134

Datagram size, 124
DNS, 123
Extended commands, 124
Host Name, 123
show ip arp, 134
show ip traffic, 135
ICMP, 122,127
IP address, 122,130,134
IP routes, 130
IP routing protocols, 134
IP routing table, 130
OSPF, 131,132
ping Command in the Privilege Exec Mode, 124
ping Command in the User Exec Mode, 123
Protocol, 123,124
protocols, 123
Repeat count, 124
RFC, 134
routing protocol, 130
show ip access-list, 133
show ip ospf database, 131
show ip ospf interface, 132
show ip ospf neighbor, 132
show ip protocols, 134
show ip route, 130
Target IP address, 124
TCP/IP, 121,122,123,124,127,128
Timeout in seconds, 124
TTL, 127
The Access Layer, 5
The Core Laye, 6
The debug Command, 82,111
The Distribution Layer, 6
The Layered Architecture of the OSI Mode, 8
The Layered Architecture of the TCP/IP Model, 13
The Layered Troubleshooting Approach, 7
The ping Commands, 115
The ping Command, 78
The show buffers Command, 73
The show cdp neighbors Command, 77
The show Command, 68,109
The show controllers Command, 72
The show flash Command, 72
The show interfaces Command, 70
The show memory Command, 74
The show process cpu Command, 75
The Show running-config Command, 70
The show stack Command, 75
The show startup-config Command, 70
The show version Command, 69
The trace Command, 86,117
Timer Problem, 183
Timer Problem in IGRP, 221
TP0, TP1, TP2, TP3, and TP4, 29
trace, 127
Traffic Director RMON, 63
Traffic Generators, 108
Transport Layer, 10,14
Triggered Updates, 194
Troubleshooting Problems on Data-Link Layer, 17
Troubleshooting Problems on the Application Layer, 19
Troubleshooting Problems on the Network Layer, 17
Troubleshooting Problems on the Physical Layer, 15
Troubleshooting Problems on the Transport Layer, 18
Troubleshooting Techniques, 138
TTL, 86,87
2-Way State, 283
ABR, 268
ABRs, 290
Adjacencies, 268
ASBR, 268
Autonomous Systems, 268
Backup Designated Router, 268
BDR, 279
BR, 268
broadcast, 268
broadcast network, 286
Designated Router, 268
DR, 279
EIGRP, 294
EXSTART, 283
FE0/0, 279
Flooding, 267
Hello packets, 282
hello timers, 286
IGP, 291
INIT, 282
IP address, 287
IP subnet, 287
IR, 268
Link State Advertisement, 267
LSA, 286
LSAs, 286
MTU, 283
Multi-access/broadcast networks, 268
NBMA, 287
Nonbroadcast multi-access networks, 268
Not So Stubby Area, 268
OSPF, 267
OSPF Routers, 268
OSPF routing, 279
OSPF routing protocols, 291
packet level, 275
protocol, 267
RIP, 291
routing loops, 291
routing protocols, 290
routing table, 267
Single area, 268
Stub area, 268
subnet, 287
subnets, 297
terminology, 267
Totally Stubby Area, 268
Type 2 service of IEEE 802.2, 29
Types of Connectionless and Connection-oriented Protocols, 28

U

UDP, 29,52,86
Undeclared Networks, 350
Using the Layered Approach, 15

V

VLAN Director Switch Management, 65

W

WAN Manager, 65